Lecture Notes in Mathematics

Edited by A. Dold and B. Eckmann

Subseries: Department of Mathematics, University of Maryland
Adviser: M. Zedek

1277

Carlos A. Berenstein (Ed.)

Complex Analysis III

Proceedings of the Special Year
held at the University of Maryland, College Park, 1985–86

Springer-Verlag
Berlin Heidelberg New York London Paris Tokyo

Editor

Carlos A. Berenstein
Department of Mathematics, University of Maryland
College Park, MD 20742, USA

Mathematics Subject Classification (1980): 32-06

ISBN 3-540-18355-8 Springer-Verlag Berlin Heidelberg New York
ISBN 0-387-18355-8 Springer-Verlag New York Berlin Heidelberg

Printing and binding: Druckhaus Beltz, Hemsbach/Bergstr.
2146/3140-543210

INTRODUCTION

The past several years have witnessed a striking number of important developments in complex analysis of both one and several variables. Through these advances the essential unity of these two previously rather separate branches of function theory has become increasingly apparent. More and more, ideas and constructs that first arose in function theory of one variable are playing an important role in the several variables theory. At the same time, techniques developed originally for use in several variables have found fruitful applications to problems in classical function theory. Examples of the former phenomenon include the development of a capacity theory for the Monge-Ampère operator and recent extensions of the Henkin-Ramirez representation formulas and their application to interpolation problems in $\mathbb{C}^n$. In the second category, the systematic use of the inhomogeneous Cauchy-Riemann equation has led to important developments in the theory of H^∞, as well as other Banach algebras on the unit disk. It has also inspired the consideration of many new questions about open Riemann surfaces. Finally the brilliant solution of the Bieberbach Conjecture by Louis de Branges offers irrefutable testimony (as if any were needed) to the continued vitality of classical ideas and approaches..

It is in this context that the Department of Mathematics of the University of Maryland decided to dedicate its sixteenth Special Year to the subject of Complex Analysis. The objective of these Special Years has been to provide a forum for the exchange of ideas between the members of the department, long-term visitors and the other participants. (In spite of the electronic mail and other gadgets, a lot of mathematics is still done on a face to face basis!) This time we have had over one hundred and fifty mathematicians from several different countries, who participated in the Special Year from July, 1985 to December, 1986. This participation has been made possible by a substantial contribution from our department as well as the very generous support of the National Science Foundation and the Argonne Universities Association. The Organizing Committee of the Special Year, D. Hamilton, J.A. Hummel, L. Zalcman and myself, would like to express their appreciation for this support to Dr. J. Polking, Dr. A.I. Thaler and Dr. J.V. Ryff of the NSF, Chancellor George A. Russell

of the AUA and Professors J. Osborn and N. Markley, past and present chairmen of our department.

It is clear that this Special Year has been a very successful one and the three volumes of Proceedings are just a token proof of it. On behalf of the Organizing Committee, I would also like to thank all the participants who made this organizational effort worthwhile. Last, but not least, our heartfelt thanks to the administrative personnel of our department, among them D. Kennedy, D. Forbes, S. Matthews, and B.L. Smith, and to the very skillful S. Smith, V. Sauber, J. Nagendra, K. Aho, and I. Grove who did a magnificent job at typing these Proceedings. Finally, there is one person who, in our opinion and the opinion expressed in many letters of our visitors, merits a special round of applause, since without her immense dedication and generosity we seriously doubt we would have fared so well, we refer to Mrs. N. Lindley.

To conclude, a word about these three volumes. The reader will find here both surveys of different areas of Complex Analysis as well as many new results and insights. We had asked for manuscripts that were accessible not only to specialists in the subject but to a broader audience; we think that the authors have responded to that request and these Proceedings will prove to be a very useful source of reference and problems in Complex Analysis.

Carlos A. Berenstein

Table of Contents

SPECIAL YEAR PAPERS -- VOLUME I

SPECIAL YEAR PAPERS -- VOLUME II

SOME RECENT SUCCESSES IN VALUE-DISTRIBUTION THEORY

D. Drasin†

Let $f(z)$ be nonconstant meromorphic in $|z| < \infty$, and let $n(r,a)$ [resp. $\bar{n}(r,a)$] be the number of solutions to the equation $f(z) = a$ in $|z| < r$ with [resp. without] regard to multiplicity. Classical Nevanlinna theory (= value-distribution theory) studies the asymptotic behavior as $r \to \infty$ of these and related functionals, and how they are influenced by more apparent properties of $f(z)$. The standard references are [17], [16], [24], or [25]; I will assume the reader is familiar with the usual notations $N(r,a)$ [$= \int^r n(t,a)t^{-1}dt$], $\bar{N}(r,a)$, $m(r,a)$, $T(r) = T(r,f)$ [$= \frac{1}{2\pi}\int_0^{2\pi} N(r,e^{i\theta})d\theta + O(1)$], the Nevanlinna densities

$$(1)\qquad \begin{aligned} \delta(a) &= \liminf_{r\to\infty} \frac{m(r,a)}{T(r)}, \\ \theta(a) &= \liminf_{r\to\infty} \frac{N(r,a) - \bar{N}(r,a)}{T(r)} \end{aligned}$$

as well as the definition of the order λ,

$$(2)\qquad \lambda = \limsup_{r\to\infty} \frac{\log T(r)}{\log r}.$$

In this lecture, I will discuss three rather recent developments.[1] The first two concern refinements of Nevanlinna's defect relation

$$\sum \delta(a,f) \le 2.$$

In (3), the sum is taken over all extended complex numbers, but as early as 1929, Nevanlinna [24] asked if we could extend (3) to a sum

†Partially supported by N.S.F.

[1] At the time of this lecture (11/85), the works [29] and [12] had not appeared.

over all "small functions" $a(z)$; i.e., those for which

$$T(r,a(z)) = o(T(r,f)) \quad (r \to \infty).$$

Following earlier work of C. Chuang [4], Le Yang [28], G. Frank-G. Weissenborn [13] and Ch. Osgood [26], N. Steinmetz [29] has now given a most compelling proof of the fact

$$\sum_{a(z)} \delta(a(z),f) \le 2, \tag{5}$$

using the Wronskian determinant, and I shall discuss this in § 1. The first proof of (5) is due to Osgood in [26].

In § 2, I return to the classical relation (3), and in § 3 ask when equality is possible. If λ, as defined in (2), is infinite, there is not much to say (except then that also $\liminf[(\log T(r))/(\log r)] = \infty$), but when $\lambda < \infty$, I have proved [7] that 2λ is an integer ≥ 2, and all numbers $\delta(a,f)$ are integral multiples of λ^{-1}.

One of the most interesting problems which remains in the subject is: how large can the left side of (3) be when the sum is taken over all entire, or meromorphic, functions of a fixed order $\lambda > 1$? For $\lambda < 1$, the answer has been known for some time (cf. [11] for entire functions and [8], using [1], for general meromorphic functions), but aside from the isolated values of λ for which $\sum\delta(a) = 2$ is possible, this is still wide open when $\lambda > 1$. Some interesting special cases are considered in [15], [20], and [22]. (See also [31].)

Finally, in § 4, I mention a most elegant recent result of A.E. Eremenko and M.L. Sodin [12], which takes a substantial step toward establishing a classical conjecture of J.E. Littlewood. Littlewood considers the behavior, for large N, of

$$\phi(N) = \sup_{P\in \mathcal{P}(N)} \left\{ \iint_{|z|<1} \rho_P(z)\,dxdy \right\}.$$

where ρ_P is the spherical derivative of the polynomial $P : \rho_P(z) = |P'(z)|/(1+|P(z)|^2)$, and $P(N)$ is the class of polynomials of degree N. Clearly

$$\phi(N) = \left\{\iint_{|z|<1} dxdy \iint_{|z|<1} \rho_f^2\, dxdy\right\}^{\frac{1}{2}} \le \pi N^{1/2} \tag{6}$$

and Littlewood conjectured in [21] that there exist positive constants A and α such that

$$\phi(N) \le AN^{\frac{1}{2}-\alpha}. \tag{7}$$

What Eremenko and Sodin have shown is that

$$\phi(N) = o(N^{\frac{1}{2}}) \quad (N \to \infty). \tag{8}$$

This weak form of (7) was sufficient to establish a fundamental result on value-distribution which involves no Nevanlinna theory:

> *If* $f(z)$ *is entire, and its maximum modulus function* $\log M(r)$ *satisfies* $\log M(r) = O(r^\lambda)$ *for some* λ, *then there is a set* S *in the plane of zero planar density such that for all complex numbers* a,

$$n(r,a) = n(r,a,S) + o(r^\lambda) \quad (r \to \infty) \tag{9}$$

> *where* $n(r,a,S)$ *is the number of solutions to* $w = f(z)$ *with* $z \in S$.

(Littlewood observed that a proof of (7) would allow similar conclusions involving more general notions of smallness of S or of the error term in (9). As noted by Hayman [18], elliptic functions, being doubly-periodic, show that no such phenomena will hold for general meromorphic functions.)

There are several excellent collections of problems/progress reports in this area, beginning with the Cornell collection [5]. Hayman's pamphlet [19] has had a significant impact in the past twenty

years, and it has been updated several times (see [2] and references therein).

1. The theorem on small functions. Nevanlinna's classical proof of (3) depended on a study of the relation between a meromorphic function f and its derivative f'. By analogous methods Nevanlinna did prove (5) when the sum is over three small functions $a_j(z)$, but all subsequent progress depended on replacing $f'(z)$ by a more general (homogeneous) differential operator $P[f]$.

We follow the analysis of Steinmetz. Let $a_1(z),\dots,a_q(z)$ be any finite set of the small functions. Then $P[f]$ is to satisfy four properties; this is motivated so that $P[f]$ has a similar effect on the $a_j(z)$ as $f'(z)$ has on constants. Thus we require at once that

$$P[f] \not\equiv 0, \tag{1.1}$$

$$P[f-a_j] = P[f] \tag{1.2}$$

($a_j = a_j(t)$, $j = 1,\dots,q$). Moreover, it is necessary to relate the growth of $P[f]$ to that of f itself; this is easiest if P has the properties

$$P[f] = f^n Q\left[\frac{f'}{f}\right], \tag{1.3}$$

where n is some fixed integer which depends on P, and Q is a polynomial with coefficients in the field generated by $a_1(z),\dots,a_q(z)$ over $\mathbb{C}$; this was already convenient in the proof of (3), since $m(r,\frac{f'}{f})$ is an error term. Finally, we need that (for almost all large r)

$$T(r,P[f]) \le nm(r,f) + (n+k)N(r,f) + o(T(r,f)), \tag{1.4}$$

for some integer $k = k(a_1,\dots,a_q)$.

Once such a P is constructed, it is possible to follow the general lines of Nevanlinna's classical of proof of (3) to obtain that

$$(1.5) \qquad m(r,f) + \sum_j^q m(r,a_j(z)) \le n^{-1}(n+k)T(r,f) + S(r,f),$$

so (5) depends on knowing that we may choose n and k tending to ∞ with $k \le (1+o(1))n$.

Steinmetz's construction of suitable P's is simple to recount. For $s \ge 0$, let $L(s)$ be the vector-space spanned by the functions $a_1^{n_1} \dots a_q^{n_q}$, with all $n_j \ge 0$, $\sum n_j = s$. For a given s let $\beta_1, \dots, \beta_n$ be a basis of $L(s)$ and $b_1, \dots, b_k$ a basis of $L(s+1)$. Then we set (W = Wronskian)

$$P[f] = W(b_1, \dots, b_k, \beta_1 f, \dots, \beta_n f).$$

It is routine to check that (1.1)-(1.4), and so (1.5) hold; for the most elegant proof that as $s \to \infty$ we have $\liminf_{s\to\infty} k/n \le 1$, we refer to § 5 of [29].

Some history: Chuang first brought the Wronskian to this problem [4]; in our notation, he takes $s = 0$, and the estimates so obtained gave (5) only when f was entire or, somewhat more generally, when $\bar{N}(r,f) = o(T(r,f))$. Frank and Weissenborn [13] obtained (5) when all the $a_j(z)$ were rational by using another choice of W.

Priority for (5) belongs to the number theorist C.F. Osgood in [26], a paper which is notable for also being one of the first to link Nevanlinna's theory with the number-theoretic work of K.F. Roth on diophantine approximation by rational numbers. The paper [26] is very hard reading, but the basic properties (1.1), (1.2) and differential polynomials are apparent. While Steinmetz's argument is so simple that it is certain to be the standard one, Osgood ([26, p. 388) considers some more general situations and his approach lends hope for Nevanlinna theory having further connections with other areas.

Steinmetz's proof also illustrates a long-standing relation between value-distribution theory and the theory of differential equations in the complex domain. Other useful contributions have been

achieved, for example, by R. Kaufman, S. Bank, I. Laine, G. Frank, E. Mues, G. Gunderson, Li Shen, and others.

Very recently, W.H.J. Fuchs and C.J. Dai have given partial solutions to the problem "inverse" to (5): given a_j's and δ_j's, find f with $\delta(a_j(z),f) = \delta_j$, and $\delta(a,f) \equiv 0$ otherwise.

2. Remarks on the deficiency relation. In fact, Nevanlinna proved more than (3): that indeed

$$\sum \delta(a,f) + \theta(a,f) \le 2, \tag{2.1}$$

where the θ's are as in (1). This more general relation may be viewed as a transcendental form of the Riemann-Hurwitz relation ([25], Ch. 12).

Here are the essential ideas: If R is a rational function of detree N, then the sum of orders of branch-points p satisfies

$$\sum_p (m-1) = 2N - 2, \tag{2.2}$$

where the left side of (2.3) is summed over all branch-points on the sphere. For each $a \in \mathbb{C}^*$, let us group those terms on the left side of (2.2) for which $R(z) = a$. If we divide by N, then (2.2) becomes

$$\sum_a r(a) = 2 - 2N^{-2}, \tag{2.3}$$

where $r(a)$ is the total ramification over a. In particular, if R is rational of degree N, it is easy to check that $\sum \delta(a) + \theta(a) = 2-N^{-1}$, which is "almost" (2.1) with equality.

Now suppose $f(z)$ is transcendental, but can be well approximated by a sequence of rational functions $R_N(z)$; for example $e^z = (1+zN^{-1})^N$. It is natural to then take a limit on the left side of (2.3), as applied to each R_N. There are several obstacles in general: for example, each limit $r_N(a)$ may fail to exist as $N \to \infty$, or for an infinite set A of a we may have $\limsup_N \sum_A r_N(a) > 0$,

but yet $r_N(a) \to 0$ for each $a \in A$. However, it is not surprising that examples are legion for which $D(a) = \lim_N r_N(a)$ exists for each a, $\sum D(a) = 2$, and $D(a) = \delta(a) + \theta(a)$; the terms in $\sum N^{-1}(m-1)$ which converge to $\delta(a)$ will have $m \to \infty$ in (2.2), while when m is bounded, the preimages in the z-plane will converge to algebraic branch points of f.

This discussion shows that equality in (2.1) will hold for many classes of functions; in particular, there exist functions of any order λ or more general rates of growth for which equality holds.

3. Equality in the deficiency relation. We have seen that the inequality (2.1) tends to become an equality for functions that can be well-approximated by rational functions. However, equality in the weaker relation (3) is a different matter. Of course, if

$$\sum \delta(a,f) = 2 \qquad (3.1)$$

this will impose great restrictions on a sequence R_N of rational functions as described in § 2; all branch points will have to advance to ∞.

A. Pfluger, A. Edrei and W.H.J. Fuchs studied (3.1) for entire functions many years ago [27], [9], [10] and proved that if $\lambda < \infty$, then λ is a positive integer, and each nonzero number $\delta(a)$ satisfies

$$\lambda\delta(a) \text{ is a positive integer.} \qquad (3.2)$$

Here is the principle of the proof. At once, (3.1) and (2.1) will show that f has few branch points:

$$\delta(0,f') = 1. \qquad (3.3)$$

Hence, we may think of f' as having no poles and (modulo some technicalities, which are controlled because of (3.3)) no zeros. Since f' has finite order λ, a Liouville-type argument will give that

$\log f'(z) = P_\lambda(z)$, a polynomial of degree λ (so that λ is an integer), and the asymptotic behavior of f follows on integration.

The analogue of the Pfluger-Edrei-Fuchs theorem for meromorphic functions is of more recent vintage [6], [7]; now if $\lambda < \infty$,

$$2\lambda \text{ is an integer } \geq 2$$

and (3.2) still holds. Very old examples of F. Nevanlinna [23] show that all these possibilities do occur.

Here is an instructive example. There exists a meromorphic function $f(z)$ of order $\frac{3}{2}$ which satisfies (3.1); it can be obtained as the quotient of two linearly independent solutions of the Airy equation

$$w'' + (Az+B)w = 0 \qquad (3.5)$$

for any choice of A,B with $A \neq 0$. The asymptotic behavior of f is simple to describe: the plane separates into three congruent sectors I_j $(1 \leq j \leq 3)$, in each of which $f(z)$ tends to a complex number a_j (the three a_j are distinct) in such a way that $\delta(a_j) = \frac{2}{3}$. (More details may be found in [14]; to achieve $f(z)$ of order $\frac{1}{2}n$, we replace $(Az+B)$ in (3.5) by a polynomial of degree $(n-2)$.)

The idea of [6] and [7] is to make a "quasi-conformal modification" of f, to obtain a function $G(z)$ to which the Pfluger-Edrei-Fuchs arguments may be applied. To do this, we "deform" the deficient values to all lie over 0 and ∞. Once this is achieved, for example, since G will have few zeros or poles, we can hope to study $\log G$. For the specific function f which was just constructed using (3.5), this is achieved as follows: Let $\mathcal{A}$ be any annulus $\{r_1 < |z| < r_2\}$ with r_1 large. Then $\mathcal{A}$ is divided into three congruent sectors, I_j, each of which joins the inner and outer boundary components, and (as soon as we are a bit away from the boundary) in each of which f is close to one of the a_j.

Now if $z = \zeta^2$, these three I_j correspond to six sectors I'_j in the ζ-plane where we may assume that I'_j is between I'_{j-1} and I'_{j+1} $(1 \le j \le 6)$. We then introduce a modification of $f(\zeta^2)$, which we call $g(\zeta)$. This is introduced in [6], and studied more extensively in [7]. The function g is chosen to satisfy an equation of the form $(z = \zeta^2 \in \mathcal{A})$

$$g(\zeta) = \omega(f(\zeta^2)) \tag{3.6}$$

where ω is a (rather rigid) quasi-conformal mapping defined on the Riemannian image of $f(z)$, and so chosen that, for example,

$$\log|g(\zeta)| \sim \log\left|\frac{1}{f(\zeta^2)-a_j}\right| \quad (\zeta \in I'_j,\ j \text{ even}) \tag{3.7}$$

and

$$\log|g(\zeta)| \sim \log|f(\zeta^2)-a_j| \quad (\zeta \in I_j,\ j \text{ odd}). \tag{3.8}$$

To do this, let $\mathcal{B}$ be the preimage of $\mathcal{A}$ under the map $\zeta^2 = z$. Let $I'_j = \{\alpha_j < \theta < \beta_j\}$, and for a fixed $\varepsilon > 0$ and $1 \le j \le 6$, let

$$\mathcal{B}(j) = \{z \in \mathcal{B} \cap I'_j;\ \alpha_j+\varepsilon < \arg z < \beta_j-\varepsilon\}.$$

We then define $g(\zeta) = f(\zeta^2) - a_j$ $(\zeta \in \mathcal{B}(j),\ j \text{ odd})$ and $g(\zeta) = (f(\zeta^2) - a_j)^{-1}$ $(\zeta \in \mathcal{B}(j),\ j \text{ even})$. Then g is defined in all of $\mathcal{B}$ as in (3.6), where ω is used to make g continuous on all of $\mathcal{B}$. In order to define ω, it is essential that each component of $(\partial\mathcal{B}(j)) \cap \mathcal{B}$ can be connected to exactly one $\partial\mathcal{B}(j')$ in $\{\mathcal{B} - \bigcup_j \mathcal{B}(j)\}$.

Note that g being alternatively large and small in the I'_j is possible only if there are an even number of I'_j, which is guaranteed by using ζ^2 in (3.6). Because we have introduced ω, the factorization (3.6) will destroy g being meromorphic, but in fact (3.6) is possible with ω having arbitrarily small maximum dilation $\|\omega_{\overline{W}}/\omega_W\|_\infty (W = f(z))$. Hence by basic quasi-conformal theory, we may write $\zeta = \phi(Z)$, with ϕ a quasi-conformal homeomorphism which is

very close to the identity, and such that

$$G(Z) = g(\phi(Z)) \tag{3.9}$$

is meromorphic, and then adapt the arguments of [9], [10], and [27] to G. Thus, for example G has integral order, which with (3.6) and (3.9), explains why g must have order equal to $\frac{1}{2}N$, N and integer.

To make this work in general requires some additional considerations, since we do not know in advance that the components I_j in which f is close to the various deficient values, have any special properties. Since $\lambda < \infty$, however, there always exist annuli

$$\mathcal{A} = \{A_n^{-1} r_n \le |z| \le A_n r_n\}, \quad \text{with} \quad A_n \to \infty,\ r_n \to \infty$$

and

$$T(r) \le \{1+o(1)\}\left(\frac{r}{r_n}\right)^{\lambda} T(r_n) \quad (A_n^{-1} r_n \le r \le A_n r_n) \tag{3.10}$$

(Polya peak annuli). In these annuli, a weak form of the situation (3.5) holds. Thus, there are $k \le 2\lambda$ components I_j, each of which reach the outer boundary of $\mathcal{A}$, and in each of which $|f(z)-a|$ is small for precisely one $a \in \{a_j\}$; moreover, near $\{|z| = r_n\}$, $m(r,a)$ may be asymptotically computed by integrating $\log^+(1/|f(re^{i\theta})-a|)$ on $\{|z| = r\} \cap \{\cup I_j(a)\}$, where $\{I_j(a)\}$ consists of the appropriate components I_j on which f is close to a.

In order to obtain g as in (3.6), it is necessary to divide $\mathcal{A}$ into concentric subannuli $\mathcal{A}(\ell)$, such that relative to each $\mathcal{A}(\ell)$, each non-empty component $I_j \cap \mathcal{A}(\ell)$ joins both boundary components of $\partial\mathcal{A}(\ell)$. This then gives a family $g_\ell(\zeta^2)$ in various ζ-annuli, as described in (3.6) and (3.7). In order to transfer analytic information through the various g_ℓ, we exploit the fortunate situation the various g's are locally nearly-Moebius transformations of each other. In particular, the expression $(G_\ell'/G_\ell)^2$ is essentially independent of ℓ, so we essentially have a single analytic object in all of $\mathcal{A}$, which is of polynomial growth.

(I have been informed that A. Eremenko has now obtained an alternate proof of this theorem, by relying more on potential theory and avoiding quasiconformal modifications.)

It seems like that that (3.1), and so the conclusions of [7], are in fact consequences of (3.3) along, at least if $\delta(\infty) = 0$. If $\lambda < \frac{1}{2}$, this has recently been obtained by D.F. Shea [28]. As Shea observes, such a result would have applications to differential equations: if f and g are entire, linearly independent and of order $\lambda < \infty$, then the Wronskian of f and g, $W \equiv fg' - gf'$ also has order λ unless perhaps 2λ is a positive integer. I don't even have a conjecture for the analogous situation of three or more entire functions.

4. Littlewood's problem. We have stated Littlewood's conjecture above as (7); the literature on this problem consists of [21], [18], [3], and now [12].

The proof of (8) depends on measure theory and an elementary but essential

Lemma. Let $u \geq 0$ be subharmonic in $D(1) = \{|z| < 1\}$ with μ its Riesz mass. Then there exist Borel sets E and L such that

$$Z \equiv \{u = 0\} \subset E \cup L, \tag{4.1}$$

with

$$\mu(E) = |L| = 0$$

($|\ |$ is planar measure).

Not only does (8) follow from this Lemma, but the authors observe that (7), in strong form, would follow from a more precise (but now only hypothetical) version: let $0 \leq u \leq 1$ be subharmonic in $D(1)$. Then there exist absolute positive constants A and β such that for each $\varepsilon > 0$, $\{z;\ u(z) < \varepsilon\} = L_\varepsilon \cup E_\varepsilon$ with

$$\mu(E_\varepsilon) \le A\varepsilon^\beta, \quad |L_\varepsilon| < A\varepsilon^\beta.$$

The proof of the Lemma is elementary: one merely chooses as E the Lebesgue set of Z, so $L = Z-E$. Estimate (8) then follows by compactness arguments applied to the family of subharmonic functions $v_n(z) \equiv m^{-1} \log \sqrt{1+|p_n(z)|^2}$, where p_n is a polynomial of degree m. Since $\Delta v_n = 2m^{-1}\rho^2_{p_n}$, the Riesz measure of each v_n is a mass on the sphere of total mass 1.

The application to entire functions described in (9) follows by considering the family

$$v_{n,a}(z) \equiv 2^{-2n\lambda} \log|f(2^n z)-a|$$

where a is chosen so that, say, $m(r,a) = O(\log r)$ and, for convenience, $a \ne f(0)$.

References

[1] A. Baernstein, Proof of Edrei's spread conjecture, Proc. London Math. Soc. (3)26 (1973), 418-434.

[2] K.F. Barth, D.A. Brannan, and W.K. Hayman, Research problems in complex analysis, Bull. London Math. Soc 16 (1984), 490-517.

[3] Y.M. Chen and M.C. Liu, On Littlewood's conjectural inequalities, J. London Math. Soc. (2)1 (1969), 385-397.

[4] C.T. Chuang, Une généralisation d'un inégalité de Nevanlinna, Sci. Sinica 13 (1964), 887-895.

[5] Classical function theory problems, Bull. Amer. Math. Soc. 68 (1962), 21-24.

[6] D. Drasin, Quasiconformal modifications of functions having deficiency sum two, Annals of Math 114 (1981), 493-518.

[7] ———, Proof of a conjecture of F. Nevanlinna concerning functions with deficiency sum two, to appear in Acta Math.

[8] A. Edrei, Solution of the deficiency problem for functions of small lower order, Proc. Lond. Math. Soc. (3)26 (1973), 435-445.

[9] A. Edrei and W.H.J. Fuchs, On the growth of meromorphic functions with several deficient values, Trans. Amer. Math. Soc. 93 (1959), 292-328.

[10] ———————, Valeurs déficientes et valeurs asymptotiques des fonctions méromorphes, Comment. Math. Helv. 33 (1959), 258-295.

[11] A. Edrei and A. Weitsman, Asymptotic behavior of meromorphic functions with small deficiencies, Bull, Amer. Math. Soc. 74 (1968), 140-144.

[12] A.E. Eremenko and M.L. Sodin, On a hypothesis of Littlewood and the distribution of values of entire functions, (in Russian), Funktzional. anal. i ego priloženia 20 (1986), 7-22.

[13] G. Frank and G. Weissenborn, Rational deficient functions of meromorphic functions, Bull. London Math. Soc. (2)18 (1986), 29-33.

[14] W.H.J. Fuchs, Topics in Nevanlinna Theory, Proc. NRL Conf. on Classical Function Theory, Naval Research Laboratory, Washington (1970).

[15] A.A. Goldberg, The deficient values for entire functions of completely regular growth, Teor. Funkcii Funkcional Anal. i Priloženia 14 (1971) 88-101.

[16] A.A. Goldberg and I.V. Ostrowskii, The Distribution of Values of Meromorphic Functions, (in Russian), Nauka, Moscow (1970).

[17] W.K. Hayman, Meromorphic Functions, Oxford (1964).

[18] ———————, On a conjecture of Littlewood, Journal d'Analyse Math 36 (1979), 75-95.

[19] ———————, Research Problems in Function Theory, London 1967.

[20] S. Hellerstein and J. Williamson, Entire functions with negative zeros and a problem of R. Nevanlinna, J. Analyse Math 22 (1969), 233-267.

[21] J.E. Littlewood, On some conjectural inequalities with applications to the theory of integral functions, J. London Math. Soc. 27 (1952), 387-393.

[22] J. Miles and D.F. Shea, An extremal problem in value distribution theory, Quart. J. Math (2)24 (1973), 373-383.

[23] F. Nevanlinna, Über eine klasse meromorpher Funktionen, C.R. 7e Congr. Math. Scad. Oslo (1929), 81-83.

[24] R. Nevanlinna, Le théoreme de Picard-Borel et la théorie des fonctions méromorphes, Paris (1929).

[25] ———————, Analytic Functions, Springer.

[26] C.F. Osgood, Sometimes effective Thue-Siegel-Roth-Schmidt-Nevanlinna bounds or better, J. Number Th. 21 (1985), 347-389.

[27] A. Pfluger, Zur Defektrelation ganzer Funktionen endlicher Ordnung, Comment. Math. Helv. 19 (1946), 91-104.

[28] D.F. Shea, On the frequency of multiples values of a meromorphic function of small order, Mich. Math. J. 32 (1985), 109-116.

[29] N. Steinmetz, Eine Verallgemeinerung des zweiten Nevanlinnaschen Hauptsatzes, J. Reine Angew. Math. 368 (1986), 134-141.

[30] Yang Le, Deficient functions of meromorphic functions, Sci. Sinica 24 (1981), 1179-89.

[31] M. Ozawa, A method to a problem of R. Nevanlinna, Kodai Math. J. 8 (1985), 14-24.

Addendum July 1987: A few weeks ago J. Lewis and J.M. Wu proved Littlewood's conjecture (7). It is possible that the bound in (7) may be as strong as A log N; examples in [18] show that the latter bound would be sharp.

BERGMAN - SZEGŐ TYPE THEORY FOR CR STRUCTURES

Roman DWILEWICZ*

Contents

ABSTRACT

In this paper some directions of generalization of the Bergman theory and the Szegő theory for CR structures are given.

---

* Research supported by NSERC of Canada and le Ministère de l'Education du Québec.

INTRODUCTION

For domains in the complex vector space $\mathbf{C}^n$ and holomorphic functions defined on these domains, there is the well-known Bergman theory, i.e. the theory of the Bergman projection and the Bergman kernel function. For real hypersurfaces (at least for boundaries of domains in $\mathbf{C}^n$), and Cauchy-Riemann (CR) functions defined on the hypersurfaces, the corresponding theory is the Szegö theory.

The question of how far these theories can be extended for CR structures is a natural one. What type of difficulties and obstacles are encountered, and if the theory works for some CR manifolds, then what are the properties of the projection and the projection kernel.

The present paper does not pretend to answer all of the above questions. The purpose is to indicate some directions of generalizations and difficulties which appear.

For the Bergman theory, the projective space, i.e. the space $H^2(D)$, $D \subset \mathbf{C}^n$, is very easy to describe. This is the space of all square integrable holomorphic functions on D. For the Szegö projection, the corresponding space $H^2(\partial D)$ is the closure in $L^2(\partial D)$ of the subset $A(\partial D) \subset L^2(\partial D)$ of continuous boundary values of holomorphic functions in D. Every functions from $H^2(\partial D)$ has a holomorphic extension to D using the Poisson formula. For an arbitrary CR manifold M, as $H^2(M)$ it is natural to take the closure in $L^2(M)$ of the set of all square integrable CR functions or, if M is embedded in $\mathbf{C}^n$, the closure of the continuous boundary values of holomorphic functions. In general, the space $H^2(M)$ is very difficult to describe because this problem is related to the solvability of a suitable system of pdes or in the embedded case, to the geometric properties of M.

The next step is to examine properties of the orthogonal projection $\pi : L^2(M) \to H^2(M)$ and the projection kernel, which seems not to be trivial. Even in the case of compact 3-dimensional abstract CR manifolds many difficulties appear and the Szegö projection has "bad" properties (see for example the paper by Burns [B]). If a hypersurface bounds a relatively compact set in $\mathbf{C}^2$, then Boutet de Monvel and Sjöstrand [BMS] show that the Szegö projection is a pseudolocal Fourier integral operator, and hence maps $C^\infty(M)$ continuously into itself in the C^∞ topology. It is related to the global solvability of the ∂_b operator in the sense that the operator has a closed range. For more details see the papers of Boas and Shaw [BS] and Burns [B].

This paper is organized as follows. In §1 we give some notation, definitions, and integral formulas on holomorphic extensions of CR functions. In §2 we examine CR manifolds on which the Bergman type theory works. In §3 and §4 the same is done for

Szegö type projections.

AKNOWLEDGMENTS

First I would like to thank Professor Paul M. Gauthier of the Université de Montréal and Sandy Gauthier for their excellent hospitality during the academic year 1985/86. I am very grateful to the Centre de recherches mathématiques and the Département de Mathématiques et de Statistiques of the Université de Montréal for making possible my stay in the academic year 1985/86 during which this research was done.

§1. Definitions and notation

(a) Definition of CR manifolds and functions

Throughout this paper M denotes a C^∞ paracompact, separable manifold of real dimension m. Let $T(M)$ be the real tangent bundle and $\mathbf{C}T(M) = T(M) \otimes \mathbf{C}$ its complexification. By a CR structure of type ℓ on M we mean an ℓ - dimensional complex subbundle H of $\mathbf{C}T(M)$ which satisfies the following conditions:

(i) $H \cap \overline{H} = 0$, where the bar denotes the complex conjugation and 0 the zero section;

(ii) H is involutive, i.e. $[P, Q]$ is a section of H whenever P and Q are sections of H.

The manifold M together with a CR structure H is called a CR manifold or abstract CR manifold.

Let X be a complex manifold of complex dimension n. Assume that M is a C^∞ submanifold embedded into X which locally, in a coordinate neighborhood U, is given by

(1.1) $M \cap U = \{z \in U;\ \rho_1(z) = \dots = \rho_k(z) = 0\}$, $k = \operatorname{codim} M$, $k = 2n - m \leq n$.

If the set

(1.2) $\{z \in M \cap U;\ \bar\partial\rho_1 \wedge \dots \wedge \bar\partial\rho_k|_z = 0\}$

is empty for all U then the complex structure on X naturally generates a CR structure

on M, namely

$$(1.3)\quad H_p(M)=\{a_1\frac{\partial}{\partial\bar z_1}+\dots+a_n\frac{\partial}{\partial\bar z_n}\Big|_p;\ a_1\frac{\partial\rho_\alpha}{\partial\bar z_1}(p)+\dots+a_n\frac{\partial\rho_\alpha}{\partial\bar z_n}(p)=0,$$

$$\alpha=1,\dots,k\},$$

and we say that M is a <u>CR submanifold of X</u> (actually generically embedded CR submanifold of X). If the set (1.2) is nowhere dense in $M\cap U$ for all U, then we say that M is an <u>almost CR submanifold of X</u>.

By a <u>CR function f</u> on a CR or almost CR submanifold M on X we mean a locally integrable function such that for an arbitrary form $\varphi\in C^\infty{}_{n,\,n-k-1}(M)$ with compact support, the following equality holds

$$(1.4)\quad \int_M f\wedge\bar\partial\varphi=0$$

If M is a CR manifold with a CR structure H of type ℓ, then a CR function f on M is locally annihilated by a system $\{L_\alpha\}$ $\alpha=1,\dots,\ell$ of independent sections of H

$$(1.5)\quad L_\alpha f=\sum_{\beta=1}^{n}a_{\alpha\beta}(x)\frac{\partial f}{\partial x_\beta}(x)=0,\ \alpha=1,\dots,\ell,\ \ell=\dim_{\mathbf C}H,$$

which becomes the system of tangential Cauchy-Riemann equations for embedded CR submanifolds and in this case it is equivalent to (1.4).

(b) <u>Definition of q-concave CR manifolds</u>

Let M be a CR manifold generically embedded into $\mathbf{C}^n$ of real codimension k locally given by (1.1) Take a point $p\in M$. By $N_p(M)$ denote the normal space to M at p with the natural inner product in $\mathbf{R}^{2n}\simeq\mathbf{C}^n$. Without loss of generality we can assume that the system $\{\text{grad}\,\rho_\alpha\}_{\alpha=1,\dots,k}$ forms an orthonormal basis in $N_p(M)$ for $p\in M\cap U$.

By the <u>Levi form of the CR manifold M at a point $p\in M$</u> we mean the quadratic form $L_p(M)$ on the space $H_p(M)$ with the values in $N_p(M)$, given by the expression

$$L_p(M)(\zeta)=-\sum_{\gamma=1}^{n}\Big(\sum_{\alpha,\beta=1}^{n}\frac{\partial^2\rho_\gamma}{\partial z_\alpha\partial\bar z_\beta}(p)\,\zeta_\alpha\,\bar\zeta_\beta\Big)\,\text{grad}\,\rho_\gamma,$$

where

$$\zeta = \zeta_1 \frac{\partial}{\partial z_1} + \ldots + \zeta_n \frac{\partial}{\partial z_n} \in H_p(M).$$

By the <u>Levi form of the CR manifold M at $p \in M$ in the direction x</u>, $x = x_1 \operatorname{grad} \rho_1 + \ldots + x_k \operatorname{grad} \rho_k$, or simply $x = (x_1, \ldots, x_k)$, because ρ_α are fixed, we mean a scalar quadratic form on $H_p(M)$

$$L_p^x(M)(\zeta) = - \sum_{\gamma=1}^{k} x_\gamma \Big(\sum_{\alpha,\beta=1}^{n} \frac{\partial^2 \rho_\gamma}{\partial z_\alpha \partial \bar{z}_\beta}(p)\, \zeta_\alpha \bar{\zeta}_\beta \Big)$$

A CR manifold M is called <u>q-concave</u> (<u>weakly q-concave</u>) <u>in the direction x at a point $p \in M$</u> if the Levi form $L_p^x(M)$ has at least q negative (q non-positive) eigenvalues on $H_p(M)$. The CR manifold M is called <u>q-concave</u> (<u>weakly q-concave</u>) at the point $p \in M$ if it is q-concave (weakly q-concave) in all directions.

We say that M is <u>q-convex</u> (<u>weakly q-convex</u>) at the point p in the direction x if it is q-concave (weakly q-concave) in the direction $-x$.

(c) <u>Barrier functions for q-concave manifolds</u>

Following Airapetyan and Henkin [AH, §3.3], a vector-valued function $P = P(\zeta, z)$, $P \in C^1(\Omega \times \Omega)$, is called a <u>strongly regular barrier for the level surface of the real function</u> $\rho \in C^1(\Omega)$ if there exists a constant $\gamma > 0$ such that

$$P_j(\zeta, z) = \frac{\partial \rho(z)}{\partial \zeta_j} + O(|\zeta - z|)$$

where

$$|O(|\zeta - z|)| \leq \gamma |\zeta - z|$$

and

$$2 \operatorname{Re} \langle P(\zeta, z), \zeta - z \rangle \geq \rho(\zeta) - \rho(z) + \gamma |\zeta - z|^2$$

for $(\zeta, z) \in \Omega \times \Omega$, and where

$$\langle P(\zeta, z), \zeta - z \rangle = P_1(\zeta, z)(\zeta_1 - z_1) + \ldots + P_n(\zeta, z)(\zeta_n - z_n).$$

Denote by S the unit sphere in $\mathbf{R}^k$ ($k = \operatorname{codim}_{\mathbf{R}} M$) and by S_p the unit sphere in

the normal space $N_p(M)$. By the assumption, the system $\{\operatorname{grad} \rho_\alpha\}_{\alpha = 1, \dots, k}$ is orthonormal. Thus the sphere S_p can be written as

$$S_p = \{x_1 \operatorname{grad} \rho_1(p) + \dots + x_k \operatorname{grad} \rho_k(p);\ \ x = (x_1, \dots, x_k) \in S\}.$$

Put $\rho_x = x_1\rho_1 + \dots + x_k\rho_k$.

<u>Lemma 1.1</u> ([AH, Lemma 3.1.1]). Let M be a smooth CR manifold of the form (1.1) which is q-concave at $p \in M$ in all directions x of a closed set $Q \subset S$. Put

$$(1.6) \qquad \tilde{\rho}_x = \rho_x + A(\rho_1^2 + \dots + \rho_k^2)$$

Then there exists a constant $A > 0$ such that for all $x \in Q$ the form

$$-\sum_{\alpha,\beta=1}^{n} \frac{\partial^2 \tilde{\rho}_x(p)}{\partial z_\alpha \partial \bar{z}_\beta} \zeta_\alpha \bar{\zeta}_\beta, \quad \zeta = \zeta_1 \frac{\partial}{\partial z_1} + \dots + \zeta_n \frac{\partial}{\partial z_n} \in H_p(M),$$

has at least $q + k$ negative eigenvalues smaller than some $c < 0$.

(d) <u>Integral formulas for holomorphic extension of CR functions</u>

There are many papers devoted to holomorphic extension of CR functions (for a review of the results see for example [AH], [H_1]). Here we formulate a theorem which was first obtained by Boggess and Polking [BP] but we give a version which uses integral formulas and which was proved by Airapetyan and Henkin [AH, Th. 5.2.1].

For a smooth vector-valued function $\eta = \eta(\tau, z, t) = (\eta_1(\tau, z, t), \dots, \eta_n(\tau, z, t))$, depending on the variables $\tau, z \in \mathbf{C}^n$ and $t \in \mathbf{R}^k$, define the Leray form by

$$\omega'(\eta) = \sum_{\gamma=1}^{n} (-1)^{\gamma-1} \eta_\gamma \bigwedge_{j \neq \gamma} (d_t\eta_j + \bar{\partial}_\tau \eta_j + \bar{\partial}_z \eta_j)$$

Also we define the n-form $\omega(z)$ by

$$\omega(z) = dz_1 \wedge \dots \wedge dz_n.$$

Let M be a CR manifold given in (1.1). Fix a point $p_0 \in M$ and take a small strictly convex neighborhood D^0 of p_0 of the form

$$(1.7) \qquad D^0 = \{\zeta \in \Omega;\ \rho_0 < 0\}, \quad \rho_0 \in C^\infty(\Omega).$$

For a number $q \in \{0, 1, ..., n-k\}$ denote by $S_{p_o, q}$ the set of those points x of the unit sphere $S \subset \mathbf{R}^k$ for which the Levi form $L^x_{p_o}(M)(\zeta)$, $\zeta \in H_{p_o}(M)$, has less than q negative eigenvalues.

By a spherical segment in S we mean the subset

$$S(a, b) = \{x \in S;\ a_1 x_1 + ... + a_k x_k > b\},$$

where $a = (a_1, ..., a_k) \in \mathbf{R}^k$ and $b > 0$.

Now we are in a position to formulate the holomorphic extension theorem for CR functions.

Theorem 1.2. ([AH, Th. 5.2.1]). Let M be a smooth CR submanifold of $\mathbf{C}^n$ given by (1.1) and let $\mathcal{U}(S_{p_o, 1})$ be a simply connected neighborhood of $S_{p_o, 1}$ entirely contained in a spherical segment $S(a, b)$. Let D^o be a sufficiently small neighborhood of p_o of the form (1.7) on which there exists a strongly regular vector-valued barrier function $P_x = P(\zeta, z, x)$ for the level surface of ρ_x (ρ_x given by (1.6)) of class C^∞ with respect to $\zeta, z \in D^o$ and C^1 with respect to x.

Then every CR function $f \in C^\infty(M \cap D^o)$ can be represented in the form

$$\frac{(2\pi i)^n}{(n-1)!} f = F$$

where

$$(1.8) \quad F = K_{\mathcal{U}} f + K_o f$$

$$(1.9) \quad K_{\mathcal{U}} f(z) = \int_{(M \cap D^o) \times \mathcal{U}} f(\zeta)\, \omega'\left(\frac{P(\zeta, z, x)}{<P(\zeta, z, x), \zeta - z>}\right) \wedge \omega(\zeta),$$

$$(1.10) \quad K_o f(z) = \int_{\partial(M \cap D^o) \times \tilde{S}} f(\zeta)\, \omega'\left(t \frac{P_o(\zeta, z)}{<P_o(\zeta, z), \zeta - z>} + (1-t) \frac{P(\zeta, z, x)}{<P(\zeta, z, x), \zeta - z>}\right) \wedge \omega(\zeta),$$

$P_o(\zeta, z) = (\partial/\partial\zeta)\, \rho_o(\zeta)$, $\tilde{S} = \{(t, (1-t)x);\ 0 \le t \le 1,\ x \in S\}$.

The function F is holomorphic in the domain

$$D_{\mathcal{U}} = \{\zeta \in \Omega;\ \tilde{\rho}_x(\zeta) > 0,\ \zeta \in U\} \cap \tilde{D}, \tag{1.11}$$

where $\tilde{D}$ is a small neighborhood of $M \cap D^o$, and $F \in C^\infty(D_{\mathcal{U}} \cup (M \cap D^o))$.

As a corollary of the above theorem we obtain

<u>Theorem 1.3</u>. ([AH, Th. 5.2.2]). Suppose that under the conditions of Th. 1.2 the set $S_{p_o, 1}$ is empty, that is, M is 1-concave at the point p_o. Then every CR function $f \in C^\infty(\overline{M \cap D^o})$ can be holomorphically extended to a neighborhood $\tilde{D}$ of $M \cap D^o$ by the formula

$$\frac{(2\pi i)^n}{(n-1)!} f(z) = K_o f(z), \tag{1.12}$$

where $K_o f(z)$ is given in (1.10).

<u>Remark 1.4</u>. Theorems 1.2 and 1.3 can be generalized for integrable CR functions; see Airapetyan and Henkin [AH], remark after Th. 5.2.2.

<u>§2. Bergman type theory for CR manifolds</u>

(a) <u>Bergman and Szegö projections</u>

The Bergman and the Szegö theories for domains of the complex vector space $\mathbf{C}^n$ are well-known. We recall them in a few words.

Let Ω be a domain in $\mathbf{C}^n$ and assume, for simplicity, that Ω is compact with C^1 boundary $\partial\Omega$. Denote by $L^2(\Omega)$, $L^2(\partial\Omega)$ the Hilbert spaces of square integrable functions on Ω and $\partial\Omega$, respectively with the natural inner products. Let $H^2(\Omega)$ be the subspace of $L^2(\Omega)$ of square integrable holomorphic functions on Ω. This subspace is closed. By $H^2(\partial\Omega)$ denote the closure in $L^2(\partial\Omega)$ of the subset of continuous bundary values on $\partial\Omega$ of holomorphic functions in Ω. Each element f of $H^2(\partial\Omega)$ has a holomorphic extension F to Ω given by its Poisson integral (see [K]). Consider the following functionals

$$H^2(\Omega) \ni F \to F(z), \quad z \in \Omega, \tag{2.1}$$

$$H^2(\partial\Omega) \ni f \to F(z), \quad z \in \Omega, \tag{2.2}$$

where F in (2.2) is the holomorphic extension of f. Both functionals are continuous because of the obvious inequalities

$$(2.3) \quad |F(z)| \le C_z \left(\int_\Omega |F(\zeta)|^2 dV_\zeta \right)^{1/2}, \; z \in \Omega$$

$$(2.4) \quad |F(z)| \le \tilde{C}_z \left(\int_{\partial\Omega} |f(\zeta)|^2 d\sigma_\zeta \right)^{1/2}, \; z \in \Omega$$

where the constants C_z, $\tilde{C}_z$ depend only on z. Notice that the inequality (2.4) is not true, in general, for $z \in \partial\Omega$.

Using the standard methods of the Bergman and the Szegö theories, the orthogonal projections

$$B : L^2(\Omega) \to \quad H^2(\Omega)$$

$$S : L^2(\partial\Omega) \to \quad H^2(\Omega)$$

can be written using integral formulas.

(b) Continuity of some functionals

Let (M, H) be an abstract CR manifold. Take an arbitrary volume element on M, say σ, which is a global function on M such that, in local coordinates $x = (x_1, ..., x_m)$ on $V \subset M$, σ can be written

$$\sigma(x) = \sigma_o(x) \mid dx_1 \wedge ... \wedge dx_m \mid, \quad \sigma_o > 0$$

i.e. for any m-tuple of tangent vectors at x to M, say $v_1, ..., v_m$,

$$\sigma(x)(v_1, ..., v_m) = \sigma_o(x) \mid dx_1 \wedge ... \wedge dx_m (v_1, ..., v_m) \mid,$$

where σ_o is a smooth positive function on V. Such smooth volume elements always exist on an arbitrary paracompact manifold.

Let us fix a volume element σ on M. Let $\{U_i\}_{i \in I}$ be a locally finite covering of M by coordinate neighborhoods and assume that $\{\varphi_i\}_{i \in I}$ is a partition of unity corresponding to the covering. Take a complex valued function f defined on M and put

$$\int_M f d\sigma = \sum_{i \in I} \int_{\mathcal{U}_i} \varphi_i f \sigma_i = \sum_{i \in I} \int_{\tilde{\mathcal{U}}_i} \varphi_i f \sigma_i^{o} \, dx_1 \ldots dx_m$$

where

$$\sigma_i = \sigma_i^{o} \, | dx_1^i \wedge \ldots \wedge dx_m^i | , \; \sigma_i^{o} > 0$$

and $\tilde{\mathcal{U}}_i$ is the image of $\mathcal{U}_i$ under the coordinates $(x_1, \ldots, x_m)$. Of coursethe integral $\int f d\sigma$ is well defined for functions f such that $\varphi_i f \sigma_i^{o}$ is measurable on $\tilde{\mathcal{U}}_i$, $i \in I$, and if the sums of the above equalities make sense for arbitrary coverings $\{\mathcal{U}_i\}_{i \in I}$ and partitions of unity $\{\varphi_i\}_{i \in I}$.

Define the space $L^2(M)$ as the space of all measurable functions f such that $\int_M |f|^2 d\sigma < \infty$ with the inner product

$$<f, g> = \int_M f \bar{g} \, d\sigma, \quad f, g \in L^2(M)$$

and the corresponding norm

$$\| f \| = \left(\int_M |f|^2 d\sigma \right)^{\frac{1}{2}}$$

This inner product makes $L^2(M)$ a Hilbert space. Notice that, in general, the space depends on the volume element σ.

Let

$CR_\infty(M)$ be the space of smooth CR functions on M;

$CR^2_\infty(M)$ be the space of smooth square-integrable CR functions on M;

$CR^2(M)$ be the closure of $CR^2_\infty(M)$ in the square norm.

We may define the following projection

$$\pi_{CR} : L^2(M) \to CR^2(M).$$

In such a general situation it is very difficult to study properties of this projection, because the space $CR^2(M)$ is rather difficult to describe. The subspace of smooth elements of

$CR^2(M)$ depends on solvability (or non-solvability) of a suitable system of linear first order pdes (1.5). In general, little is known about solvability of such systems (see for example $[N_1]$, $[N_2]$, $[T_1]$, $[T_2]$, $[T_3]$).

In the extremal situations the space $CR^2(M)$ is equal to $L^2(M)$ or the space of constant functions (if the volume of M is finite) or even to the space consisting of only the zero function (if the volume of M is infinite). In the first case the projection π_{CR} is the identity or in the integral form is given by the Dirac delta

$$\pi_{CR} f(z) = \int_M \delta_z f \, d\sigma$$

In the second case if we project $L^2(M)$ onto the space of constant functions, then the integral form is

$$\pi_{CR} f(z) = \frac{1}{\mathrm{vol}(M)} \int_M f \, d\sigma$$

and the kernel function $P(z, \bullet)$ is the constant function equal to $\frac{1}{\mathrm{vol}(M)}$. If the $\mathrm{vol}(M) = \infty$ then the function P is the zero function.

In the first case the kernel function does not belong to $\pi_{CR}(L^2(M))$; in the second case yes. The properties of P in the extremal cases are completely different. For general CR manifolds, one can expect that the properties of the kernel function "lie between the properties of the extremal cases".

In this subsection we examine more carefuly the situation when the functionals

$$(2.5) \quad CR^2_\infty(M) \ni f \to f(p), \; p \in M,$$

are continuous in the square norm.

From the continuity of the functionals (2.5) the following inequality holds

$$(2.6) \quad |f(p)| \le C_p \Big(\int_M |f(\zeta)|^2 \, d\sigma \Big)^{1/2}, \quad f \in CR^2_\infty(M), \quad p \in M,$$

where C_p is a constant depending only on the point p and independent of the functions f.

Notice that from the inequlaity (2.6) the maximum principle for smooth square integrable CR functions follows, for if not, then we can find a function $g \in CR^2_\infty(M)$ such that

$$g(p_o) = 1 \quad \text{and} \quad |g(p)| < 1 \quad \text{for } p \neq p_o.$$

Taking the sequence of smooth CR functions defined by

$$h_k(z) = g^k(z)$$

we see that

$$1 = |g^k(p_o)| \leq C_{p_o} \left(\int_M |g|^{2k} d\sigma\right)^{1/2} \underset{k \to \infty}{\longrightarrow} 0,$$

which gives the contradiction.

Def. 2.1. Let M be a CR manifold. We say that CR functions on M satisfy the local weak (resp. strong) maximum modulus principle iff: given any open connected set V in M and any differentiable CR non-constant function u on V, then $|u|$ cannot have a weak (resp. strong) local maximum at any point of V.

For CR manifolds embedded into $\mathbf{C}^n$, there is a nice characterization of such CR manifolds for which CR functions satisfy the local maximum modulus principle. The characterization was conjectured by Ellis, Hill and Seabury [EHS] and they proved the necessary condition. The sufficiency of the condition was proved by Iordan [I]. Also there is a nice paper by Sibony [S]. Of course the local maximum principle implies the global one.

We formulate a weaker version which will be useful in our considerations.

Theorem 2.2. (see [EHS], [I], [S]). Let M be a CR manifold embedded into $\mathbf{C}^n$. Then

(i) If CR functions on M satisfy the local weak maximum modulus principle, then M is weakly 1-concave.

(ii) If M is 1- concave, then CR functions on M satisfy the local strong maximum principle.

In the next subsection we examine more precisely the Bergman type theory for

1-concave CR manifolds.

(c) The case of 1–concave CR manifolds

From the considerations in subsection (b) it follows that there is some hope that the Bergman type theory for embedded CR manifolds works when the manifolds are 1-concave. This class of CR manifolds has interesting properties. The main property was given in §1 (d), namely any CR function on M can be holomorphically extended to a neighborhood of M in $\mathbf{C}^n$. The local extension of CR functions is given by the formula (1.12).

Let M be a 1-concave CR submanifold of $\mathbf{C}^n$. Take an arbitrary point $p \in M$. Using the integral formula (1.12) we obtain immediately the inequality

$$(2.7) \quad |F(p)| \le \tilde{C}_p \int_M |f(\zeta)|\, d\sigma, \; f \in CR_\infty(M) \cap L^1(M),$$

where the constant $\tilde{C}_p$ does not depend on f and F is the holomorphic extension of f to a small neighborhood Ω of M in $\mathbf{C}^n$ and $p \in \Omega$.

If the volume of M is finite and $f \in CR^2_\infty(M)$ then the Hölder inequality gives

$$(2.8) \quad |F(p)| \le C_p\Big(\int_M |f(\zeta)|^2 d\omega_\zeta\Big)^{1/2}, \quad f \in CR^2_\infty(M)$$

or in other words the functionals

$$(2.9) \quad CR^2_\infty(M) \ni f \longrightarrow F(p)$$

are continuous on the space $CR^2_\infty(M)$, for all $p \in \Omega$. In this special case the space $CR^2(M)$ can be easily described, namely

$$(2.10) \quad CR^2(M) = CR^2_\infty(M)$$

Using the standard procedure of the Bergman theory we obtain the reproducing kernel and the integral formula for square integrable CR functions:

$$(2.11) \quad f(p) = \int_M P(p,\zeta) f(\zeta)\, d\sigma_\zeta ,$$

where the volume element σ is generated by Lebesgue measure on $\mathbf{C}^n \simeq R^{2n}$. Of course the right-hand side of (2.11) makes sense for arbitrary square integrable functions on M. The projection π_{CR} can be written in the form

$$\pi_{CR}\, g(p) = \int_M P(p,\zeta)\, g(\zeta)\, d\sigma_\zeta\,, \quad g \in L^2(M).$$

We list the main properties of the projection π_{CR} and the projection kernel $P = P(p, \zeta)$.

1. For an arbitrary function $g \in L^2(M)$ we have $\pi_{CR} g \in CR^2(M)$ and consequently $\pi_{CR} g$ is the restriction to M of a holomorphic function defined in a neighborhood of M.

2. The functionals in (2.9) are continuous for all $p \in \Omega$, where Ω is a neighborhood of M in $\mathbf{C}^n$. Thus the function

$$M \ni \zeta \to \overline{P(p,\zeta)}$$

belongs to $CR^2(M)$ for all $p \in \Omega$ and consequently is holomorphically extendable to a neighborhood of M.

3. The kernel $P(p, \zeta)$ is conjugate symmetric $P(p, \zeta) = \overline{P(\zeta, p)}$, because

$$\int_M P(p,\zeta)\, \overline{P(q,\zeta)}\, d\sigma_\zeta = \overline{P(q,p)}$$

On the other hand

$$\int_M P(p,\zeta)\, \overline{P(q,\zeta)}\, d\sigma_\zeta = \int_M \overline{P(q,\zeta)\, \overline{P(p,\zeta)}}\, d\sigma_\zeta =$$

$$= \overline{\overline{P(p,q)}} = P(p,q).$$

4. The kernel $P = P(p, \zeta)$ is uniquely determined by the properties that it is an element of $CR^2(M)$ in p, is conjugate symmetric, and reproduces $CR^2(M)$. Let $Q(p, \zeta)$ be another such kernel. Then

$$P(p,\zeta) = \overline{P(\zeta,p)} = \int_M Q(p,t)\,\overline{P(\zeta,t)}\,d\sigma_t =$$

$$= \overline{\int_M P(\zeta,t)\,\overline{Q(p,t)}\,d\sigma_t} = \overline{\overline{Q(p,\zeta)}} = Q(p,\zeta)$$

5. Here we show that under some assumptions a reproducing kernel for holomorphic functions can be obtained from a family of reproducing kernels for CR functions.

Let $\{M_t\}_{t \in T}$ be a family of smooth 1-concave CR submanifolds of $\mathbf{C}^n$ with the following properties

(i) T is a domain in $\mathbf{R}^k$, $k = \operatorname{codim}_{\mathbf{R}} M_t$;

(ii) $M_t \cap M_s = \emptyset$ for $t \neq s$;

(iii) $G = \bigcup_{t \in T} M_t$ is a domain in $\mathbf{C}^n$;

(iv) any CR function f defined on M_t can be holomorphically extended onto G;

(v) the family $\{M_t\}_{t \in T}$ is given locally by the equations

$$S_1(\zeta, t) = 0, \ldots, S_k(\zeta, t) = 0$$

where S_α are smooth real-valued functions and such that

$$\operatorname{rank}\left(\frac{\partial S_\alpha}{\partial t_\beta}\right)_{\alpha,\beta=1,\ldots,k} = k \text{ at each } (\zeta, t) \in G \times T;$$

(vi) the functionals $CR^2(M_t) \ni f \to f(p)$, $p \in G$ are continuous.

Take a holomorphic function F on G and denote

$$f(\bullet, t) = F/M_t$$

Assume moreover that $f(\bullet, t)$ belongs to $CR^2(M_t)$ for $t \in T$. Using the above properties we can write the following formula

$$(2.12) \quad F(p) = \int_{M_t} Q(p,\zeta,t)\, f(\zeta,t)\, d\sigma_t(\zeta), \quad p \in G,$$

where $Q(p, \bullet, t) \in CR^2(M_t)$ for each $p \in G$, $t \in T$. Integrating both sides of (2.12) over T we obtain

$$(2.13) \quad \int_T F(p)\,dt = \int_T \int_{M_t} Q(p,\zeta,t)\, f(\zeta,t)\, d\sigma_t(\zeta)\, dt$$

In general, one cannot expect that $d\sigma_t(\zeta)dt$ is the volume element on G, but, from the property (v), there exists a nonvanishing function $\mu = \mu(\zeta, t)$ such that

$$\mu(\zeta, t)\, d\sigma_t(\zeta)\, dt = dV(\zeta, t),$$

where $dV(\zeta, t)$ is the volume element on $G \subset \mathbf{C}^n$. Therefore (2.13) can be written

$$(2.14) \quad F(p) = \frac{1}{|T|} \int_G Q(p,z) \frac{1}{\mu(z)} f(z)\, dV(z), \quad z \in G,\ z \sim (\zeta, t).$$

Notice that (2.14) gives a reproducing formula for such holomorphic functions F on G for which $F|_{M_t} \in CR^2(M_t)$, $t \in T$.

§3. Szegö type projections for CR manifolds

In the previous section we have seen that the Bergman theory can be transformed for a sufficiently narrow class of CR manifolds, namely 1-concave manifolds. Here we examine more carefuly the Szegö type theory - what class of CR manifolds it works for?

Let M be a generic CR manifold embedded into $\mathbf{C}^n$. Take $f \in CR^2_\infty(M)$. We are interested in continuous functionals on the space $CR^2(M)$, which involve values of the functions f or their (holomorphic) extension. If the CR manifold M is Levi flat then there is no chance that such functionals could be continuous. In this case f is not, in general, holomorphically extendable in any direction. We have to put some additonal convexity conditions on M, namely 1-concavity of M, at least in some directions (see §1 (b)). More precisely, assume that M is a smooth CR submanifold of $\mathbf{C}^n$ locally given by (1.1) and such that for each point $p_0 \in M$ the set $S_{p_0, 1}$ (see §1 (d)) is entirely contained in a spherical segment $S(a, b)$. Then, using Th. 1.2, there exists an open subset Ω_M of $\mathbf{C}^n$ such that the boundary $\partial\Omega_M$ contains M and any smooth CR function on M extends holomorphically to Ω_M and such that the extended function is smooth on $M \cup \Omega_M$.

Using (1.8) - (1.10), the properties of the barrier function

$P = P(\zeta, z, x)$ and 1-concavity of M in suitable directions, the following estimation can be obtained immediatey

$$(3.1)\quad |F(z)| \le \tilde{C}_z \int_M |f(\zeta)|\, d\sigma_\zeta\,,\ f \in CR_\infty(M) \cap L^1(M),\quad z \in \Omega_M,$$

where F holomorphically extends f to Ω_M and the constant $\tilde{C}_z$ does not depend on f. If moreover the volume of M is finite then for $f \in CR_\infty(M)$ the above inequality gives

$$(3.2)\quad |F(z)| \le C_z \left(\int_M |f(\zeta)|^2\, d\sigma_\zeta\right)^{1/2}$$

Using Remark 1.4 inequality (3.2) remains true for $f \in CR^2(M)$. Under the above notation, take the following continuous functional $CR^2(M) \ni f \to F(z)$, $z \in \Omega_M$. This functional can be written in the form

$$F(z) = \int_M P(z, \zeta) f(\zeta)\, d\sigma_\zeta\,,\ f \in CR^2(M),$$

where the function

$$M \ni \zeta \to \overline{P(z, \zeta)},\ z \in \Omega_M,$$

belongs to the class $CR^2(M)$ for each $z \in \Omega_M$. Denote

$$(3.3)\quad (Pg)(z) = \int_M P(z, \zeta) g(\zeta)\, d\sigma_\zeta,\quad g \in L^2(M),\quad z \in \Omega_M\,.$$

Now we list some properties of the projection P and the projection kernel.

1. For fixed $z \in \Omega_M$ the function $\overline{P(z, \bullet)}$ belongs to the class $CR^2(M)$ and therefore is holomorphically extendable to the domain Ω_M.

2. The kernel $P(z, \zeta)$ is conjugate symmetric i.e. $P(z, \zeta) = \overline{P(\zeta, z)}$ for $(z, \zeta) \in \Omega_M \times \Omega_M$.

3. The kernel $P = P(z, \zeta)$ is uniquely determined by the properties that it is an element of $CR^2(M)$ in ζ, is conjugate symmetric (as in 2) and reproduces $CR^2(M)$, (the proof is standard).

The projection given in (3.3) is a generalization of the Szegö type projection for

boundaries of domains in $\mathbf{C}^n$, which are 1-concave in the outward direction.

§4. Szegö type projection for boundaries of domains in 1-concave CR manifolds

In the previous section we have shown a way of generalization of Szegö type theory for CR submanifolds which are convex in some directions. These submanifolds are not necessarily the boundaries of higher dimensional manifolds. It is natural to ask if there is another way of generalization, namely, for boundaries of domains in CR manifolds. The answer is in part positive.

Let M be a smooth CR submaniflod of $\mathbf{C}^n$, $\dim_{\mathbf{R}} M = m$. Take an open domain D in M with smooth boundary ∂D. In general, ∂D is not a CR submanifold of $\mathbf{C}^n$, but it is natural to assume that ∂D is an almost CR submanifold; for the definition see §1 (a). Define $B(\overline{D})$ to be the space of functions continuous on $\overline{D}$ which are CR in D;

$CR^2_B(\partial D)$ to be the $L^2(\partial D)$ closure of the elements of $B(\overline{D})$ restricted to ∂D.

Put $\pi_{CR} : L^2(\partial D) \to CR^2_B(\partial D)$ the orthogonal projection of $L^2(\partial D)$ onto $CR^2_B(\partial D)$.

In the general case it is very hard to give a more exact description of the space $CR^2_B(\partial D)$ or even $B(\overline{D})$. But for domains of 1-concave CR manifolds, the Bochner-Hartogs phenomenon occurs and, as a consequence, it is much easier to describe the space $B(\overline{D})$.

Now we formulate the theorem which was proved by Henkin in 1984.

<u>Theorem 4.1</u> ([H_1, Th. 7.12]). Let Ω be a pseudoconvex domain in $\mathbf{C}^n$ and $\rho_1, \ldots, \rho_k$ real valued smooth functions on Ω such that $\overline{\partial}\rho_1 \wedge, \ldots, \wedge \overline{\partial}\rho_k \neq 0$ on Ω and such that all CR manifolds $M_\varepsilon = \{z \in \Omega;\ \rho_1(z) = \varepsilon_1, \ldots, \rho_k(z) = \varepsilon_k\}$, $\varepsilon = (\varepsilon_1, \ldots, \varepsilon_k)$ are 1-concave.

Let U be a relatively compact domain of Ω

$$U = \{z \in \Omega;\ \rho(z) < 0\}$$

given by a smooth function ρ and such that for some $\varepsilon^o = (\varepsilon^o_1, \ldots, \varepsilon^o_k)$ the set $D = M_{\varepsilon^o} \cap U$ has connected complement in M_{ε^o} with the boundary ∂D being an almost CR manifold. Then any continuous CR function f on ∂D can be extended to a continuous CR function F on D.

If the domain D in the above theorem is sufficiently small, then the extension of CR functions from the boundary is given explicitely by the formula (1.12). Moreover the formula works for square integrable CR functions on ∂D, and consequently the space $CR^2_B(\partial D)$ is known.

The Bochner-Hartogs effect for domains in CR manifolds is described in greater detail in the papers by Henkin {H_1, §7.4], [H_2] and Airepetyan and Henkin [AH].

Using standard methods of the Szegö theory, the corresponding theory for boundaries of domains, which satisfy assumptions of Th. 4.1., in 1-concave CR manifolds can be given. In the general case, when the space $CR^2_B(\partial D)$ is very hard to describe, it seems much more difficult to study properties of the orthogonal projection.

REFERENCES

[AH] R.A. Airapetyan and G.M. Henkin, Integral representations of differential forms on CR-manifolds and the theory of CR-functions. Uspekhi Mat. Nauk vol. 39:3 (1984), 39-106; English transl.: Russian Math. Surveys 39:3 (1984), 41-118.

[B] D.M. Burns, Jr., Global behavior of some tangential Cauchy-Riemann equations. In: Partial differential equations and geometry. Proc Park City Conf., pp. 51-56. New York: Dekker 1979.

[BMS] L. Boutet de Monvel and J. Sjöstrand, Sur la singularité des noyaux de Bergman et de Szegö. Soc. Math. Fr. Astérisque 34-35 (1976), 123-164.

[BP] A. Boggess and T.C. Polking, Holomorphic extension of CR-functions. Duke Math. J. 49 (1982), 757-784.

[BS] H.P. Boas and M.-C. Shaw, Sobolev estimates for the Lewy operator on weakly pseudo-convex boundaries. Math. Ann. 274 (1986), 221-231.

[D] R. Dwilewicz, Some remarks on maximum type principles for solutions of linear pdes in two and three variables. (to appear).

[EHS] D. Ellis, C.D. Hill and C. Seabury, The maximum modulus principle I. Necessary conditions. Indiana Univ. Math. J. 25 (1976), 709-715.

[H_1] G.M. Henkin, Methods of integral representations in complex analysis (Russian).

In: Current problems in mathematics - "Complex analysis - several variables" I, vol. 7 (1985), 23-124. (English translation in preparation).

[H_2] G.M. Henkin, Hartogs-Bochner effect on CR manifolds. (Russian). Dokl. Akad. Nauk 274:3 (1984), 553-558.

[I] A. Iordan, The maximum modulus principle for CR functions. Proc. A.M.S. 96 no. 3 (1986), 465-469.

[K] S.G. Krantz, Function theory of several complex variables. John Wiley, New York 1982.

[N_1] L. Nirenberg, Lectures on linear partial differential equations. Regional Conference Series in Mathematics, no. 17, A.M.S. (1973).

[N_2] L. Nirenberg, Ob odnom zadache Hans Lewy. (Russian). Upsekhi Math. Nauk 29:2 (1974), 241-251; English transl.: Russian Math. Surveys 29:2 (1974), 251-262.

[S] N. Sibony, Principe du maximum sur une variété C.R. et équations de Monge-Ampere complexes. Séminaire P. Lelong 1975/76, Lecture Notes in Math. 578 (1977), 14-27.

[T_1] F. Treves, Lectures on linear pdes. Proceedings of the 1980 Beijing Symposium on Differential Geometry and Differential Equations. Science Press, Beijing, China 1982. Gordon and Breach, Science Publishers, Inc., New York, vol. II, pp. 861-928.

[T_2] F. Treves, Remarks about certain first-order linear pde in two variables. Comm. in Partial Diff. Equat. 5(4), (1980), 381-425.

[T_3] F. Treves, On the local solvability and the local integrability of systems of vector fields. Acta Math. 151 (1983), 1-48.

Institute of Mathematics, Warsaw University, PKiN 1Xp, 00-901 Warsaw, Poland and Centre de recherches mathématiques, Université de Montréal, C.P. 6128, Succ. A, Montréal, QC, H3C 3J7, Canada.

1980 Mathematics Subject Classification (1985) Revision. Primary 32F25, 32H10; Secondary 32F10, 35F05.

Key words and phrases. Bergman theory, Szegö theory, CR structures.

REGULAR COMPLEX GEODESICS FOR THE DOMAIN $D_n = \{(z_1, \ldots, z_n) \in \mathbb{C}^n : |z_1| + \ldots + |z_n| < 1\}$.

by
Graziano Gentili

1. Introduction.

Let $Hol(\Delta,D)$ be the set of all holomorphic maps from the open unit disc Δ of $\mathbb{C}$ into a domain $D \subset \mathbb{C}^n$. A map $f \in Hol(\Delta,D)$ is a complex geodesic of D for the Carathéodory (or Kobayashi) pseudo-distance, if f is an isometry with respect to the Poincaré distance on Δ and the Carathéodory (or Kobayashi) pseudo-distance on D, [V1].

Several results have been obtained in the study of existence and uniqueness of complex geodesics joining two given points of a (linearly) convex domain, [V2], [L], [R-W], on which the Carathéodory and the Kobayashi pseudo-distances coincide [L], [D-T-V].

In particular, if D is bounded and convex, then for any two given points of D there exists at least one complex geodesic whose range contains both of them, and, if D is bounded and strictly convex, then the complex geodesic joining any two given points is unique (up to a Moebius transformation). Moreover, the character of the non-uniqueness has been investigated [G1], [G2].

The complex geodesics of a convex domain D have been characterized as the solutions of an extremum problem [L], [R-W]. In the case of a bounded comvex domain D for which a normal unit vector $\nu(z)$ is defined for $z \in \partial D$, this characterization turns out to be the following:

f is extremal, if there exists a positive function $p : \partial\Delta \to \mathbb{R}^+$ such that the map $\partial\Delta \to \mathbb{C}^n$

$$\xi \mapsto \xi p(\xi)\overline{\nu(f(\xi))}$$

can be extended to a holomorphic map of Δ into $\mathbb{C}^n$.

Notwithstanding the above results and the fact that a direct connection has been established between the complex geodesics of D and the foliation associated to a solution of the complex Monge-Ampère equation for the domain D, [L], [P], it is not clear in general how to determine explictly the family of all complex geodesics of D, and in fact the results in this direction are very few [V2], [V1], [G1].

In this paper, as an application of the extremal characterization of Lempert and Royden-Wong [L], [R-W], we explictly determine all regular (i.e., continuous up to the boundary of Δ) complex geodesics for the domain $D_n = \{(z_1,\dots,z_n) \in \mathbb{C}^n : |z_1| + \dots + |z_n| < 1\}$. The explicit study of the complex geodesics of the domain $D_n^{\alpha} = \{(z_1,\dots,z_n) \in \mathbb{C}^n : |z_1|^{\alpha} + \dots + |z_n|^{\alpha} < 1\}$, for $\alpha > 1$, has been carried out by Polekii [POL], from a different point of view in which smoothness of ∂D is required.

2. Preliminaries.

Let $\Delta = \{z \in \mathbb{C} : |z| < 1\}$. For $p = 1,2,\dots,\infty$ the symbol $H^p(\Delta)$ will denote, as usual, the Hardy space of all holomorphic functions $f \in \mathrm{Hol}(\Delta,\mathbb{C})$ such that $\|f\|_p < \infty$.

If D is a domain in $\mathbb{C}^n$, K_D and C_D will denote, respectively, the Kobayashi and Carathéodory pseudo-distances and k_D, γ_D the Kobayashi and Carathéodory pseudo-differential metrics associated to D (cf. e.g. [F-V]). If D is convex it turns out that

$$C_D \equiv K_D \ , \ k_D \equiv \gamma_D$$

(see e.g. [L], [D-T-V]).

Let $f : \Delta \to D$ be a holomorphic map. Then f is a complex geodesic for C_D if, and only if, there exist $\xi_0 \neq \xi_1 \in \Delta$ such that

$$C_D(f(\xi_0) \ , \ f(\xi_1)) = \omega(\xi_0,\xi_1),$$

or there exists $\xi_0 \in \Delta$ such that

$$\gamma_D(f(\xi_0);f'(\xi_0)) = \langle 1 \rangle_{\xi_0}$$

where ω and $\langle \ \rangle$ denote the Poincaré distance and the Poincaré differential metric of Δ.

Any two complex geodesics f and g (for C_D or K_D) have the same range if, and only if, there exists a Moebius transformation m of Δ such that $f = g \circ m$ (see e.g. [V2]). Therefore, the complex geodesics f and g will be said to be equal if they differ by a Moebius transformation.

Let us consider now the domain $D = \{(z_1,\dots,z_n) \in \mathbb{C}^n : |z_1| + \dots + |z_n| < 1\}$. The boundary of D consists of all the points $(z_1,\dots,z_n) \in \mathbb{C}^n$ such that $|z_1| + \dots + |z_n| = 1$. Setting

$$z_1 = r_1 e^{i\theta_1}$$
$$\cdots$$
$$\cdots$$
$$z_n = r_n e^{i\theta_n}$$

with $r_1, \dots r_n \in \mathbb{R}^+$ and $\theta_1, \dots, \theta_n \in \mathbb{R}$, we obtain

$$\partial D = \{(r_1 e^{i\theta_1}, \dots, r_n e^{i\theta_n}) : 0 \le r_i \le 1, (i = 1, \dots, n), \sum_{i=1}^{n} r_i = 1\}.$$

If Π is defined as the set

$$\{(r_1, \dots, r_{n-1}) \in \mathbb{R}^{n-1} : r_i \ge 0, (i = 1, \dots, n-1), \sum_{i=1}^{n-1} r_i \le 1\},$$

then the boundary of D is the image of $\Pi \times [0, 2\pi]^n$ by the map

$$(r_1, \dots, r_{n-1}, \theta_1, \dots, \theta_n) \overset{F}{\longmapsto} (r_1 e^{i\theta_1}, \dots, r_{n-1} e^{i\theta_{n-1}}, (1 - \sum_{i=1}^{n-1} r_i) e^{i\theta_n}).$$

The partial derivatives

$$\frac{\partial F}{\partial r_s}(r_1, \dots, r_{n-1}, \theta_1, \dots, \theta_n) = (0, \dots, e^{i\theta_s}, \dots, -e^{i\theta_n})(s = 1, \dots, n-1)$$
$$\frac{\partial F}{\partial \theta_s}(r_1, \dots, r_{n-1}, \theta_1, \dots, \theta_n) = (0, \dots, ir_s e^{i\theta_s}, \dots, 0)(s = 1, \dots, n-1)$$
$$\frac{\partial F}{\partial \theta_n}(r_1, \dots, r_{n-1}, \theta_1, \dots, \theta_n) = (0, \dots, 0, \dots, i(1 - \sum_{s=1}^{n-1} r_s) e^{i\theta_n}$$

are $(n-1)$ vectors tangent to ∂D at the point $(r_1 e^{i\theta_1}, \dots, r_n e^{i\theta_n})$, and therefore the 1-dimensional vector space orthogonal to ∂D at $F(r_1, \dots, r_{n-1}, \theta_1, \dots, \theta_n)$ is the space

$$N(r_1, \dots, r_{n-1}, \theta_1, \dots, \theta_n) = \{t(e^{i\theta_1}, \dots, e^{i\theta_n}) : t \in \mathbb{R}\}.$$

In conclusion:

<u>Proposition 1</u>. If $(z_1, \dots, z_n) \in \partial D$ and if $z_1 \cdots z_n \ne 0$, then a real vector orthogonal to ∂D at $(z_1, \dots, z_n)$ is given by

$$\left(\frac{z_1}{|z_1|}, \dots, \frac{z_n}{|z_n|}\right).$$

3. <u>A class of functions in $H^1(\Delta)$</u>.

For any $f \in H^1(\Delta)$, the symbol f^* will denote the boundary value of f, defined almost everywhere on $\partial\Delta$ by the radial limit

$f^*(e^{i\theta}) = \lim_{r\to 1} f(re^{i\theta})$.

The following lemma is a characterization of a class of functions which will be used in the sequel.

<u>Lemma 2</u>. Let $\phi \in H^1(\Delta)$ be such that

$$\frac{\phi^*(e^{i\theta})}{e^{i\theta}} \in \mathbb{R}$$

for a.a. $\theta \in [0,2\pi]$. Then

$$\frac{\phi(z)}{z} = \frac{\bar{a}}{z} + r + az \quad (z \in \Delta)$$

with $a \in \mathbb{C}$, $r \in \mathbb{R}$.

<u>Proof</u>. For $\bar{a} \in \mathbb{C}$ we have

$$\phi(z) = \bar{a} + f(z) \quad (z \in \Delta)$$

with $f \in H^1(\Delta)$ such that $f(0) = 0$. Therefore

$$\frac{\phi(z)}{z} = \frac{\bar{a}}{z} + \frac{f(z)}{z} \quad (z \in \Delta).$$

Since $f(0) = 0$, $\psi(z) = \frac{f(z)}{z} \in H^1(\Delta)$, and by hypothesis,

$$\frac{\bar{a}}{e^{i\theta}} + \psi^*(e^{i\theta}) = \frac{a}{e^{-i\theta}} + \overline{\psi^*(e^{i\theta})}$$

for a.a. $\theta \in [0,2\pi]$. Hence

$$e^{-i\theta}\bar{a} + \psi^*(e^{i\theta}) = e^{i\theta}a + \overline{\psi^*(e^{i\theta})}$$

a.e. and

$$e^{-i\theta}\bar{a} - \overline{\psi^*(e^{i\theta})} = e^{i\theta}a - \psi^*(e^{i\theta})$$

$$\overline{e^{i\theta}a - \psi^*(e^{i\theta})} = e^{i\theta}a - \psi^*(e^{i\theta}) \quad \text{a.e.}$$

The function $\gamma(z) = za - \psi(z)$ belongs to $H^1(\Delta)$ and hence it is the Poisson integral of

$$\gamma^*(e^{i\theta}) = \lim_{r\to 1} \gamma(re^{i\theta}) = e^{i\theta}a - \psi^*(e^{i\theta})$$

which we have proved to be real a.e. Therefore

$$\gamma(z) \in \mathbb{R} \quad \text{for all} \quad z \in \Delta$$

implying $\gamma(z) = -r \in \mathbb{R}$, for all $z \in \Delta$.

It follows $az - \frac{f(z)}{z} = -r$

$$\frac{f(z)}{z} = az + r$$

for all $z \in \Delta$, and, in conclusion

$$\frac{\phi(z)}{z} = \frac{\bar{a}}{z} + r + az$$

for all $z \in \Delta$. □

4. Some properties of the complex geodesics of D_n.

If $g = (g_1,\ldots,g_n) : \Delta \to D_n$ is a complex geodesic of D_n, then every component g_i $(i = 1,\ldots,n)$ belongs to $H^\infty(\Delta)$, since D_n is bounded. Therefore

Lemma 3. If $g = (g_1,\ldots,g_n) : \Delta \to D_n$ is a complex geodesic, then, for every $i = 1,\ldots,n$, either the set of points of $\partial\Delta$ where

$$|g_i^*(e^{i\theta})| = 0 \quad (i = 1,\ldots,n)$$

has Lebesgue measure zero, or g_i is the zero function. (See e.g. [R], th. 17.18 page 273.)

As a consequence of the extremal characterization for the complex geodesics given by Lempert [L] and Royden-Wong [R-W], we obtain from Lemma 3

Proposition 4. If $g = (g_1,\ldots,g_n) : \Delta \to D_n$ is a complex geodesic, then a positive real function

$$\tilde{r} : \partial\Delta \to \mathbb{R}^+$$

exists such that, for every $j = 1,\ldots,n$ for which $g_j \not\equiv 0$, there is a holomorphic function $G_j \in H^1(\Delta)$ with

$$G_j^*(e^{i\theta}) = \frac{e^{i\theta}\overline{g_j^*(e^{i\theta})}\tilde{r}(e^{i\theta})}{|g_j^*(e^{i\theta})|}$$

for almost all $\theta \in [o,2\pi]$.

Let $g = (g_1,\ldots,g_n) : \Delta \to D_n$ be a complex geodesic and let $g_{i_1},\ldots,g_{i_m}$ $(m < n)$ be all the identically zero components of g. By rearranging (if necessary), the order of the components we can suppose that $g_{i_1} \equiv g_1,\ldots,g_{i_m} \equiv g_m$ so that $\tilde{g} = (g_{m+1},\ldots,g_n) : \Delta \to D_n \cap \{z_1 = \ldots z_m = 0\} = D_{n-m}$ is a complex geodesic of D_{n-m} without identically zero components. It is, therefore, not restrictive to study the complex geodesics of D_n having no identically zero components. Hence by Proposition 4, we have, for all complex geodesics without identically zero components,

$$G_j^*(e^{i\theta}) = e^{i\theta}\overline{g_j^*(e^{i\theta})}r_j(e^{i\theta}) \tag{1}$$

where r_j $(j = 1,\dots,n)$ are positive real functions defined almost everywhere on $\partial\Delta$.

<u>Theorem 5</u>. If $g = (g_1,\dots,g_n) : \Delta \to D_n$ is a complex geodesic without identically zero components, then there exist $G = (G_1,\dots,G_n) \in H_n^1(\Delta)$, $(a_1,\dots,a_n) \in \mathbb{C}^n$, $(r_1,\dots,r_n) \in (\mathbb{R}^+)^n$ with $r_j \geq 2|a_j|$ for all $j = 1,\dots,n$ such that

$$g_j(z)G_j(z) = \bar{a}_j + r_j z + a_j z^2 \quad (j = 1,\dots,n) \tag{2}$$

for all $z \in \Delta$.

<u>Proof</u>. The function $\phi_j(z) = g_j(z)G_j(z)$ $(j = 1,\dots,n)$ belongs to $H^1(\Delta)$, and, by (1)

$$\phi_j^*(e^{i\theta}) = g_j^*(e^{i\theta})G_j^*(e^{i\theta}) = e^{i\theta}|g_j^*(e^{i\theta})|^2 r_j(e^{i\theta}) \quad (j = 1,\dots,n).$$

Therefore we have, almost everywhere

$$\frac{\phi_j^*(e^{i\theta})}{e^{i\theta}} = |g_j^*(e^{i\theta})|^2 r_j(e^{i\theta}) \quad (j = 1,\dots,n). \tag{3}$$

Lemma 2 implies now the existence of $a_j \in \mathbb{C}$, $r_j \in \mathbb{R}$ $(j = 1,\dots,n)$ such that

$$\frac{\phi_j(z)}{z} = \frac{\bar{a}}{z} + r_j + a_j z \quad (j = 1,\dots,n)$$

for all $z \in \Delta$. By (3)

$$\bar{a}_j e^{-i\theta} + r_j + a_j e^{i\theta} = r_j + 2\mathrm{Re}(a_j e^{i\theta}) \geq 0$$

for all $\theta \in [0,2\pi]$, i.e.

$$r_j \geq 2|a_j| \quad (j = 1,\dots,n). \qquad \square$$

Let us consider now for $a \in \mathbb{C}$ and $r \in \mathbb{R}$ with $r \geq 2|a|$, the polynomial

$$P(z) = \bar{a} + rz + az^2 \quad (z \in \mathbb{C}).$$

If $a = 0$, the only (simple) root of the equation $P(z) = 0$ is the point $z = 0$ (unless $r = 0$). If $a \neq 0$, we obtain for the same equation

$$\bar{a} + rz + az^2 = 0 \tag{4}$$

with $(r \geq 2|a|)$ the two roots

$$\alpha_0 = \frac{-r+\sqrt{r^2-4|a|^2}}{2a}$$

$$\alpha_1 = \frac{-r-\sqrt{r^2-4|a|^2}}{2a}$$

which satisfy the condition

$$\alpha_0\bar{\alpha}_1 = 1.$$

Moreover $|\alpha_0| \leq 1$, and hence $|\alpha_1| \geq 1$. In fact

$$\left|\frac{-r+\sqrt{r^2-4|a|^2}}{2a}\right| \leq 1$$

is equivalent to

$$|-r+\sqrt{r^2-4|a|^2}\,| \leq 2|a|$$

$$r-\sqrt{r^2-4|a|^2} \leq 2|a|$$

$$r-2|a| \leq \sqrt{r^2-4|a|^2}$$

$$r^2 + 4|a|^2 - 4r|a| \leq r^2 - r|a|^2$$

$$2|a|^2 - r|a| \leq 0$$

$$r \geq 2|a|.$$

If $r = 2|a|$, then $\alpha_0 = \alpha_1$ is a double root of (4) having modulus 1.

Conversely, given any $\alpha_0 \in \bar{\Delta}$, there exists a polynomial equation of type (4) such that α_0 and $1/\bar{\alpha}_0$ are its only roots: consider

$$P_{\alpha_0}(z) = -\alpha_0 + (|\alpha_0|^2+1)z - \bar{\alpha}_0 z^2 = (z-\alpha_0)(1-\bar{\alpha}_0 z).$$

5. <u>Regular geodesics for D_n</u>.

A complex geodesic $g = (g_1,\ldots,g_n) : \Delta \to D_n$ such that g_j extends continuously to $\bar{\Delta}$ for $j = 1,\ldots,n$ will be called a <u>regular geodesic</u>.

In the following $g = (g_1,\ldots,g_n)$ will always be a regular geodesic of D_n without identically zero components. According to Theorem 5, there exist $\alpha_1,\ldots,\alpha_n \in \bar{\Delta}$, $G_1,\ldots,G_n \in H^1(\Delta)$ and $t_1,\ldots,t_n \in \mathbb{R}^+_*$ such that

$$g_j(z)G_j(z) = t_j(z-\alpha_j)(1-\bar{\alpha}_j z) \quad (j = 1,\dots,n),$$

for all $z \in \Delta$.

Suppose g_n has no zeros in $\bar{\Delta}$. If, for $j \neq n$, $g_j(z) \neq 0$ for all $z \in \bar{\Delta}$ we obtain

$$\frac{G_n(z)}{G_j(z)} = \frac{t_n(z-\alpha_n)(1-\bar{\alpha}_n z)}{t_j(z-\alpha_j)(1-\bar{\alpha}_j z)} \cdot \frac{g_j(z)}{g_n(z)} .$$

The function

$$\frac{(z-\alpha_j)}{(1-\bar{\alpha}_j z)} \cdot \frac{G_n(z)}{G_j(z)} = \frac{t_n(z-\alpha_n)(1-\bar{\alpha}_n z)}{t_j(1-\bar{\alpha}_j z)^2} \cdot \frac{g_j(z)}{g_n(z)}$$

is inner, continuous on $\bar{\Delta}$ and has a simple zero at α_n. Therefore

$$\frac{t_n(z-\alpha_n)(1-\bar{\alpha}_n z)}{t_j(1-\bar{\alpha}_j z)^2} \cdot \frac{g_j(z)}{g_n(z)} = \exp(i\phi_j)\frac{(z-\alpha_n)}{(1-\bar{\alpha}_n z)}$$

(5) $$g_j(z) = \exp(i\phi_j)\frac{t_j(1-\bar{\alpha}_j z)^2}{t_n(1-\bar{\alpha}_n z)^2} \cdot g_n(z)$$

On the other hand, if, for $k \neq n$, g_k has a zero (at α_k) on $\bar{\Delta}$, the function

$$\frac{G_n(z)}{G_k(z)} = \frac{t_n(z-\alpha_n)(1-\bar{\alpha}_n z)}{t_k(z-\alpha_k)(1-\bar{\alpha}_k z)} \cdot \frac{g_k(z)}{g_n(z)}$$

is inner, continuous on $\bar{\Delta}$ and with a simple zero at α_n. Hence

$$\frac{t_n(z-\alpha_n)(1-\bar{\alpha}_n z)}{t_k(z-\alpha_k)(1-\bar{\alpha}_k z)} \cdot \frac{g_k(z)}{g_n(z)} = \exp(i\phi_k)\frac{(z-\alpha_n)}{(1-\bar{\alpha}_n z)}$$

i.e.,

(6) $$g_k(z) = \exp(i\phi_k)\frac{t_k}{t_n} \cdot \frac{(z-\alpha_k)(1-\bar{\alpha}_k z)}{(1-\bar{\alpha}_n z)^2} \cdot g_n(z).$$

By rearranging the order of the components of g, if necessary, we can suppose that, of all components of g, only the first m $(0 \le m < n)$ have a zero on $\bar{\Delta}$. By hypothesis we have

$$|g_1(e^{i\theta})| + \dots + |g_m(e^{i\theta})| + |g_{m+1}(e^{i\theta})| + \dots + |g_n(e^{i\theta})| = 1$$

i.e., by (5) and (6)

$$|g_n(e^{i\theta})|\left[\sum_{k=1}^{m}\frac{t_k|e^{i\theta}-\alpha_k||1-\bar\alpha_k e^{i\theta}|}{t_n|1-\bar\alpha_n e^{i\theta}|^2}+\sum_{j=m+1}^{n-1}\frac{t_j|1-\bar\alpha_j e^{i\theta}|^2}{t_n|1-\bar\alpha_n e^{i\theta}|^2}+1\right]=1.$$

Now, since for any $s = 1,\dots,n$

$$\begin{aligned}|1-\bar\alpha_s e^{i\theta}|^2 &= |1-\bar\alpha_s e^{i\theta}||1-\bar\alpha_s e^{i\theta}| \\ &= |(1-\alpha_s e^{i\theta})(1-\bar\alpha_s e^{i\theta}| = |e^{i\theta}-\alpha_s||1-\bar\alpha_s e^{i\theta}| \\ &= |\alpha_s|^2+1-2\mathrm{Re}(\alpha_s e^{i\theta}) \ge 0\end{aligned}$$

we obtain

$$(7)\qquad |g_n(e^{i\theta})| = \frac{t_n(|\alpha_n|^2+1-2\mathrm{Re}(\alpha_n e^{-i\theta}))}{\sum_{s=1}^{n} t_s(|\alpha_s|^2+1-2\mathrm{Re}(\alpha_s e^{-i\theta}))}.$$

The polynomial

$$(8)\qquad -\left[\sum_{s=1}^{n} t_s\bar\alpha_s\right]z^2+\left[\sum_{s=1}^{n} t_s(|\alpha_s|^2+1)\right]z-\left[\sum_{s=1}^{n} t_s\alpha_s\right]$$

can be written as

$$r(z-\gamma)(1-\bar\gamma z),$$

for suitable $r > 0$ and $\gamma \in \bar\Delta$. Furthermore

$$\sum_{s=1}^{n} t_s(|\alpha_s|^2+1) \ge \sum_{s=1}^{n} t_s 2|\alpha_s| \ge 2\left|\sum_{s=1}^{n} t_s\alpha_s\right|.$$

The function

$$F(z) = \frac{t_n(1-\bar\alpha_n z)^2}{r(1-\bar\gamma z)^2}$$

is holomorphic on Δ, without zeros and such that

$$|F(e^{i\theta})| = |g_n(e^{i\theta})|$$

for all θ. Hence $g_n(z) = \exp(i\delta_n)F(z)$. Now, using (5) and (6):

$$g_k(z) = \exp(i\psi_k)\,\frac{t_k(z-\alpha_k)(1-\bar\alpha_k z)}{r(1-\bar\gamma z)^2}\quad (1 \le k \le m)$$

$$g_j(z) = \exp(i\psi_j)\,\frac{t_k(1-\bar\alpha_j z)^2}{r(1-\bar\gamma z)^2}\quad (m+1 \le j \le n).$$

An analogous argument can be carried out in the case in which g_n has a zero in $\bar\Delta$. In conclusion we are led to the following characterization.

<u>Theorem 6</u>. Let $g = (g_1,\dots,g_n) : \Delta \to D_n$ be a regular geodesic of D_n. Then there exist:

i) K, J, S non-negative integers such that $K+J+S = n$,

ii) $\alpha_1,\dots\alpha_{K+J} \in \bar{\Delta}$, not all coinciding with the same $\alpha \in \partial\Delta$,

iii) $t_1,\dots,t_{K+J} \in \mathbb{R}^+_*$,

iv) $\psi_1,\dots,\psi_{K+J} \in \mathbb{R}$,

such that, up to a permutation of the components

$$g_k(z) = \exp(i\psi_k)\frac{t_k(z-\alpha_k)(1-\bar{\alpha}_n z)}{r(1-\bar{\gamma}z)^2} \quad \text{for} \quad 1 \le k \le K$$

$$g_j(z) = \exp(i\psi_j)\frac{t_j(1-\bar{\alpha}_j z)^2}{r(1-\bar{\gamma}z)^2} \quad \text{for} \quad K < j \le K+J$$

$$g_s(z) \equiv 0 \quad \text{for} \quad K+J < s \le n,$$

where $r \in \mathbb{R}^+_*$ and $\gamma \in \Delta$ are such that

$$r(z-\gamma)(1-\bar{\gamma}z) = -\left(\sum_{j=1}^{K+J} t_j\bar{\alpha}_j\right)z^2 + \left(\sum_{j=1}^{K+J} t_j(|\alpha_j|^2+1)\right)z - \left(\sum_{j=1}^{K+J} t_j\alpha_j\right).$$

<u>Proof</u>. What is left to prove is that condition ii) holds. If that is not the case, i.e., if $\alpha_1 = \dots = \alpha_{K+J} = \alpha \in \partial\Delta$, then (7) yields

$$|g_{K+J}(e^{i\theta})| = \frac{t_{K+J}}{\sum_{j=1}^{K+J} t_s} \quad \text{for all} \quad \theta \in [0,2\pi].$$

Therefore, being $t_{K+J} \neq 0$,

$$g_{K+J}(z) = \exp(i\psi_{K+j})\,\frac{t_{K+J}}{\sum_{s=1}^{K+J} t_s},$$

and by (5) or (6), $g_\ell(z) = \exp(i\psi_{K+J})\,\dfrac{t_{K+J}}{\sum_{s=1}^{K+J} t_s}$ $(\ell = 1,\dots,K+J)$.

Hence g is constant, contradicting the fact that g is a complex geodesic. □

By construction (see, e.g. [L], [R-W]) every non constant holomorphic map $g = (g_1,\dots g_n) : \Delta \to D_n$, whose components are as in Theorem 6, is a complex geodesic of D_n. Remark also that by taking $\alpha_1 = \dots = \alpha_n = \alpha \in \Delta$ and $K \ge 1$ (with the notations of Theorem 6) we obtain the obvious "linear" complex geodesics whose components are

$$g_k(z) = \exp(i\psi_k)\left(\frac{z-\alpha}{1-\alpha\bar{z}}\right)^{\frac{t_k}{\sum_{s=1}^{K+J} t_s}} \quad \text{for} \quad 1 \le k \le K,$$

$$g_j(z) = \exp(i\psi_j)\frac{t_j}{\sum_{s=1}^{K+J} t_s} \quad \text{for} \quad k < j \le K+J$$

$$g_\ell(z) \equiv 0 \quad \text{for} \quad K+J < \ell \le n.$$

References

[D-T-V] S. Dineen, R.M. Timoney, J.P. Vigué, Pseudodistances Invariantes sur les Domaines d'un Espace Localment Convexe, Ann. Scuola Norm. Sup., Pisa (4), 12 (1985), 515-530.

[G1] G. Gentili, On Complex Geodesics of Balanced Convex Domains, Annali di Mat. Pura e Appl., to appear.

[G2] G. Gentili, On non-Uniqueness of Complex Geodesics in Convex Bounded Domains, Rend. Acc. Naz. dei Lincei, to appear.

[L] L. Lempert, La métrique de Kobayashi et la répresentation des domaines sur la boule, Bull Soc. Math. Franc, 109 (1981), 427-474.

[P] G. Patrizio, The Kobayashi Metric and the Homogeneous Complex Monge-Ampére Equation, to appear.

[POL] E.A. Polekii, The Euler-Lagrange Equations for Extremal Holomorphic Mappings of the Unit Disk, Michigan Math. J. 30 (1983), 317-333.

[R] W. Rudin, Real and Complex Analysis, McGraw-Hill.

[R-W] H. Royden, Pit-Mann Wong, Carathéodory and Kobayashi Metric on Convex Domains, to appear.

[V1] E. Vesentini, Complex Geodesics, Compositio Mathematica 44 (1981), 375-394.

[V2] E. Vesentini, Complex Geodesics and Holomorphic Maps, Symposia Mathematica 26 (1982), 211-230.

Scuola Normale Superiore
Pisa, Italy

SUBELLIPTIC, SECOND ORDER DIFFERENTIAL OPERATORS

David Jerison[1] and Antonio Sánchez-Calle[2]

1. Introduction

Let Ω be an open subset of $\mathbf{R}^n$. In this article we will be discussing second order partial differential operators of the form

$$L = \sum_{i,j=1}^{n} (1/h(x))(\partial/\partial x_i)(h(x)a_{ij}(x)\partial/\partial x_j),$$

where the coefficients a_{ij} and h are C^∞ real-valued functions on $\overline{\Omega}$, h is positive, and the matrix $A(x) = (a_{ij}(x))$ is symmetric positive semidefinite for every $x \in \overline{\Omega}$. This class as it stands is too general. We will also suppose that L satisfies a *subelliptic* estimate: There is a constant C and a number $\epsilon > 0$ such that all $u \in C_0^\infty(\Omega)$ satisfy

$$(1.1) \qquad \|u\|_{2\epsilon} \le C(\|Lu\| + \|u\|).$$

(Here, $\|u\|_s = \left[\int |\hat{u}(\xi)|^2(1 + |\xi|^2)^s d\xi\right]^{1/2}$ denotes the standard Sobolev norm of order s, $\hat{u}$ is the Fourier transform of u, $\|u\| = \left[\int |u(x)|^2 d\mu(x)\right]^{1/2}$ and $d\mu(x) = h(x)dx$.) Because L is positive and selfadjoint on $L^2(d\mu)$ condition (1.1) is equivalent to

$$(1.2) \qquad \|u\|_\epsilon^2 \le C'\left[\int \sum_{i,j=1}^{n} a_{ij}(x)\frac{\partial u}{\partial x_i}(x)\frac{\partial u}{\partial x_j}(x)h(x)dx + \|u\|^2\right].$$

Clearly an elliptic operator, that is, one for which the matrix A is positive definite, satisfies (1.1) and (1.2) with $\epsilon = 1$. What is interesting is that there are many examples other than elliptic operators. The simplest is an operator on $\mathbf{R}^2$

$$(1.3) \qquad \partial^2/\partial x_1^2 + x_1^{2k}\partial^2/\partial x_2^2$$

which satisfies (1.1) and (1.2) with $\epsilon = 1/(k+1)$, but is not elliptic where $x_1 = 0$. (See Grushin [14].) Despite the possible failure of ellipticity, operators satisfying (1.1) and (1.2) share many properties with elliptic operators. These include hypoellipticity [16], the

---

[1]Research supported in part by NSF grant DMS 8514341 and a Presidential Young Investigator Award. The author is an Alfred P. Sloan Research Fellow.

[2]Research supported in part by NSF grant DMS 8602944.

strong maximum principle and Harnack's inequality [3], estimates on the size of Green's function and fundamental solutions for the corresponding heat and wave equations [4,8,23,29,30,31,32,40] and on the number of eigenvalues [7].

Our purpose here is to describe this far-reaching analogy with elliptic operators and to add a few further links with elliptic theory. In the second section we will describe how to identify subelliptic operators and state several fundamental estimates. In the third section we will study in detail the analogue for subelliptic operators of dilation. We will also prove that definitions used by various people of the metric associated to the operator L are identical. The fourth section will be devoted to detailed estimates for the heat kernel. In the fifth section, we will prove a lower bound on the lowest eigenvalue in the Neumann problem, or equivalently the Poincaré inequality, and describe its application to potential theory. Examination of the Poincaré inequality for subelliptic operators suggests a conjecture concerning the ordinary Poincaré inequality in planar domains. We will explain that conjecture in the last section, which can be read independently of the rest of the paper. In the other sections we will suggest several other problems, formulated in terms of the metric defined by L.

The motivation for studying these subelliptic operators comes from function theory in several complex variables. The induced Cauchy-Riemann equations on a submanifold of $\mathbf{C}^k$ give rise to a system of operators denoted $\bar\partial_b$ by Kohn and Rossi [28]. Hodge theory for the $\bar\partial_b$-complex requires the study of the corresponding Laplacian $\Box_b = \bar\partial_b\bar\partial_b^* + \bar\partial_b^*\bar\partial_b$ on the boundary of a suitably non-degenerate pseudoconvex domain in $\mathbf{C}^k$. The principal part of $-\Box_b$ is an operator of the type of L (with $n = 2k-1$) satisfying (1.1) and (1.2). Indeed, one can choose real C^∞ vector fields X_j, $j = 1,\dots,2k$, spanning the real and imaginary parts of holomorphic vector fields tangent to the boundary in such a way that $L = -\sum_{j=1}^{2k} X_j^* X_j$, where X_j^* is the formal adjoint of X_j. (See [12,24,39].) The simplest example is the operator on the boundary of the unit ball in $\mathbf{C}^2$. After a change of variable to $\mathbf{R}^3$ the principal part of $-\Box_b$ is

(1.4) $$(\partial/\partial x_1 + 2x_2\partial/\partial x_3)^2 + (\partial/\partial x_2 - 2x_1\partial/\partial x_3)^2.$$

Note that the rank of the coefficient matrix A is two; thus A is singular at every point. Nevertheless, this operator satisfies (1.1) and (1.2) with $\epsilon = 1/2$ [24,25]. The operator can be dealt with in a

very explicit way, much like the ordinary Laplace operator on Euclidean space because it is invariant under left translation by Heisenberg group multiplication:

$$(x_1,x_2,x_3)\cdot(v_1,v_2,v_3) = (x_1+v_1,x_2+v_2,x_3+v_3+2x_2v_1-2x_1v_2).$$

In connection with (1.4) it is also worthwhile to recall the example of H. Lewy of an operator ($\bar{\partial}_b$ as it acts on scalar functions) that is not locally solvable. In these coordinates it is $(\partial/\partial x_1 + 2x_2\partial/\partial x_3) + i(\partial/\partial x_2 - 2x_1\partial/\partial x_3)$. A full understanding of $\bar{\partial}_b$ and $\Box_b$ requires far more than just an understanding of the purely real operators we are considering. We will not attempt to describe the considerable progress in that direction, but instead refer the reader to the literature. (See [26] for recent results.)

2. Background

How does one recognize whether an operator satisfies the subelliptic estimate (1.1) or (1.2)? We will describe three related ways to do so. The first of these, due to Hormander, applies only to the case in which L can be written smoothly as a sum of squares of vector fields.

An arbitrary non-negative matrix $A = (a_{ij}(x))$ can be factored as the matrix product $A = BB^t$, where $B = (b_{ij}(x))$ is an $n\times m$ matrix. (The matrix B and the choice of m are not unique.) However, even if the entries of A are C^∞ functions the entries of B are in general only Lipschitz functions of x [38]. If $X_j = \sum_{i=1}^{n} b_{ij}(x)\partial/\partial x_i$, then

$$L = \sum_{j=1}^{m} (1/h(x))(\partial/\partial x_i)(h(x)a_{ij}(x)\partial/\partial x_j) = -\sum_{j=1}^{m} X_j^* X_j = \sum_{j=1}^{m} X_j^2 + X_0,$$

where X_j^* is the formal adjoint of X_j in $L^2(d\mu)$. Even though a smooth factorization is not available in general, the case in which the vector fields X_j can be chosen to be C^∞ is the most important special case arising as it does in function theory on pseudoconvex domains.

In 1967 Hormander [16] proved that given C^∞ vector fields X_0, $X_1,\dots,X_m$ on $\mathbb{R}^n$, operators of the form $L = \sum_{j=1}^{m} X_j^2 + X_0$ satisfy a subelliptic estimate of the form (1.1) for some $\epsilon>0$ if and only if there exists an integer k such that the vector fields and their commutators up to length k,

$$X_0,X_1,\dots,X_m,[X_{i_1},X_{i_2}],\dots,[X_{i_1},[X_{i_2},[\dots,X_{i_k}]]\dots],\quad i_j= 0,\dots,m,$$

span the tangent space at each point. We will say the vector fields $X_0, \ldots, X_m$ satisfy *Hormander's condition (of order k)* if this property holds. Hormander also proved as a corollary that the operator L is *hypoelliptic*, that is, if $Lu = f$ then u is C^∞ in any open set where f is C^∞.

Notice that when L is selfadjoint with respect to $L^2(d\mu)$, then necessarily $L = -\Sigma X_j^* X_j$, where $X_j^* = -X_j + f_j$ is the formal adjoint of X_j in $L^2(d\mu)$. It follows that X_0 belongs to the span of $X_1, \ldots, X_m$. Thus in the case of interest to us the vector field X_0 is superfluous in Hormander's condition. We will refer to operators of the form $-\Sigma X_j^* X_j$ for which the X_j's may be chosen to be C^∞ as *Hormander type* operators and to the more general selfadjoint operators we are considering as *subelliptic* operators. (The reader should beware that the word subelliptic is used elsewhere in connection with operators that are neither selfadjoint nor second order nor real [see 17].)

In 1969, J.-M. Bony showed that for operators $L = \Sigma X_j^2 + X_0$, where $X_0, \ldots, X_m$ satisfy Hormander's condition, the weak maximum principle holds: $Lu=0$ in Ω and u continuous in $\bar{\Omega}$ implies $\max_\Omega u \le \max_{\partial\Omega} u$. This is the best one can expect since Hormander's general class includes not only the ordinary Laplace operator, but also the heat operator $\partial/\partial t - \partial^2/\partial x^2$. However if one considers the special case in which the vector fields $X_1, \ldots, X_m$ and their commutators up to order k suffice to span the tangent space at each point, Bony proved the strong maximum principle: $Lu = 0$ in Ω and u is continuous in $\bar{\Omega}$, then $u(x) < \max_{\partial\Omega} u$ for all $x \in \Omega$, unless u is identically constant. Bony's result is the first indication that our operators will resemble elliptic operators.

A more precise version of Hormander's theorem was proved by Rothschild and Stein in 1976

Theorem 2.1[39] *Suppose that* $L = -\Sigma X_j^* X_j$ *for some* C^∞ *vector fields* $X_1, \ldots, X_m$. *The vector fields satisfy Hormander's condition of order k, if and only if the operator* L *satisfies (1.1) and (1.2) with* $\epsilon = 1/k$.

In order to characterize operators satisfying (1.1) and (1.2) in the general case we cannot rely on derivatives of the coefficients of the vector fields X_j. Our next characterization will overcome that difficulty at the expense of some precision.

Theorem 2.2 (Oleinik and Radkevic [37]) *Consider the operators* $Y_j = \sum_{i=1}^{n} a_{ij}(x)\partial/\partial x_i$, $j = 1,\dots,n$. *If* $Y_1,\dots,Y_n$ *and their commutators up to some finite order* k *span the tangent space at each point, then the associated operator* L *will satisfy* (1.1) *and* (1.2) *for some* $\epsilon>0$.

In Section 5 of this paper we will prove a converse to this theorem:

Proposition 2.1 *If the operator* L *satisfies* (1.1) *or* (1.2), *then there exists an integer* k *for which the vector fields* $Y_1,\dots,Y_n$ *and their commutators up to order* k *span the tangent space at each point.*

For example, if we take the Grushin operator (1.3) we have $Y_1 = \partial/\partial x_1$ and $Y_2 = x_1^{2k}\partial/\partial x_2$, while we can choose $X_1 = \partial/\partial x_1$ and $X_2 = x_1^{k}\partial/\partial x_2$. The vector fields Y_1 and Y_2 and their commutators up to order 2k+1 span the tangent space, while the vector fields X_1 and X_2 and their commutators up to order k+1 span the tangent space. Whereas the order of commutators of the vector fields X_j needed to span determines exactly the order of subellipticity, there is no known simple relationship between the corresponding order for the vector fields Y_j and the order of subellipticity.

The third characterization is formulated in terms of a fundamental notion of distance associated to L defined as follows.

We say that a vector $v = (v_1,\dots,v_n)$ is *subunit* for L at x if $|v\cdot\xi|^2 \le \sum_{i,j=1}^{n} a_{ij}(x)\xi_i\xi_j$ for all $\xi \in \mathbf{R}^n$. Notice that for any choice of X_j for which $L = -\sum X_j^* X_j$, v is subunit for L at x if and only if $\sum_{i=1}^{n} v_i\partial/\partial x_i = \sum_{j=1}^{m} a_j X_j(x)$, with $\sum_{j=1}^{m} a_j^2 \le 1$. We say that a path γ is *subunit for L* if $\gamma: [0,b] \to \mathbf{R}^n$ is a Lipschitz function and $\gamma'(t)$ is subunit for L at $\gamma(t)$ for a. e. $t \in [0,b]$. The metric d_L is defined by

$$d_L(x,y) = \inf\{b: \ \gamma \text{ is a subunit path for } L,\ \gamma(0)=x \text{ and } \gamma(b)=y\}$$

We denote the ball of radius r by $B_L(x;r) = \{y \in \mathbf{R}^n: \ d_L(x,y) < r\}$. Ordinary Euclidean distance is just $d_\Delta(x,y)$ for the standard Laplace operator $\Delta = \sum \partial^2/\partial x_j^2$.

Theorem 2.3 (C. Fefferman and D. Phong [7]) *The operator* L *satisfies* (1.1) *and* (1.2) *for some* $\epsilon>0$ *if and only if there are constants* $r_0>0$ *and* C *such that for all* x *and* r

$$B_\Delta(x,r) \subset B_L(x,Cr^\epsilon) \quad \text{whenever} \quad 0 < r < r_0.$$

In the case of the Grushin operator, the ball $B_L((0,0),r)$ is comparable to a rectangle of the form $[-r,r]\times[-r^{k+1},r^{k+1}]$. In the general case this long thin strip may bend wildly. It is perhaps surprising that the presence inside of a tiny pea (namely the Euclidean ball of radius r^{k+1}) suffices for subelliptic estimates. Fefferman and Phong also showed how to deduce the theorem of Rothschild and Stein from theirs, but it should be emphasized that it is not necessarily easy to confirm the hypothesis in Theorem 2.3 on the balls in the metric d_L. For instance in our proof of a lower bound for the first eigenvalue in the Neumann problem we will have to use Proposition 2.1 instead. We would also like to remind the reader of the implicit hypothesis in Theorem 2.3 that the coefficients a_{ij} are C^∞. (See [44] for a discussion of the case of C^2 coefficients in two dimensions.)

The notion of distance is crucial to the formulation of most of the rest of the results we will describe. Let $G(x,y)$ denote a fundamental solution for L, that is, for some open set Ω and every function $f \in C_0^\infty(\Omega)$, $L\int G(x,y)f(y)d\mu(y) = f(x)$. Then for $n > 2$,

$$|G(x,y)| \leq Cr^2/\mu(B_L(x,r)), \quad \text{where} \quad r = d_L(x,y)$$

This result was proved independently by Nagel, Stein, and Wainger [36] and Sanchez-Calle [40] in the case of Hormander type operators and by Fefferman and Sanchez-Calle [8] in the general case. At the same time one also has bounds on the derivatives of G of the form

$$|X_{i_1}X_{i_2}\dots X_{i_p}G(x,y)| \leq C_p r^{2-p}/\mu(B_L(x,r))$$

in the case of Hormander type operators and

$$|Y_{i_1}Y_{i_2}\dots Y_{i_p}G(x,y)| \leq C_p r^{2-p}/\mu(B_L(x,r))$$

in the general case. These estimates are best in terms of order of magnitude. For instance we also have the lower bound

$$|G(x,y)| \geq cr^2/\mu(B_L(x,r)).$$

It would be interesting to know if there is a reasonable asymptotic formula for G as r tends to 0. The formula cannot be expressed in terms of the distance $d_L(x,y)$ alone. For instance, the fundamental solution for (1.4) was calculated by Folland [10], and it depends on a comparable, but different distance.

Let us now examine the heat kernel. Consider a subelliptic operator L on Ω. It is easy to see that $\partial/\partial t - L$ is also subelliptic in the sense of (1.1). In fact, if $\varphi \in C_0^\infty(\mathbb{R}\times\Omega)$, and we denote by $\| \ \|$ and $\langle \ , \ \rangle$ the L^2-norm and inner product with

respect to the measure $dtd\mu$ we have (abbreviating $\partial_t = \partial/\partial t$)

$$\|\partial_t\varphi\|^2 = \langle\partial_t\varphi - L\varphi,\partial_t\varphi\rangle + \langle L\varphi,\partial_t\varphi\rangle = \langle\partial_t\varphi - L\varphi,\partial_t\varphi\rangle \leq$$
$$\leq \|(\partial_t - L)\varphi\| \, \|\varphi\|, \quad \text{because}$$

$\langle L\varphi,\partial_t\varphi\rangle = -\langle\partial_t L\varphi,\varphi\rangle = -\langle\partial_t\varphi,L\varphi\rangle$ shows that $\langle L\varphi,\partial_t\varphi\rangle = 0$. Thus

$$\|\partial_t\varphi\| \leq \|(\partial_t - L)\varphi\|.$$

Since the derivatives in the "spatial" or Ω directions can be controlled by $\langle L\varphi,\varphi\rangle = \langle(\partial_t - L)\varphi,\varphi\rangle \leq (\|(\partial_t - L)\varphi\| + \|\varphi\|)^2$, it follows that $\partial_t - L$ is subelliptic:

$\|\varphi\|_\epsilon \leq C(\|(\partial_t - L)\varphi\| + \|\varphi\|)$, for the same ϵ for which the property (1.2) holds for L. Therefore, $\partial_t - L$ is hypoelliptic and it is not hard to prove the existence of the heat kernel, that is, a function $h(t,x,y)$ satisfying $h \in C^\infty(\mathbb{R}\times M\times M - \{(0,x,x): x \in M\})$, $h(t,x,y) = 0$ if $t < 0$ and such that

$$(\partial_t - L)u = 0 \quad \text{for } t > 0 \quad \text{and} \quad u(t,x) \to f(x) \quad \text{as } t \to 0^+$$

where $u(t,x) = \int h(t,x,y)f(y)d\mu(y)$ and $f \in C_0^\infty(\Omega)$.

The heat kernel for the standard Laplace operator is, of course,

$$h(t,x,y) = c_n t^{-n/2} e^{-|x - y|^2/4}.$$

It is well-known that the heat kernel for a general elliptic operator has upper and lower bounds of this type (at least locally) with the Euclidean distance $|x - y|$ replaced by the distance associated to (a_{ij}) defined above. (This is the same as the geodesic distance for the metric $g_{ij}dx_i\otimes dx_j$ where the matrix (g_{ij}) is the inverse matrix to (a_{ij})). An explicit formula for the heat kernel can be calculated rather rarely. In the case of (1.4) a formula is given by Gaveau [13]. We will discuss bounds for the heat kernel (and wave kernel) in the subelliptic case in some detail in Section 4. For now we would like to state just one limiting result

$$\lim_{t\to 0+} 4t \log h(t,x,y) \quad \text{exists and equals} \quad -d_L(x,y)^2.$$

This result was proved by Leandre using probablistic methods. (See [31] and also the work of Kusuoka and Stroock [30], where a simpler proof is given.) Along similar lines, there is a result of Melrose [32] that the support of the forward fundamental solution to the wave equation is supported in the set $d_L(x,y) \leq t$. One could speculate that its singular support is the set $d_L(x,y) = t$.

Next, we would like to comment briefly on boundary value problems. The size estimates on the fundamental solution for L make it possible to develop a theory of capacity and a Wiener test. It is easy to see that some domains with C^∞ boundaries are not regular for

the Dirichlet problem in the sense that solutions to the problem with continuous boundary data need not be continuous at some points. (See [18],[43].) The points where things sometimes go wrong are the points x of the boundary where $\Sigma a_{ij}(x)\nu_i\nu_j = 0$ where $\nu = (\nu_1,\dots,\nu_n)$ is the normal to the boundary at x. Such points don't exist in the elliptic case. At such so-called characteristic boundary points the solution to the Dirichlet problem is typically not C^∞ even when the boundary data are C^∞. (See [5],[9],[20],[27].)

To conclude this section, we would like to describe briefly the methods used in proving these theorems. Hormander's approach and the technique used by Kohn [25] were based on integration by parts. The technique used by Folland and Stein [12], Folland [11], and culminating in the work of Rothschild and Stein was to consider first a model case of a nilpotent Lie algebra generated by the vector fields $X_1,\dots,X_m$ with a group of dilations acting on the Lie algebra in such a way that L is homogeneous of degree -2. In example (1.4) we have given generators for the Lie algebra of the Heisenberg group. The group of dilations in this case is $(x,y,z) \to (sx,sy,s^2z)$ for $s \in \mathbf{R}_+$. The advantages of the group situation are as follows. First every point is equivalent after translation by the group action to every other point--thus a fundamental solution is given by convolution. Second the homogeneity makes it possible to select a dilation invariant fundamental solution. The translation and dilation invariance completely determine its properties. General Hormander type operators are well approximated by such operators only in the "constant rank" case, that is, the case in which $\dim V_k(x)$ is independent of x for all k, where $V_k(x)$ is the subspace of the tangent space to Ω at x spanned by the commutators of the X_j up to order k at x. However, one can still obtain estimates for all Hormander type operators by lifting the vector fields to a higher dimensional space and reversing the process by a quotient operation. The simplest example is (1.3) with $k = 1$, in which we can add a new variable z and consider the vector fields

$X_1 = \partial/\partial x_1$ and $X_2 = \partial/\partial z + x_1\partial/\partial x_2$. After a linear change of variables, this is the same as example (1.4). We obtain information about the original problem (1.3) by restricting our attention to functions that are independent of z.

By contrast, the methods used by Fefferman and Phong involve reduction of the number of variables. One can always change variables so that L is written as $\partial^2/\partial x_1^2$ + a suitable remainder. It then turns out that one can average the operator in the x_1 direction and

reduce the number of variables by one. In both the Lie group case and the more general one, scale invariance according to the correct notion of scale given by the distance $d_L(x,y)$ is of central importance.

3. Rescaling of subelliptic operators and elementary properties of the distance function

When we speak of elementary properties we mean properties that make no reference to solutions of partial differential equations. Such properties are often hard to prove and serve as excellent preparation for proofs of the "non-elementary" properties of solutions to the Poisson, heat, and wave equations associated to the subelliptic operator L. In this section we will describe a few such properties and show that several natural definitions of distance are equivalent.

One very useful property of the distance function is the so-called *doubling condition* which says that

$$(3.1)\quad \mu(B_L(x,2r)) \leq C\mu(B_L(x,r)) \quad \text{for all} \quad x \in Q \quad \text{and } 0 < r < 1.$$

(We will suppose for the rest of this section that Q is a unit cube and that L is defined in a concentric cube of side-length 3. This is merely a convenient normalization.)

One of the immediate consequence of the doubling property is

$$(3.2)\quad \mu(B_L(x,R\delta)) \leq CR^m\mu(B_L(x,\delta)), \text{ for } |x| \leq 1/2 \text{ and } R\delta < r_0.$$

(We obtain (3.2) with $C = 2^m$, by applying (3.1) at radii $2^k\delta$.)

There are far more detailed descriptions of balls $B_L(x,r)$ from which it is easy to deduce (3.1). In the Hormander-type case Nagel, Stein, and Wainger have proved that the ball $B_L(x,r)$ is comparable to the image under the exponential mapping of an appropriate cube. More precisely, there is a collection of vector fields $Z_1,\dots,Z_n$ such that Z_j is a commutator of the X_i's of order k_j, the Z_j's form a basis for the tangent space at x, and if we denote

$$\phi(t) = \exp_x\Big(\sum_{j=1}^{n} t_jZ_j\Big), \quad \text{for} \quad t \in \mathbf{R}^n,$$

then there are positive constants c and C independent of r such that

$$\phi(Q(cr)) \subset B_L(x,r) \subset \phi(Q(Cr)),$$

where $Q(r) = [-r^{k_1},r^{k_1}]\times[-r^{k_2},r^{k_2}]\times\dots\times[-r^{k_n},r^{k_n}]$.

In addition to this sort of approximate description of balls there is a powerful notion of rescaling. Let us first recall what happens with uniformly elliptic operators. Let $Q_\rho(x_0)$ denote the

cube in $\mathbf{R}^n$ sides parallel to the axes, center x_o, and side length 2ρ, and denote by ψ the dilation to unit size: $\psi\colon Q_1(0) \longrightarrow Q_\rho(x_0)$; $\psi(y) = x = x_0 + \rho y$. If $L = \sum_{i,j=1}^n a_{ij}(x)\partial^2/\partial x_i \partial x_j$ + (lower order terms) is an elliptic operator in the x-coordinates, and we take the pull-back by ψ (i.e., we write L in the y-coordinates), then we find that $L = \rho^{-2} L_{\rho,x_0}$, where $L_{\rho,x_0} = \sum_{i,j=1}^n a_{ij}(x_0 + \rho y)\partial^2/\partial y_i \partial y_j$ + (lower order terms). The family $\{L_{\rho,x_0}\}_{\rho,x_0}$ is uniformly elliptic: the ellipticity constant λ for L serves equally well for each L_{ρ,x_0} — we have the matrix inequality $(\lambda\delta_{ij}) \le (a_{ij}(x_0 + \rho y)) \le (\lambda^{-1}\delta_{ij})$, independent of ρ and x_0. Moreover, if the coefficients of L are C^∞, then there are uniform bounds on all derivatives in the y variable of the coefficients of the family L_{ρ,x_0} independent of x_0 and of ρ for $\rho \le 1$. Many of the properties of elliptic operators can be deduced from these two observations. For general subelliptic operators Fefferman, Phong, and Sanchez-Calle [7,8,40] have given a description of the balls and the change of variable analogous to dilation. We will recall that result here.

We will describe in detail below some "standard cubes" which we will denote $Q_L(x_0,\rho)$. These "standard cubes" are equivalent to the metric balls in the sense that, for some positive constant K, $B_L(x_0,\rho/K) \subset Q_L(x_0,\rho) \subset B_L(x_0,K\rho)$. In order to state the rescaling theorem, we define a notion of *subellipticity constants*. If L is a subelliptic operator defined in a neighborhood of the closed Euclidean unit cube $\overline{Q_1(0)}$, there are constants c_0, and C_α such that $h(x) \ge c_0 > 0$ and $|(\partial^\alpha/\partial x^\alpha)a_{ij}(x)| + |(\partial^\alpha/\partial x^\alpha)h(x)| \le C_\alpha$ for all $x \in \overline{Q_1(0)}$. We will call the *subellipticity constants* of L the constants ϵ and C of (1.1), c_0, the dimension n, and C_α for all $|\alpha| \le N$ for an integer N depending only on ϵ, C, and n.

<u>Theorem 3.1</u> [8] *Let L be a subelliptic operator defined on the unit cube $Q_1(0)$. There are constants $C, c, \rho_0 > 0$ depending only on the subellipticity constants of L such that for $|x| \le 1/2$ and $0 < \rho < \rho_0$, there is a diffeomorphism $\phi = \phi_{\rho,x}\colon Q_1(0) \longrightarrow Q_L(x,\rho)$ such that the Jacobian $J\phi$ has the property that*

$$c\cdot\mu(Q_L(x,\rho)) \le |J\phi_{\rho,x}| \le C\cdot\mu(Q_L(x,\rho)).$$

Moreover, if we take the pull-back by $\phi_{\rho,x}$ of L to $Q_1(0)$ we get an operator $\rho^{-2}L_{\rho,x}$, and the subellipticity constants of $L_{\rho,x}$ depend only on those of L, independent of ρ and x.

By definition, $Q_L(x_0,\rho) = Q_{\rho^2 L}(x_0,1)$, and we can assume that $x_0 = 0$, so we only have to describe $Q_L(0,1)$ for $L = \sum_{i,j=1}^{n} a_{ij}(x)\partial^2/\partial x_i \partial x_j$ + (lower order terms). Choose $\delta = 2^{-k}$ for the first integer k such that

$$\max_{1\le i\le n} \max_{x\in Q_\delta(0)} |a_{ii}(x)| \ge R\delta^2,$$

where R is a constant much larger than a bound for the a_{ij}'s and their derivatives up to order 2. Observe that since $(a_{ij}) \ge 0$, $|a_{ij}|^2 \le a_{ii}a_{jj}$. Hence, $|a_{ij}(x)| \le 4R\delta^2$ for $x \in Q_\delta(0)$. We can assume that the index for which the maximum is attained is i = 1. It follows that $a_{11}(x) \ge R\delta^2/4$ for all $x \in Q_\delta(0)$. In fact, we know that we have four times that lower bound at some point of $Q_\delta(0)$, and the inequality for all points is a consequence of the following remark: *If f is a C^2 non-negative function and $|f''| \le 1$, then $(1/2)(f(0) - t^2) \le f(t) \le 2f(0) + t^2$.*

If we now change to y-coordinates defined by $x = \delta y/10$, we see that L has the form $\sum_{i,j=1}^{n} \tilde{a}_{ij}(y)\partial^2/\partial y_i \partial y_j + \dots$, with $\tilde{a}_{ij}(y) = 100\delta^{-2}a_{ij}(\delta y/10)$. The bounds on the $\tilde{a}_{ij}$'s and their derivatives with respect to y depend only on the corresponding bounds for the a_{ij}'s and not δ. Next, we change variables to $(t,z) = (t,z_1,\dots,z_{n-1})$ by setting $y = \varphi(t,z)$, where $\varphi(\cdot,z)$ is the integral curve of the vector field $Y = \partial/\partial y_1 + \sum_{j=2}^{n} \frac{\tilde{a}_{1j}(y)}{\tilde{a}_{11}(y)}\partial/\partial y_j$ starting at $(0,z)$, that is, $\varphi(0,z) = (0,z)$ and $(d/dt)\varphi(t,z) = Y|_{\varphi(t,z)}$. (Observe that $|\tilde{a}_{1j}(y)| \le 8\tilde{a}_{11}(y)$ if $y \in Q_{10}(0)$, so $\varphi(t,z) \in Q_{10}(0)$ for $|t| \le 1$, $|z_1| \le 1,\dots,|z_{n-1}| \le 1$) In these new coordinates,

$$L = \tilde{a}_{11}(\partial^2/\partial t^2 + \sum_{i,j=1}^{n-1} b_{ij}(t,z)\partial^2/\partial z_i \partial z_j) + \dots$$

and we set $Q_L(0,1) = (-1,1)\times Q_{\bar{L}}(0,1)$, where

$$\bar{L} = \sum_{i,j=1}^{n-1} \bar{b}_{ij}(z)\partial^2/\partial z_i \partial z_j + \dots, \quad \bar{b}_{ij} = \frac{1}{2}\int_{-1}^{1} b_{ij}(t,z)dt.$$

We now use induction on the dimension n to complete the definition of a standard cube.

After composing all the changes of variables we get the mapping $\Phi = \Phi_{\rho,x_0} : Q_1(0) \longrightarrow Q_L(x_0,\rho)$ of Theorem 3.1.

So far we have only discussed order of magnitude properties of the metric. In order to compare results of various authors that depend on the exact metric we will need to show that several notions

of distance are equivalent. For a subelliptic operator $L = -\Sigma X_j^* X_j$ as above, we will denote

$$C(x,y) = \{\sigma : \sigma \text{ is a Lipschitz curve, } \sigma:[0,1] \longrightarrow \mathbf{R}^n,\ \sigma(0) = x,\ \sigma(1) = y, \text{ and } \sigma'(t) = \sum_{j=1}^{m} a_j(t) X_j(\sigma(t)) \text{ for a.e. } t\}$$

$$a_\sigma(t) = (a_1(t),\dots,a_m(t)) \text{ and } |a_\sigma(t)| = \left[\sum_{j=1}^{m} a_j(t)^2\right]^{1/2},$$

where the a_j's are from the equation for σ' above.

The length of a_σ does not depend on the particular choice of the X_j's. In fact, as we mentioned implicitly in Section 2, $|a_\sigma(t)|^2 = \sup_{\xi \in \mathbf{R}^n} \langle \sigma'(t), \xi \rangle^2 / \langle A(\sigma(t))\xi, \xi \rangle$, provided that $a_\sigma(t)$ is chosen minimally, i.e., orthogonal to all vectors v such that $\sum_{j=1}^{m} v_j X_j(\sigma(t)) = 0$. Next, denote $\|f\|_{L^p[0,1]} = \left[\int_0^1 |f(t)|^p dt\right]^{1/2}$, $1 \le p < \infty$, and $\|f\|_{L^\infty[0,1]} = \operatorname*{ess\,sup}_{t \in [0,1]} |f(t)|$.

We define several metrics as follows:

$$d_{L,p}(x,y) = \inf \{\|a_\sigma\|_{L^p[0,1]} : \sigma \in C(x,y)\}, \quad 1 \le p \le \infty.$$

$$\alpha_L(x,y) = \lim_{\epsilon \to 0+} d_{L+\epsilon\Delta}(x,y)$$

$$\beta_L(x,y) = \sup \{\psi(x) - \psi(y) : \psi \in C^\infty, \ \langle A(z)\nabla\psi(z), \nabla\psi(z) \rangle \le 1 \text{ for all } z\}$$

We obviously have $d_L(x,y) = d_{L,\infty}(x,y)$, by the reparametrization $\gamma(t) = \sigma(t/b)$, where $b = \|a_\sigma\|_{L^\infty[0,1]}$. The limit defining α_L exists because $d_{L+\epsilon\Delta}(x,y)$ is a monotone function of ϵ.

<u>Proposition 3.1</u> *The distances $d_L(x,y)$, $d_{L,p}(x,y)$, $\alpha_L(x,y)$, and $\beta_L(x,y)$ are all equal.*

<u>Proof</u>: Denote $L_k = L + (1/k^2)\Delta = \sum_{j=1}^{m} -X_j^* X_j + (1/k^2)\Delta$, for $k = 1,2,\dots$. As we observed in Section 2, we can suppose that the coefficients of X_j are Lipschitz continuous. The operator L_k is also in sum of squares form for the vector fields $X_1,\dots,X_m,(1/k)\partial/\partial x_1,\dots,(1/k)\partial/\partial x_n$. It is obvious that $d_{L_k}(x,y)$ increases with k and that $d_{L_k}(x,y) \le d_L(x,y)$. Therefore, in order to prove that α_L is the same as d_L, it suffices to prove that

$$d_L(x,y) \le \sup_k d_{L_k}(x,y).$$

We abbreviate $d_k = d_{L_k}(x,y)$. Since L_k is elliptic, we can choose a C^∞ curve $\sigma_k\colon [0,1] \longrightarrow \mathbf{R}^n$ with $\sigma_k(0) = x$, $\sigma_k(1) = y$, and

$$\sigma_k'(t) = \sum_{j=1}^{m} a_j^k(t)X_j(\sigma_k(t)) + \sum_{i=1}^{n} b_i^k(t)(1/k)\partial/\partial x_i, \quad \text{where}$$

$$\sum_{j=1}^{m}(a_j^k(t))^2 + \sum_{i=1}^{n}(b_i^k(t))^2 \le d_k^2.$$

Clearly, $|\sigma_k'(t)| \le C$, so that the functions σ_k are uniformly Lipschitz. Therefore there exists a subsequence which we continue to denote σ_k such that σ_k tends to a limit σ uniformly in $[0,1]$. It follows that σ is Lipschitz. In particular, $\sigma'(t)$ exists for almost every t. Next, there is a subsequence of the coefficients a_j^k and b_i^k converging weakly (say in $L^2[0,1]$) and pointwise to functions a_j, $j=1,\dots,m$ and b_i, $i=1,\dots n$. We also know that $\Sigma\, a_j(t)^2 + \Sigma\, b_i(t)^2 \le \sup_k d_k^2$ for almost every t.

It remains to show that $\sigma'(t) = \sum_{j=1}^{m} a_j(t)X_j(\sigma(t))$ for a.e. t. Once we have done so, the curve $\gamma(t) = \sigma(t/b)$ defined on $[0,b]$ is subunit for L if $b = \sup_k d_k^2$; thus $d_L(x,y) \le b$. In fact, σ_k satisfies the integral equation

$$\sigma_k(t) = x + \int_0^t \left[\sum_{j=1}^{m} a_j^k(s)X_j(\sigma_k(s)) + b^k(s)/k\right]ds,$$

where b^k is the vector in $\mathbf{R}^n$ $(b_1^k,\dots,b_n^k)$. If $f_k(s)$ tends to $f(s)$ uniformly for s in $[0,t]$, then f_k certainly tends to f in L^2. Thus, since $a_j^k(s)$ is uniformly bounded (in all variables),

$$\int_0^t a_j^k(s)f_k(s)ds = \int_0^t a_j^k(s)f(s)ds + \int_0^t a_j^k(s)[f_k(s) - f(s)]ds$$

tends to $\int_0^t a_j(s)f(s)ds$ as k tends to infinity. Recall also that the functions $b_i^k(s)$ are uniformly bounded, so that $b_i^k(s)/k$ tends uniformly to 0. It follows that

$$\sigma(t) = x + \int_0^t \sum_{j=1}^{m} a_j(s)X_j(\sigma(s))ds,$$

and so $\sigma'(t)$ has the desired formula.

Next, we consider the distance $\beta_L(x,y)$. Suppose that ψ is a function satisfying the constraint in the definition of β_L. If γ is a subunit path on $[0,b]$ with $\gamma(0) = x$ and $\gamma(b) = y$, then

$$|\psi(y) - \psi(x)| = \left|\int_0^b \langle\nabla\psi(\gamma(t)),\gamma'(t)\rangle dt\right| \le$$

$$\le \int_0^b \sqrt{\langle A(\gamma(t))\nabla\psi(\gamma(t)),\nabla\psi(\gamma(t))\rangle}\,dt \le b.$$

Hence, $\beta_L(x,y) \le d_L(x,y)$. Our previous discussion shows that to prove the opposite inequality, it suffices to show that $d_{L_k}(x,y) \le \beta_L(x,y)$, for all k. In fact denote $A_k = A + (1/k^2)I$, the matrix of coefficients for L_k, and define $\eta(y) = d_{L_k}(x,y)$. Then $\langle A_k\nabla\eta,\nabla\eta\rangle \le 1$. Because L_k is elliptic, one can use an approximate identity to regularize η, that is, to find C^∞ functions η_ϵ such that $|\eta(y) - \eta_\epsilon(y)| \le \epsilon$ for all y and $\langle A_k\nabla\eta_\epsilon,\nabla\eta_\epsilon\rangle \le 1$. If we use these functions η_ϵ for ψ in the definition of $\beta_{L_k}(x,y)$, we find that $d_{L_k}(x,y) \le \beta_{L_k}(x,y)$. But it is trivial from the definition that $\beta_{L_k}(x,y) \le \beta_L(x,y)$, so we are done.

Finally, we show that $d_{L,p}$ is the same as d_L. Actually, the infimum is a minimum for each of the $d_{L,p}$ and a minimizing σ can be taken such that $|a_\sigma(t)|$ = constant a.e. t. Since $\|a_\sigma\|_{L^1} \le \|a_\sigma\|_{L^p} \le \|a_\sigma\|_{L^\infty}$, $1 \le p \le \infty$, it clearly suffices to show that $d_{L,\infty} \le d_{L,1}$. Assume that $\sigma \in C(x,y)$ is such that $\|a_\sigma\|_{L^1[0,1]} = \rho < \infty$. Let $u(t) = (1/\rho)\int_0^t |a_\sigma(s)|ds$. u is a monotone increasing function from [0,1] onto [0,1]. We define what is essentially its inverse function by $\bar{t}(s) = \inf\{t\colon\ u(t) = s\}$. Let $\tilde{\sigma}(s) = \sigma(\bar{t}(s))$, and $\tilde{a}_j(s) = \rho a_j(\bar{t}(s))/|a_\sigma(\bar{t}(s))|$. We will check presently that

$$\tilde{\sigma}(s) = x + \int_0^{\bar{t}(s)} \sum_{j=1}^m a_j(t)X_j(\sigma(t))dt = \tag{3.3}$$

$$x + \int_0^s \sum_{j=1}^m \tilde{a}_j(v)X_j(\tilde{\sigma}(v))dv.$$

Note that the definition of $\tilde{a}_j$ then implies that $\|a_{\tilde{\sigma}}\|_{L^\infty} = \rho$. Thus $\inf\{\|a_\sigma\|_{L^\infty[0,1]}\colon\ \sigma \in C(x,y)\} \le \rho$ and we have $d_{L,\infty} \le d_{L,1}$. The first equality in (3.3) follows from the definition of σ. For the second, the only difficulty is the set of points where u is not strictly increasing. The set $\{t\in[0,1]\colon\ \bar{t}(u(t)) \ne t\} = \bigcup_{j=1}^\infty (a_j,b_j]$, a disjoint union of half-open intervals. The points a_j and b_j satisfy $u(a_j) = u(b_j)$. Thus on the interval $(a_j,b_j]$, $|a_\sigma(t)|$ is almost everywhere equal to zero. Hence,

$$\int_0^s f(\bar{t}(v))dv = \int_0^{\bar{t}(s)} f(\bar{t}(u(t)))\frac{|a_\sigma(t)|}{\rho}dt = \int_0^{\bar{t}(s)} f(t)\frac{|a_\sigma(t)|}{\rho}dt.$$

(The first equation is the result of the change of variable $v = u(t)$, and the second equation is justified by the fact that $|a_\sigma(t)|$ is zero where $\bar{t}(u(t)) \ne t$.) We obtain the desired equation by applying this

formula for $f(v)$ equal to the second integrand in (3.3). Notice that $\tilde{a}_j(v)$ is defined because $|a_\sigma(\bar{t}(v))|$ is non-zero for almost every v.

4. Estimates for the heat kernel

Sanchez-Calle showed in [40] that for a Hormander type operator L the heat kernel satisfies

$$(4.1)\quad |\partial_t^k X_{j_1} \ldots X_{j_p} h(t,x,y)| \le C_{N,k,p} t^{-k-p/2} \mu(B(x,t^{1/2}))^{-1} (1 + d(x,y)^2/t)^{-N},$$

for $0 < t < 1$, $x,y \in \Omega$ and all nonnegative integers N and k. (Henceforth we will frequently drop the subscript L from d_L and B_L. We will also abbreviate $\partial_t = \partial/\partial t$.) This was improved in [23] to

$$h(t,x,y) \le A\mu(B(x,t^{1/2}))^{-1} e^{-d(x,y)^2/\gamma t}$$

$$(4.2)\quad h(t,x,y) \ge A'\mu(B(x,t^{1/2}))^{-1} e^{-d(x,y)/\gamma' t}$$

$$|\partial_t^k X_{j_1} \ldots X_{j_p} h(t,x,y)| \le C_{p,k} t^{-k-p/2} \mu(B(x,t^{1/2}))^{-1} e^{-d(x,y)^2/\gamma t},$$

for some positive constants A, A', γ, γ', $C_{p,k}$, and all $x,y \in \Omega$ $0 < t < 1$. In both cases, the basic tool is the approximation by nilpotent Lie groups method developed by Folland, Rothschild and Stein [11,12,39]. Similar results were obtained using probabilistic methods by Kusuoka and Stroock [29,30]; they also discuss estimates as t tends to infinity under a uniform global subelliptic hypothesis. Furthermore, Davies [4] used logarithmic Sobolev inequalities to prove estimates of the form

$$(4.3)\quad h(t,x,y) \le C_\delta t^{-q} e^{-d(x,y)^2/4(1+\delta)t}, \quad \text{for all } \delta > 0.$$

(Davies used the distance $\beta_L(x,y)$, which we showed was the same as $d_L(x,y)$ in the last section. Another proof of Davies's result was given independently at the same time by Melrose; see the discussion of the wave equation below.)

The exponent q in Davies's result is only sharp in the constant rank case, referred to at the end of Section 2. On the other hand, the exponential term is essentially sharp as indicated by the elliptic case and as we shall now see in more detail. As a consequence of (4.3) one obtains

$$\limsup_{t \to 0+} 4t \log h(t,x,y) \le -d(x,y)^2.$$

As we mentioned in Section 2, Leandre proved that

$$\lim_{t \to 0+} 4t \log h(t,x,y) \text{ exists and equals } -d(x,y)^2.$$

Thus Davies's result is quite sharp when x and y are unequal and fixed and t tends to zero.

However the upper bound (4.1) is the only one of the numbered inequalities that is known to be correct up to order of magnitude, and this is only the case in the range of values of x, y, and t for which $d(x,y)^2 < Ct$. The estimate (4.2) is an improvement on the other two provided that $1 \ll d(x,y)^2/t \ll \log 1/t$.

For general subelliptic operators, estimate (4.1) was proved by Fefferman and Sanchez-Calle [8]. Also the result of Davies holds in this general setting. However, the existence of the Leandre limit is not yet known. The proof of (4.2) for general subelliptic operators will be presented later in this section. Another proof can be found in [29]. Both of these proofs rely on [8].

It would by nice to know the size of the heat kernel more exactly even in the Hormander-type case. For instance, is there an asymptotic expansion for $h(t,x,y)$ as $t \to 0+$ that is uniform in x and y? A less ambitious question is whether such an asymptotic expansion exists for $h(t,x,x)$. The hope is to find a main term expressed in terms of some average of $\mu(B(x,r))$.

Suppose that L is a subelliptic operator on a compact manifold M, that is, on each coordinate patch L has a formula like the one given above on Ω. Let $N(\lambda,L)$ denote the number of eigenvalues of L of modulus less than λ. Fefferman and Phong proved that for large λ, $N(\lambda,L)$ is bounded above and below by constant multiples of

$$\tilde{N}(\lambda,L) = \int_M \mu(B(x,\lambda^{-1/2})^{-1} d\mu(x)$$

Understanding more exactly the asymptotic behavior of the trace of the heat kernel $\int_M h(t,x,x)d\mu(x)$ would lead to a more precise knowledge of the asymptotics of $N(\lambda,L)$. We cannot expect an expansion in terms of powers of t. In the case of Hormander-type operators, there is an asymptotic expansion of the form $h(t,x,x) = c_0(x)t^{-q} + \dots$, but q depends on the point x. It is only constant in the constant rank case mentioned above--in that case the remainder term in the expansion is estimated uniformly in x. The key difficulty in understanding the trace is to have some uniform control on the remainder. (See [33], where the constant rank condition was first considered.)

Next, let us look at the wave operator $\partial_t^2 - L$. Melrose [32] proved that the support of the forward fundamental solution (the one supported in $t>0$) of the wave equation is contained in the set where $d(x,y) \leq t$. He deduced (4.3) from this. Actually, the distance defined by Melrose was given in terms of bicharacteristic curves. For

elliptic operators this is well known to be equivalent to the distance we have defined. Since in his proof the distance is obtained as a limit of the case $L + \epsilon\Delta$ as $\epsilon \to 0+$, Proposition 3.1 shows that his result can be formulated in terms of our metric. One can also expect to deduce asymptotics of $N(\lambda,L)$ from information about solutions to the wave equation.

We will conclude this section by proving (4.2) in the general subelliptic case. The lower bound for $h(t,x,y)$ follows exactly as in [23], with the added estimates of [8]. The upper bound will follow from the rescaling described in Section 3 and the following uniform Gevrey regularity result.

Lemma 4.1 *Let L be a subelliptic operator defined in the Euclidean unit cube $Q_1(0)$. Then there is an $R > 0$ that depends only on the subellipticity constants of L such that if $(\partial/\partial t - L)u(t,x) = 0$ and $|u(t,x)| \le 1$ for $|t| \le 1$, $|x| \le 1$, then $|(\partial/\partial t)^k u(t,x)| \le R^k(k!)^2$ whenever $|t| \le 1/2$ and $|x| \le 1/2$, $k = 0,1,2,\dots$.*

Proof: We will use the following notations. C and D will denote constants depending only on c_0, C_α, n and the subelliptic constants, which may change from place to place. Let

$$I(r) = \{(t,x)\in\mathbf{R}\times\mathbf{R}^n: \ |t| < r,\ |x| < r\}.$$

A classical argument [25] shows that if $\varphi, \psi \in C_0^\infty(I(1))$, $\psi = 1$ on $\operatorname{supp}\varphi$, then

$$\|\varphi v\|_{s+\epsilon} \le C_{s,\varphi,\psi}(\|\psi(\partial_t - L)v\|_s + \|\psi v\|) \quad \text{for all} \quad v \in C^\infty(I(1)).$$

(The norms are defined in (1.1).) Applying this to $v = \partial_t^k u$ with $\varphi = 1$ in $I(1/2)$ and $\operatorname{supp}\psi \subset I(2/3)$, $s = (n+1)/2$ and using the Sobolev embedding we get

$$\sup_{I(1/2)} |\partial_t^k u(t,x)| \le C\|\partial_t^k u\|_{L^2(I(2/3))},$$

where the L^2 norm is with respect to the measure $dtd\mu$. (Observe that $(\partial_t - L)v = 0$.) It is therefore enough to estimate $\|\partial_t^k u\|_{L^2(I(2/3))}$. We will prove by induction that

$$\|\partial_t^k u\|_{L^2(I(1-2k\delta))} \le D^k\delta^{-2k}.$$

The theorem will then follow by taking $\delta = 1/6k$.

Choose $\varphi_{k,\delta}$ and $\psi_{k,\delta}$, smooth functions satisfying

$$\varphi_{k-1,\delta} = 1 \ \text{on} \ I(1-2k\delta) \ \text{and} \ \varphi_{k-1,\delta} = 0 \ \text{outside} \ I(1-2k\delta+\delta),$$

$$\psi_{k-1,\delta} = 1 \ \text{on} \ I(1-2k\delta+\delta) \ \text{and} \ \psi_{k-1,\delta} = 0 \ \text{outside} \ I(1-2k\delta+2\delta),$$

$$|\nabla^j\varphi_{k,\delta}| + |\nabla^j\psi_{k,\delta}| \le C\delta^{-j} \quad \text{for} \quad j = 0,1,2.$$

Assume that $\|\partial_t^k u\|_{L^2(I(1-2k\delta))} \le D^k\delta^{-2k}$. (Observe that this is true for $k = 0$ since $|u| \le 1$.) Next, we have

$$\|\partial_t^{k+1}u\|_{L^2(I(1-2(k+1)\delta))} \le \|\partial_t(\varphi_{k,\delta}\partial_t^k u)\| \le \|(\partial_t-L)(\varphi_{k,\delta}\partial_t^k u)\|.$$

The latter inequality was proved in the discussion of the heat kernel in Section 2. Since $(\partial_t-L)\partial_t^k u = 0$, we have

$$(\partial_t-L)(\varphi_{k,\delta}\partial_t^k u) = (\partial_t-L)(\varphi_{k,\delta})\partial_t^k u - 2\sum_{i=1}^n(\partial\varphi_{k,\delta}/\partial x_i)Y_i\partial_t^k u,$$

where as before we denote $Y_i = \sum_{j=1}^n a_{ij}(x)\partial/\partial x_j$. Therefore,

$$\|\partial_t^{k+1}u\|_{L^2(I(1-2(k+1)\delta))} \le C\delta^{-2}D^k\delta^{-2k} + C\delta^{-1}\left[\sum_{i=1}^n\|Y_i(\psi_{k,\delta}\partial_t^k u)\|^2\right]^{1/2}.$$

But $\sum_{i=1}^n\left[\sum_{j=1}^n a_{ij}\xi_j\right]^2 \le C\sum_{i,j=1}^n a_{ij}\xi_i\xi_j$, so

$$\sum_{i=1}^n\|Y_i(\psi_{k,\delta}\partial_t^k u)\|^2 \le C|\langle L(\psi_{k,\delta}\partial_t^k u),\psi_{k,\delta}\partial_t^k u\rangle|.$$

Now, $L(\psi_{k,\delta}\partial_t^k u) = L(\psi_{k,\delta})\partial_t^k u + \psi_{k,\delta}L\partial_t^k u + 2\sum_{i=1}^n(\partial\psi_{k,\delta}/\partial x_i)Y_i\partial_t^k u$, and $L\partial_t^k u = \partial_t^{k+1}u$; so $\sum_{i=1}^n\|Y_i(\psi_{k,\delta}\partial_t^k u)\|^2 \le$

$$C\{\delta^{-2}(D^k\delta^{-2k})^2 + |\langle\psi_{k,\delta}^2\partial_t^k u,\partial_t^{k+1}u\rangle| + \sum_{i=1}^n|\langle\psi_{k,\delta}Y_i\partial_t^k u,(\partial\psi_{k,\delta}/\partial x_i)\partial_t^k u\rangle|\}.$$

Integration by parts gives

$$\langle\psi_{k,\delta}^2\partial_t^k u,\partial_t^{k+1}u\rangle = -(1/2)\langle\partial_t(\psi_{k,\delta}^2)\partial_t^k u,\partial_t^k u\rangle.$$

Also, $\langle\psi_{k,\delta}Y_i\partial_t^k u,(\partial\psi_{k,\delta}/\partial x_i)\partial_t^k u\rangle =$

$$= \langle Y_i(\psi_{k,\delta}\partial_t^k u),(\partial\psi_{k,\delta}/\partial x_i)\partial_t^k u\rangle - \langle Y_i(\psi_{k,\delta})\partial_t^k u,(\partial\psi_{k,\delta}/\partial x_i)\partial_t^k u\rangle.$$

Thus

$$\sum_{i=1}^n\|Y(\psi_{k,\delta}\partial_t^k u\|^2 \le CD^{2k}\delta^{-4k-2} + CD^k\delta^{-2k-1}\left[\sum_{i=1}^n\|Y_i(\psi_{k,\delta}\partial_t^k u)\|^2\right]^{1/2},$$

and therefore

$$\left[\sum_{i=1}^n\|Y_i(\psi_{k,\delta}\partial_t^k u)\|^2\right]^{1/2} \le CD^k\delta^{-2k-1}.$$

Finally, combining this with our earlier estimate for $\partial_t^{k+1}u$,

$$\|\partial_t^{k+1}u\|_{L^2(I(1-2(k+1)\delta))} \le CD^k\delta^{-2(k+1)} \le D^{k+1}\delta^{-2(k+1)},$$

provided we choose $D \ge C$. This concludes the proof of Lemma 3.1.

An application of Taylor's formula [23] then yields

<u>Lemma 4.2</u> *Under the assumptions of Lemma 3.1, if in addition* $\partial_t^k u(0,x) = 0$ *for all* k, *then* $|u(t,x)| \le Ce^{-1/\gamma t}$ *for* $|t| \le 1/2$ *and* $|x| \le 1/2$, *where* C *and* γ *depend only on the subellipticity constants of* L.

We are now in a position to prove the upper bounds in (4.2). First of all, the estimate (4.1) (proved in the general subelliptic case in [8]) gives in particular $h(t,x,y) \le C/\mu(B(x,\sqrt{t}))$. This is the correct upper bound if $d(x,y)^2 \le t$. If $d(x,y)^2 > t$, let ϕ be

the rescaling transformation $\phi_{\rho,x}$ defined in Section 3, with $\rho = d(x,y)/2K$ for K as in Section 3. The function $u(s,z) = h(\rho^2 s,\phi(z),y)$ satisfies the equation $(\partial_s - \hat{L})u = 0$ for $|s| \le 1$ and $|z| \le 1$, where $\hat{L}$ is the pullback under the change of variables ϕ of L, denoted $L_{\rho,x}$ in Section 3. Also $u(s,z) = 0$ if $s \le 0$. We would now like to check that $u(s,z) \le C/\mu(B(x,\sqrt{t}))$ for $|s| \le 1$ and $|z| \le 1$. In fact 4.1 implies that $u(s,z) \le C_N s^N/\mu(B(\phi(z),\rho\sqrt{s}))$ for all N. (Note that by definition of K, $d(\phi(z),y) \ge d(x,y)/2$.) Since from 3.2 we have $\mu(B(q,R\delta)) \le CR^m\mu(B(q,\delta))$, we conclude that $\mu(B(\phi(z),\rho)) \le Cs^{-m/2}\mu(B(\phi(z),\rho\sqrt{s}))$. Thus if we take $N \ge m/2$,

$$u(s,z) \le C/\mu(B(\phi(z),\rho)) \le C/\mu(B(x,\rho)) \le C/\mu(B(x,\sqrt{t})).$$

(The last two inequalities hold because $d(\phi(z),x) \le C\rho$, $\rho^2 \ge t/4K^2$, and because of the doubling property 3.1.)

The proof of estimates for the derivatives of $h(t,x,y)$ is analogous. The main point is that a subunit vector field Y scales by ϕ to a vector field $\rho^{-1}Z$ where Z is subunit for $\hat{L}$. Also, for solutions to $(\partial_s - \hat{L})u = 0$, the C^{k+p} norm of u is controlled by the sup norm. The details are left to the reader. (See also [23]).

5. The Poincare inequality

Jerison [21] proved that for vector fields $X_1,\dots,X_m$ satisfying Hormander's condition we have a Poincare-type inequality of the form

$$\int_\Omega (f - f_\Omega)^2 d\mu \le C\rho^2 \int_\Omega \Sigma|X_j f|^2 d\mu \quad \text{for all } f \in C^\infty(\bar{\Omega})$$

where $\Omega = B_L(x,\rho)$ $(L = -\Sigma X_j^* X_j)$ and $f_\Omega = (1/\mu(\Omega))\int f d\mu$. The extremal for this inequality is, at least formally, the first nontrivial eigenfunction for the Neumann problem in Ω, that is, a function φ satisfying $L\varphi = -\lambda\varphi$ in Ω and $\Sigma\, a_{jk}\nu_j \partial\varphi/\partial x_k = 0$ on $\partial\Omega$, where $(\nu_1,\dots,\nu_n)$ is the normal to $\partial\Omega$. The bound in the Poincare inequality is equivalent to a lower bound $\lambda \ge C^{-1}\rho^{-2}$. The Poincare inequality is trivial for functions that are compactly supported in the open set Ω. In that case it is not necessary to subtract the constant f_Ω and the inequality is true for any bounded open set. This corresponds to a positive lower bound for the first eigenvalue in the Dirichlet problem in any open, bounded set. In contrast, the full inequality can fail even on C^∞ domains Ω. However, a ball in the metric $d_L(x,y)$ is always sufficiently nice for the inequality to hold. The proof of this inequality follows the

Lie group approach. It depends on elementary (but not easy) estimates for the geometry of the balls $B_L(x,r)$ due to Nagel, Stein, and Wainger[36]. We will extend this inequality to the general subelliptic case shortly.

One of the reasons for proving a Poincare inequality is that with it one can carry out the proof of the Harnack inequality due to Moser [34]. His proof was designed for the case of uniformly elliptic divergence form operators with bounded measurable coefficients. The result in our setting is

Suppose u *is a positive solution to* $Lu = 0$ *in* $B(x,2\rho)$, *then there is a constant* C *independent of* ρ, $0 < \rho < 1$, *and* x *for which*

$$\max_{B(x,\rho)} u \leq C \min_{B(x,\rho)} u.$$

With this scale invariant version of Harnack's inequality, one can, for instance, carry out much of potential theory in domains as in [1],[6],[22]. Robert Johnson proves the Harnack inequality for operators of the form $\Sigma X_i^* a_{ij}(x) X_j$ for bounded measurable coefficients a_{ij} in his forthcoming dissertation at Princeton University.

Kusuoka and Stroock have developed a "non-elementary" approach to the Poincare and Harnack inequalities based on estimates of [8] and [40] for the heat kernel.

The general Poincare inequality can be stated as follows.

<u>Theorem 5.1</u> *Let* L *be a subelliptic operator defined in the unit cube* $Q_1(0)$. *Then there are constants* C *and* $\rho_0 > 0$ *that depend only on the subellipticity constants of* L *such that if* $0 < \rho < \rho_0$ *and* $|x| \leq 1/2$,

$$\int_B |u - u_B|^2 d\mu \leq C\rho^2 \int_B \sum_{i,j=1}^n a_{ij}(\partial u/\partial x_i)(\partial u/\partial x_j) d\mu$$

where $B = B_L(x,\rho)$, $u_B = (1/\mu(B))\int u\,d\mu$.

We will deduce this inequality from the version for Hormander type operators and the following quantitative version of Proposition 2.1. Recall the notation $Y_i = \Sigma_{j=1}^n a_{ij}\partial/\partial x_j$. For a multi-index $J = (i_1, \ldots, i_s)$ we denote $|J| = s$ and $Y_J = [Y_{i_1}, \ldots, [Y_{i_{s-1}}, Y_{i_s}]\ldots] = \sum_{j=1}^n a_J^j(x)\partial/\partial x_j$. Finally, for $X = \sum_{j=1}^n b_j(x)\partial/\partial x_j$, we define the symbol $X(x;\xi) = \sum_{j=1}^n b_j(x)\xi_j$.

<u>Proposition 2.1'</u> *There is a constant* $r > 0$ *and an integer* N *depending only on the subellipticity constants of* L *such that*

$$\sum_{|J| \leq N} |Y_J(0;\xi)| \geq r|\xi|, \quad \textit{for all} \quad \xi \in \mathbf{R}^n.$$

(This is a quantitative statement of Hormander's condition at the origin.)

Covering arguments in [21] based only on the doubling property (3.1) show that in order to prove Theorem 5.1, it suffices to prove

$$\min_{a \in \mathbf{R}} \int_{B_1} (u - a)^2 d\mu \leq C\rho^2 \int_{B_2} \sum_{i,j=1}^{n} a_{ij}(\partial u/\partial x_i)(\partial u/\partial x_j) d\mu,$$

where $B_1 = B_L(x, c\rho)$ and $B_2 = B_L(x, \rho/c)$ for a positive constant c with the same dependence as C. If we change variables using the mapping $\Phi_{\rho,x}: Q_1(0) \longrightarrow Q_L(x,\rho)$ from Theorem 3.1, we can choose c so that $\Phi_{\rho,x}(Q_{1/2}(0)) \subset B_2$ and $B_1 \subset \Phi_{\rho,x}(Q_{1/4}(0))$; thus we can reduce matters to proving that

$$\min_{a \in \mathbf{R}} \int_{Q_{1/4}(0)} (u(y) - a)^2 d\mu(y) \leq C \int_{Q_{1/2}(0)} \sum_{i,j=1}^{n} \hat{a}_{ij}(y) \frac{\partial u}{\partial y_i}(y) \frac{\partial u}{\partial y_j}(y) d\mu(y)$$

where $\hat{a}_{ij}(y)$ are the coefficients of the operator $L_{\rho,x}$ of Theorem 3.1. By Proposition 2.1' (with the origin replaced by any point of $Q_{1/2}(0)$), the vector fields $\hat{Y}_i = \sum_{j=1}^{n} \hat{a}_{ij}(y)\partial/\partial y_j$ satisfy Hormander's condition uniformly for all ρ, x, and y. That condition is equivalent to the condition given in 2.3c) in [21], so we can apply the Poincare inequality for Hormander type vector fields to a finite covering of $Q_{1/4}(0)$ with balls associated to $\tilde{L} = -\Sigma_{i=1}^{n} \hat{Y}_i^* Y_i$ of a fixed small radius so that they are contained in $Q_{1/2}(0)$. The balls can be chosen so that each is connected by a chain of balls to one central ball in such a way that successive balls in the chain overlap in a significant fraction of their measure. (See [21].) It then follows that

$$\min_{a \in \mathbf{R}} \int_{Q_{1/4}(0)} (u - a)^2 d\mu \leq C \int_{Q_{1/2}(0)} \sum_{i=1}^{n} (\hat{Y}_i u)^2 d\mu,$$

with a constant C independent of ρ and x. However, if (c_{ij}) is any positive semi-definite matrix,

$$\sum_{i,j,k=1}^{n} c_{ij} c_{ik} \xi_j \xi_k \leq M \sum_{i,j=1}^{n} c_{ij} \xi_i \xi_j,$$

where M is the largest eigenvalue of (c_{ij}). (In fact, the inequality reduces to the trivial case of a diagonal matrix after an

orthogonal transformation of ξ.) The constant M depends only on an upper bound for the coefficients c_{ij}. Using this inequality with $c_{ij} = \hat{a}_{ij}$, we get $\sum_{i=1}^{n} (\hat{Y}_i u)^2 \leq M\sum_{i,j=1}^{n} \hat{a}_{ij}(\partial u/x_i)(\partial u/x_j)$, which proves the theorem.

The proof of Proposition 2.1' is rather technical. A similar procedure can be found in [37, Theorem 2.8.1]. Let L, (a_{ij}), and Y_i be as above. L is a subelliptic operator, and we will let ϵ be as in (1.2). Choose an integer $s > 10/\epsilon$. Let $r_1 > r_2 > \ldots > r_n > 0$ be numbers that we will choose later depending only on the subellipticity constants of L. We will abbreviate $\partial_j = \partial/\partial x_j$. Consider the following finite inductive procedure:

Step 1 Choose i such that $|Y_i(0)| \geq r_1$. If no such i exists then stop. If such an i does exist, then change coordinates so that $Y_i = \partial_1$.

Step 2 Let $N_1 = 1$ and $N_2 = 10(1+s)N_1$. Choose a multi-index J_2 of length $|J_2| \leq N_2$ and $(\xi_2,\ldots,\xi_n) \in \mathbf{R}^{n-1}$ of unit length such that $|Y_{J_2}(0;(0,\xi_2,\ldots,\xi_n))| \geq r_2$. If no such J exists, then stop. If it does exist, then change variables as follows. We can write $a^j_{J_2}(x) = h^j(0,x_2,\ldots,x_n) + x_1 g^j(x)$, $j=2,\ldots,n$, and make a change in the variables $(x_2,\ldots,x_n)$ so that $\sum_{j=2}^{n} h^j(0,x_2,\ldots,x_n)\partial_j$ becomes ∂_2.

Step k Let $N_k = 10(1+s)N_{k-1}$. Choose J_k such that $|J_k| \leq N_k$ and such that for some $(\xi_k,\ldots,\xi_n) \in \mathbf{R}^{n-k+1}$ of unit length, $|Y_{J_k}(0;(0,\ldots,0,\xi_k,\ldots,\xi_n))| \geq r_k$. If none such exists, then stop. If it does exist, then we make a change of the variables $(x_k,\ldots,x_n)$ so that

$$Y_i = \partial_1$$

$$Y_{J_2} = a_{J_2}(x)\partial_1 + \partial_2 + x_1\sum_{j=2}^{n} f^j(x)\partial_j$$

$$\ldots$$

$$Y_{J_k} = \sum_{j=1}^{k-1} a^j_{J_k}(x)\partial_j + \partial_k + \sum_{j=k}^{n}(x_1 f_1^{jk}(x) + \ldots + x_{k-1}f_{k-1}^{jk}(x))\partial_j.$$

(Here we have retained the notation $a^j_J(x)$ after change of variable.)

If this procedure does not stop until after Step n, then we have proved that the vector fields Y_i satisfy Hormander's condition at the origin with $N = N_n$ and $r = r_n$. If the procedure does stop, then we will prove that the subelliptic estimate (1.2) fails.

Suppose that the procedure stops after Step k. Then it is easy to see that we can write $\partial_1,\ldots,\partial_k$ in terms of Y_J as follows.

$$\partial_1 = Y_i$$

$$\partial_2 = \sum_{|J|\le N_2} a_{2,J}Y_J + x_1 Z_{12},$$

$$\ldots\ldots$$

$$\partial_k = \sum_{|J|\le N_k} a_{k,J}Y_J + \sum_{i=1}^{k-1} x_i Z_{ik},$$

for some C^∞ coefficients $a_{j,J}$ and C^∞ vector fields Z_{ij} with bounds on their derivatives depending only on the subellipticity constants of L. Moreover, for all $(\xi_{k+1},\ldots,\xi_n)$ of unit length and all J, $|J| \le N_{k+1}$, $|Y_J(0;(0,\ldots,0,\xi_{k+1},\ldots,\xi_n))| \le r_{k+1}$. (Note that this statement applies equally well in the case $k = 0$.)

Next, we claim that we can write

$$(5.1)\qquad \partial_p = \sum_{|J|\le 2(1+s)N_p} b_{p,J}Y_J + E_{p,s}, \quad 1 \le p \le k.$$

(The coefficients $b_{p,J}$ and the vector fields $E_{p,s}$ are likewise controlled by the subellipticity constants of L.) Moreover, The coefficients of $E_{p,s}$ are bounded by $C_s(|x_1| +\ldots+ |x_p|)^s$. We will explain how to do this for the vector ∂_2 first. Suppose that we already have $\partial_2 = V_{2,j} + x_1^jF_j$ with $V_{2,j}$ a linear combination of the Y_J. Observe that $[\partial_1,V_{2,j}] = [\partial_1,\partial_2 - x_1^jF_j] = -jx_1^{j-1}F_j - x_1^j[\partial_1,F_j]$. Therefore, $\partial_2 = V_{2,j} - (1/j)x_1([\partial_1,V_{2,j}] + x_1^j[\partial_1,F_j])$, which has the form $V_{2,j+1} + x_1^{j+1}F_{j+1}$, since ∂_1 is itself a linear combination of the Y_J—indeed, it is Y_i. Rather than write out the general case we will illustrate it with an example of how to use ∂_2 to increase the powers of x_2 that appear in the error term. Suppose that we already have $\partial_2 = V_{2,j} + x_1^jF_jS$ and $\partial_3 = V_{3,j} + x_1^jG_j + x_2^\ell H_\ell$, $j \ge \ell+1$, with $V_{2,j}$ and $V_{3,j}$ linear combinations of the Y_J, and suppose that we are trying to increase ℓ by one. Observe that

$$[V_{2,j},V_{3,j}] = -\ell x_2^{\ell-1}H_\ell + x_1^j[\partial_3,F_j] - x_1^j[\partial_2,G_j] - x_2^\ell[\partial_2,H_\ell] + [x_1^jF_j,x_1^jG_j] + [x_1^jF_j,x_2^\ell H_\ell].$$

Thus if we write,

$$\partial_3 = V_{3,j} - (1/\ell)x_2[V_{2,j},V_{3,j}] + x_1^jG_j + (x_2^\ell H_\ell + (1/\ell)x_2[V_{2,j},V_{3,j}]),$$

we see that ∂_3 is expressed as a linear combination of the Y_J and a term whose coefficients are bounded by $C(|x_1| + |x_2|)^{\ell+1}$.

Now, with all due apology, we have to distinguish the original variable $x = (x_1,\ldots,x_n)$ from the change of variable $\tilde{x} = (\tilde{x}_1,\ldots,\tilde{x}_k,\tilde{x}_{k+1},\ldots,\tilde{x}_n)$ in order to keep track of how the operator L has changed. We will also write $y = (y_1,\ldots,y_k) = (\tilde{x}_1,\ldots,\tilde{x}_k)$ and

$t = (\tilde{x}_{k+1},\dots,\tilde{x}_n)$. Denote $Y_i = \sum_{j=1}^{n} d_{ij}(\tilde{x})\partial/\partial\tilde{x}_j$. If $d\mu(x) = h(x)dx$ $= \tilde{h}(\tilde{x})d\tilde{x}$, then $-\int(Lu)(u)d\mu = \int\sum_{i,j=1}^{n}\tilde{a}_{ij}(\tilde{x})(\partial u/\partial\tilde{x}_i)(\partial u/\partial\tilde{x}_j)\tilde{h}(\tilde{x})d\tilde{x}$. Since $L = \sum_{i=1}^{n}(1/h(x))(\partial/\partial x_i)(h(x)Y_i)$ and $\partial/\partial x_i =$ $\sum_{j=1}^{n}(\partial\tilde{x}_j/\partial x_i)\partial/\partial\tilde{x}_j$, $\tilde{a}_{\ell j}(\tilde{x}) = \sum_{i=1}^{n} d_{ij}(\tilde{x})(\partial\tilde{x}_\ell/\partial x_i)$. It is clear that bounds on the functions $\tilde{x}_\ell$ as functions of x and upper and lower bounds for $\tilde{h}(\tilde{x})$ depend only on the subellipticity constants s and $r_1,\dots,r_k$, which in turn depend only on the subellipticity constants. We claim that there is a constant C_k depending only on these constants such that for $j = k+1,\dots,n$,

(5.2) $|(\partial^\alpha d_{ij}/\partial y^\alpha)(0)| \le C_k r_{k+1}$, for all i and $|\alpha| < s$,

(5.3) $|\tilde{a}_{\ell j}(y,t)| \le C_k(r_{k+1} + |y|^s + |t|)$ for all ℓ,

(5.4) $|\tilde{a}_{\ell j}(y,t)| \le C_k(r_{k+1} + |y|^s + |t|^2)$ for $\ell = k+1,\dots,n$.

To prove (5.2) we write

$$(\partial^\alpha d_{ij}/\partial y^\alpha)(0) = [\underbrace{\partial/\partial y_1,[\partial/\partial y_1,\dots}_{\alpha_1\text{ times}},\underbrace{[\partial/\partial y_2,\dots}_{\alpha_2\text{ times}\dots},[\partial/\partial y_k,Y_i]]..]](0;e_j)$$

where $e_j = (0,\dots,0,1,0,\dots,0)$, the unit vector in the jth slot. The formula (5.1) for each $\partial/\partial y_p$, $p = 1,\dots,k$ says that they are each linear combinations of Y_J modulo error terms that vanish to order s. Therefore, if $|\alpha| < s$, the value of the commutator at the origin is the same as if the $\partial/\partial y_p$ were replaced by a linear combination of Y_J. Thus $(\partial^\alpha d_{ij}/\partial y^\alpha)(0)$ is a linear combination of terms $Y_J(0;e_j)$ with $|J| \le N_{k+1}$, because we chose N_{k+1} much larger than N_k. The terms $Y_J(0;e_j)$ are precisely those for which we have the estimate $|Y_J(0;e_j)| \le r_{k+1}$, so we have proved (5.2). The estimate (5.3) for $\tilde{a}_{\ell j}(y,t)$ is an immediate consequence of (5.2) and Taylor's formula given the expression for $\tilde{a}_{\ell j}$ in terms of d_{ij}. Finally, inequality (5.4) is a corollary of (5.3) and the

<u>Remark</u> *Suppose that* $(a_{ij}(u))$ *is a positive semidefinite matrix and a* C^2 *function of the real variable* u. *If* $|a_{ij}(u)| \le |u| + \delta$ *and* $|d^2a_{ij}/du^2| \le 1$ *for all* i,j, *then* $|a_{ij}(u)| \le 2\delta + u^2$.

The remark follows immediately from the fact that $|a_{ij}|^2 \le a_{ii}a_{jj}$ and the corresponding property for non-negative scalar functions: $a_{ii}(u) \le 2a_{ii}(0) + u^2$. We can now conclude the proof by contradicting the subelliptic estimate (1.2). Choose fixed nonzero functions φ and $\psi \in C_0^\infty$. Denote $u(y,t) =$ $\sigma^{-k/2}\varphi(y/\sigma)\upsilon^{-(n-k)/2}\psi(t/\upsilon)$, with $\upsilon < \sigma << 1$ to be

chosen later. The change of variables from x to (y,t) is controlled by the subellipticity constants, s, and $r_1,\dots,r_k$, all of which depend only on subellipticity constants. Thus the L^2 and ϵ-Sobolev norm in the coordinates (y,t) are comparable to the ones in the x coordinates. It is important to note that these constants do not depend on r_{k+1}. Our test function is normalized so that its L^2 norm is 1. Also, (5.3) and (5.4) show that

$$\begin{aligned}\int(Lu)u\,d\mu &\le C_k(\sigma^{-2} + (r_{k+1}+\sigma^s+\upsilon^2)\upsilon^{-2} + (r_{k+1}+\sigma^s+\upsilon)\upsilon^{-1}\sigma^{-1}) \\ &\le 10C_k(\sigma^{-2} + r_{k+1}\sigma^{-s}),\end{aligned}$$

if we choose $\upsilon = \sigma^{s/2}$. On the other hand, $\|u\|_\epsilon^2 \ge C_k^{-1}\upsilon^{-2\epsilon} = C_k^{-1}\sigma^{-\epsilon s}$. Therefore, if the estimate (1.2) were true, we would have a constant C depending only on the subelliptic constants, s, and $r_1,\dots,r_k$, such that $1 \le C(\sigma^{-2} + r_{k+1}\sigma^{-s})\sigma^{\epsilon s} \le C(\sigma^{-2} + r_{k+1}\sigma^{-s})\sigma^{10}$. If we choose $\sigma^8 = 1/2C$, and then choose $r_{k+1} < \sigma^{s-10}/2C$, then we arrive at a contradiction, $1 < 1$.

6. <u>The classical Poincaré inequality</u>

As we mentioned above, one difficulty with the Poincaré inequality for subelliptic operators is that it is false for certain C^∞ domains. The property of a ball (say $B = B_L(x_0,1)$) in the metric $d_L(x,y)$ that makes it better than an arbitrary C^∞ domain is that there are constants C and $\alpha > 0$ such that for any $r > 0$, any δ, $0 < \delta < 1$, and any $y \in \partial B$,

$$(6.1)\qquad \mu(B\cap B_L(y,r)\cap\{x:\ d_L(x,\partial B) \le \delta r\}) \le C\delta^\alpha\mu(B_L(y,r)).$$

This sort of condition is obvious for a C^∞ domain in the elliptic case. In the elliptic case the domain must be very pathological in order for the Poincaré inequality to fail.

What follows is a discussion of the Poincaré inequality in the simplest case, leading up to a conjecture about how to characterize the domains for which it holds. The simplest case is that of a simply-connected open set Ω in the plane. The classical Poincaré inequality is:

$$(6.2)\qquad \int_\Omega (f - f_\Omega)^2dx \le C\int_\Omega |\nabla f|^2dx, \quad \text{for all } f \in C^\infty(\Omega).$$

(Here we interpret the inequality as valid if the right hand side is infinite. As usual, f_Ω is the average value of f on Ω. One could also ask the question for $f \in C^\infty(R^2)$.)

This inequality can fail for the trivial reason that the region Ω is not connected. If Ω has two connected components, then the

function that is 0 on one component and 1 on the other violates (6.2). This just reflects the fact that 0 is a multiple eigenvalue for the Neumann problem. On the other hand we can imagine a region made up of two fixed squares joined by an arbitrarily thin strip for which the constant C is arbitrarily large. If we rescale this picture to a small scale and string together a countable collection of such regions, we can construct a bounded, connected set for which the inequality fails. Here is an example in the form of a comb with infinitely many teeth.

$$\Omega = \{(x,y):\ 0 < x < 1,\ 0 < y < 1,\ \text{and } y < r_k \text{ for } 2^{-2k} \leq x \leq 2^{-2k+1},\ \ k = 1,2,\ldots\}$$

If the numbers r_k decrease to zero sufficiently fast, then the inequality fails. For example, it fails if $r_k = 2^{-100k}$.

Hummell [19] showed that the inequality still can fail even if we restrict to analytic functions f. D. H. Hamilton has pointed out recently that the question for analytic f and general f are equivalent. For a discussion of related function theoretic questions, see [2,41]. In particular, Stegenga has a characterization of domains satisfying (6.2) in terms of the conformal mapping to the disk.

Inequality (6.2) is readily reduced to a discrete inequality by means of a Whitney decomposition. By a Whitney decomposition we mean a family of squares Q_j such that Ω is the union of the Q_j the distance from Q_j to the boundary of Ω is comparable to the side length of Q_j and distinct squares overlap at most on an edge. The decomposition can be constructed by taking maximal dyadic squares in a dyadic grid with the comparability property. (See [42].) It is not hard to see from the Poincare inequality in a square that it suffices to look at functions that are essentially constant on Whitney squares. In this way the problem becomes discrete.

A simple sufficient condition for (6.2) which is somewhat weaker than the condition (6.1) is

<u>Proposition</u> (See [21]) *Let $\rho(x,y)$ be the quasi-hyperbolic distance between x and y, that is, the smallest number of squares in a chain of adjacent Whitney squares joining x to y. If $\int_\Omega \rho(x_0,y)dy < C < \infty$, then (6.1) holds with a comparable constant.*

In order to motivate our guess for a necessary and sufficient condition we will look at a one-dimensional model case. Consider

$$\Omega = \{(x,y):\ 0 < x < 1,\ 0 < y < w(x)\}$$

for a continuous function $0 < w(x) \le 1$ for $0 < x < 1$. We will also suppose that $w(x) = 1$ for $0 < x < 1/2$. Thus the pathology of the region is concentrated at the end where $x = 1$. Finally, to simplify our analysis even further we will concentrate on test functions f that depend only on x. The right hand side of (6.2) equals $C\int_0^1 f'(x)^2 w(x)dx$. The left hand side is $\int_0^1 (f(x) - f_\Omega)^2 w(x)dx$. Because we have assumed that $w(x) = 1$ for $x \le 1/2$, (6.2) is equivalent to the similar inequality in which we subtract $f(0)$ instead of f_Ω. Thus, if we put $h(x) = f'(x)$, we are considering the inequality

$$(6.3)\qquad \int_0^1 \Big[\int_0^x h(s)ds\Big]^2 w(x)dx \le C\int_0^1 h(x)^2 w(x)dx.$$

This is a weighted Hardy-type inequality. The "weights" $w(x)$ for which it is satisfied were characterized by Muckenhoupt as a special case of a large family of inequalities involving exponents other than 2.

Proposition [35] *A necessary and sufficient condition in order that (6.3) hold for all $h \in L^2(w(x)dx)$ is that*

$$(6.4)\qquad \sup_a \int_0^a w(x)^{-1}dx \int_a^1 w(x)dx \le C < \infty.$$

Moreover, the constants C in (6.3) and (6.4) are comparable.

Proof: The implication (6.3) implies (6.4) is trivial: take $h(s) = \chi_{[0,a]}(s)w(s)^{-1}$. For the converse, denote $H(x) = \int_0^x w(s)^{-1}ds$, $sUt = \max(s,t)$, and $I(x) = \int_x^1 w(s)ds$. By (6.4),

$$\int_0^1 \Big[\int_0^x h(s)ds\Big]^2 w(x)dx = \int_0^1\int_0^x h(s)ds\int_0^x h(t)dt\,w(x)dx =$$

$$\int_0^1\int_0^1 h(s)h(t)I(sUt)dsdt \le C\int_0^1\int_0^1 h(s)h(t)H(sUt)^{-1}dsdt.$$

Now let $g(x) = h(x)w(x)^{1/2}$. We want to show that

$$\int_0^1\int_0^1 g(s)g(t)w(s)^{-1/2}w(t)^{-1/2}H(sUt)^{-1}dsdt \le A\int_0^1 g(x)^2dx.$$

for some absolute constant A. Equivalently, we want to show for another absolute constant which we will also denote by A that

$$\Big[\int_0^1\Big[\int_0^1 w(s)^{-1/2}w(t)^{-1/2}H(sUt)^{-1}g(s)ds\Big]^2dt\Big]^{1/2} \le A\Big[\int_0^1 g(s)^2ds\Big]^{1/2}.$$

Without loss of generality we can estimate the integral on the left hand side restricted to the region $s \geq t$. Applying successively Minkowski's and Schwarz's inequalities, we have

$$\left[\int_0^1\left[\int_t^1 w(s)^{-1/2}w(t)^{-1/2}H(s\cup t)^{-1}g(s)ds\right]^2 dt\right]^{1/2} =$$

$$= \left[\int_0^1\left[\sum_{k=0}^{\infty} \int_{2^kH(t)\leq H(s)\leq 2^{k+1}H(t)} w(s)^{-1/2}w(t)^{-1/2}H(s)^{-1}g(s)ds\right]^2 dt\right]^{1/2} \leq$$

$$\leq \sum_{k=0}^{\infty}\left[\int_0^1\left[\int_{2^kH(t)\leq H(s)\leq 2^{k+1}H(t)} w(s)^{-1/2}w(t)^{-1/2}H(s)^{-1}g(s)ds\right]^2 dt\right]^{1/2} \leq$$

$$\leq \sum_{k=0}^{\infty}\left[\int_0^1 \int_{2^kH(t)\leq H(s)\leq 2^{k+1}H(t)} w(s)^{-1}w(t)^{-1}H(s)^{-2}ds \int_{2^kH(t)\leq H(s)\leq 2^{k+1}H(t)} g(s)^2ds \quad dt\right]^{1/2} =$$

$$= \sum_{k=0}^{\infty}\left[\int_0^1 2^{-k-1}w(t)^{-1}H(t)^{-1}\int_{2^kH(t)\leq H(s)\leq 2^{k+1}H(t)} g(s)^2ds \quad dt\right]^{1/2} =$$

$$= \sum_{k=0}^{\infty}\left[\int_0^1 2^{-k-1}\int_{2^{-k-1}H(s)\leq H(t)\leq 2^{-k}H(s)} w(t)^{-1}H(t)^{-1}dt \quad g(s)^2ds\right]^{1/2} =$$

$$= \sum_{k=0}^{\infty}\left[2^{-k-1}\int_0^1(\log 2)g(s)^2ds\right]^{1/2} = A\left[\int_0^1 g(s)^2ds\right]^{1/2}.$$

The argument that we have just given can also be stated in discrete form: *There is an absolute constant* A *such that any sequence of positive numbers* a_i, $i = 1,2,\ldots$ *satisfy*

$$\sum_{i,j=1}^{\infty}(a_ia_j)^{1/2}c_ic_j/(A_i+A_j) \leq A\sum_{i=1}^{\infty}c_i^2, \text{ for all square summable}$$

sequences c_i, *where* $A_j = \sum_{i=1}^{j} a_i$. In this form the inequality is due to Schur [15]. (The example $a_i = 1$ for all i is the Hilbert series.)

One can interpret condition (6.4) as a discrete condition on Whitney squares. The largest Whitney square near the boundary point $(x,0)$ has sidelength roughly $w(x)$. Thus, if we take the test function $h(x) = \chi_{[0,a]}(x)w(x)^{-1}$, the derivative $f'(x)$ is comparable to the reciprocal of the sidelength of the Whitney square, and f increases by about 1 each time we traverse a Whitney square. This suggests a lower bound for the constant in the general case (6.2), which we will now explain.

We will suppose that Ω is contained in the disk of diameter 10 and that Ω contains the disk of diameter $1/10$. There is a natural (almost) partial ordering for the Whitney decomposition of Ω

analogous to the linear ordering by x in the one-dimensional model. We fix the largest Whitney square Q_0. We say that two squares Q_1 and Q_2 are related by $Q_1 \leq Q_2$ if the shortest chain of squares from Q_2 to Q_0 passes within say 10 squares of Q_1. This is not quite a partial ordering, but it essentially is if we think of squares that are within roughly 10 squares of eachother as equivalent. Now consider an arbitrary finite collection $Q_1,\dots,Q_N$ of unrelated Whitney squares. Suppose that the length of the shortest chain from Q_j to Q_0 is k_j. Denote by M the number of Whitney squares in the *union* of the chains from each Q_j, $j = 1,\dots,N$, to Q_0. (Note that M may be much smaller than $\sum_{j=1}^{N} k_j$.) Denote by A_j the area of the union of the Whitney squares $\geq Q_j$. The test function that we can construct is 0 on the "center" square Q_0 and increases by one as we pass from one Whitney square to the next along each of the chains from Q_0 to Q_j. For Whitney cubes larger than Q_j the function will have the constant value k_j. The smoothest function f that we can construct with these properties will have the property that $|\nabla f|$ is bounded by $1/s$, where s is the sidelength of one of the M Whitney squares, and $|\nabla f| = 0$ elsewhere. Since the area of such Whitney squares is s^2, the integral of $|\nabla f|^2$ on a Whitney square is bounded by an absolute constant A. Thus $\int_\Omega |\nabla f|^2 dx \leq AM$. On the other hand, because f is 0 on a large set Q_0, there is no loss of generality in replacing the average value f_Ω by 0. Since f takes the value k_j on a set with area A_j (and these sets are disjoint for different values of j), the left hand side of (6.2) is bounded below by $\sum_{j=1}^{N} k_j^2 A_j$. Thus we have a lower bound for the constant C in (6.2) of the form $\sum_{j=1}^{N} k_j^2 A_j / AM$. The conjecture is that this lower bound is comparable to an upper bound.

The conjecture asserts something much stronger than a characterization of domains satisfying the Poincare inequality. It suggests an estimate within an order of magnitude for the size of the lowest eigenvalue in the Neumann problem, depending only on the diameter and inradius of the region. The conjecture is, of course, correct in the linearly ordered model. The problem is to prove estimates for quadratic forms on partially ordered sets.

References

[1] Ancona, A., Principe de Harnack à la frontière et théorème de Fatou pour un opérateur elliptique dans un domaine Lipschitzien, Ann. Inst. Fourier 28(4), 169-213(1978).

[2] Axler, S. and Shields, A.L., Univalent multipliers of the Dirichlet space, preprint.

[3] Bony, J.-M., Principe du maximum, inégalité de Harnack et unicité du problème de Cauchy pour les opérateurs elliptiques dégénérés, Ann. Inst. Fourier 19(1), 277-304(1969).

[4] Davies, E.B., Explicit constants for Gaussian upper bounds on heat kernels, preprint.

[5] Derridj, M., Un problème aux limites pour une classe d'opérateurs hypoelliptiques du second ordre, Ann. Inst. Fourier 21(4), 99-148(1971).

[6] Fabes, E., Jerison, D. and Kenig, C., Boundary behavior of solutions to degenerate elliptic equations, Proceedings of the Conference on Harmonic Analysis in Honor of Antoni Zygmond, Wadsworth Math. Series, 577-589(1981).

[7] Fefferman, C.L. and Phong, D.H., Subelliptic eigenvalue problems, Proceedings of the Conference on Harmonic Analysis in Honor of Antoni Zygmund, Wadsworth Math. Series, 590-606(1981).

[8] Fefferman, C.L. and Sánchez-Calle, A., Fundamental solutions for second order subelliptic operators, Ann. of Math. 124, 247-272(1986).

[9] Fichera, G., On a unified theory of boundary value problems for elliptic-parabolic equations of second order, In Boundary problems in Differential Equations, Univ. of Wisconsin Press, Madison, 97-120(1973).

[10] Folland, G.B., A fundamental solution for a subelliptic operator, Bull. Amer. Math. Soc. 79, 373-376(1973).

[11] Folland, G.B., Subelliptic estimates and function spaces on nilpotent Lie groups, Arkiv för Mat. 13, 161-207(1975).

[12] Folland, G.B. and Stein, E.M., Estimates for the $\bar{\partial}_b$-complex and analysis on the Heisenberg group, Comm. Pure Appl. Math. 27, 429-522(1974).

[13] Gaveau, B., Principe de moindre action, propagation de la chaleur et estimées sous elliptiques sur certains groupes nilpotents, Acta Math. 139, 95-153(1977).

[14] Grushin, V., On a class of hypoelliptic pseudodifferential operators degenerate on a submanifold, Math. USSR Sbornik 13, 155-185(1971).

[15] Hardy, G.H., Littlewood, J.E. and Pólya, G., Inequalities, Cambridge University Press (1952).

[16] Hörmander, L., Hypoelliptic second order differential equations, Acta Math. 119, 147-171(1967).

[17] Hörmander, L.: Subelliptic operators, Annals of Math. Studies 91, Princeton Univ. Pres, 127-208 (1978).

[18] Hueber, H. Further examples of irregular domains for some hypoelliptic operators, preprint.

[19] Hummell, J.A., Counterexamples to the Poincaré inequality, Proc. Amer. Math. Soc. 8, 207-210(1957).

[20] Jerison, D., The Dirichlet problem for the Kohn Laplacian on the Heisenberg group, I,II, Jour. of Funct. Anal. 43, 97-142, 224-257(1981).

[21] Jerison, D., The Poincaré inequality for vector fields satisfying Hörmander's condition, Duke Math. Jour. 53, 503-523(1986).

[22] Jerison, D. and Kenig, C., Boundary behavior of harmonic functions in nontangentially accessible domains, Advances in Math. 46, 80-147(1982)

[23] Jerison, D. and Sánchez-Calle, A., Estimates for the heat kernel for a sum of squares of vector fields, to appear in Indiana Univ. J. of Math.

[24] Kohn, J.J., Boundaries of complex manifolds, Proc. conf. on complex manifolds (Minneapolis), Springer-Verlag, N.Y., 81-94(1965).

[25] Kohn, J.J., Pseudodifferential operators and hypoellipticity, Proc. Symp. Pure Math. 23, 61-69(1973).

[26] Kohn, J.J.: Estimates for $\bar{\partial}_b$ on compact pseudoconvex CR manifolds, Proc. Symp. Pure Math. 43, 207-217(1985).

[27] Kohn, J.J. and Nirenberg, L., Degenerate elliptic-parabolic equations of second order, Comm. Pure Appl. Math. 20, 797-872(1967).

[28] Kohn, J.J. and Rossi, H., On the extension of holomorphic functions from the boundary of a complex manifold, Ann. Math. 81, 451-472(1965).

[29] Kusuoka, S. and Stroock, D., Applications of the Malliavin calculus, part III, preprint.

[30] Kusuoka, S. and Stroock, D., Long time estimates for the heat kernel associated with a uniformly subelliptic symmetric second order operator, preprint.

[31] Léandre, R., Estimation en temps petit de la densité d'une diffusion hypoelliptique, C.R. Acad. Sc. Paris, t.301, série I, 17, 801-804(1985).

[32] Melrose, R., Propagation for the wave group of a positive subelliptic second order differential operator, preprint.

[33] Metivier, G., Fonction spectrale et valeurs propres d'une classe d'opérateurs non elliptiques, Comm. in P.D.E. 1(5), 467-519(1976).

[34] Moser, J., On Harnack's theorem for elliptic differential equations, Comm. Pure Appl. Math. 14, 577-591(1961).

[35] Muckenhoupt, B., Hardy's unequality with weights, Studia Math. 44, 31-38(1972).

[36] Nagel, A., Stein, E.M. and Wainger, S., Balls and metrics defined by vector fields I: Basic properties, Acta Math. 155, 103-147(1985).

[37] Oleinik, O. and Radkevic, E., Second order equations with nonnegative characteristic form, Amer. Math. Society, Providence, R.I. (1973).

[38] Phillips, R.S. and Sarason, L., Elliptic-Parabolic equations of the second order, Jour. of Math. and Mech. 17, 891-917(1968).

[39] Rothschild, L. and Stein, E.M., Hypoelliptic differential operators and nilpotent Lie groups, Acta Math. 137, 247-320(1977).

[40] Sánchez-Calle, A., Fundamental solutions and geometry of the sum of squares of vector fields, Invent. Math. 78, 143-160(1984).

[41] Stegenga, D., Multipliers of the Dirichlet space, Illinois J. Math. 24, 113-139(1980).

[42] Stein, E.M., Singular integrals and differentiability properties of functions, Princeton Univ. Press (1970).

[43] Stroock, D. and Taniguchi, S., Regular points for the first boundary problem associated with degenerate elliptic equations, preprint.

[44] Xu, C.-J., Opérateurs sous-elliptiques et regularité des solutions d'équations aux derivées partielles non linéaires du second ordre en deux variables, U. of Paris-Sud, preprint.

Department of Mathematics
Massachusetts Institute of Technology
Cambridge, Massachusetts 02139

RECENT RESULTS ON HOMOGENEOUS COMPLEX MANIFOLDS

by

Karl Oeljeklaus and Wolfgang Richthofer

Introduction.

Throughout this paper a complex manifold X is called homogeneous, if there is a complex Lie group G, which acts transitively and holomorphically as a group of biholomorphic transformations on X. For a $p \in X$ let H be the isotropy group of p, i.e., $H = \{g \in G | g(p) = p\}$. The group H is a closed complex subgroup of G and the manifold X can be naturally identified with the quotient space G/H, the G-action being simply the left multiplication.

The central problem in the theory of homogeneous complex manifolds is to classify or characterize these manifolds in terms of certain additional conditions, for example:

1) the topology, e.g. $\dim X$ or X compact,
2) group theoretical conditions on G,
3) differential geometric conditions, e.g. X Kähler,
4) function theory of X, e.g. conditions on the algebra of holomorphic function $\mathcal{O}(X)$, the field of meromorphic functions $\mathcal{M}(X)$ or the set of complex analytic hypersurfaces $\mathcal{H}(X)$.

Now we mention some results, which enlighten the philosophy of the general problem stated above.

Theorem 1 (Borel-Remmert [BR], Grauert-Remmert [GR]). Let X be a compact, homogeneous, complex manifold and assume that X satisfies one of the following conditions:

i) X is Kähler
ii) X is hypersurface separable (see section 1).

Then X is isomorphic to a product $Q \times T$, where Q is a homogeneous rational manifold and T is a compact, complex torus. In case ii) T is an abelian variety.

Theorem 2 (Huckleberry-Margulis [HuM]). Let S be a semisimple complex Lie group and $I \subset S$ be a closed, complex subgroup of S. Then the following conditions are equivalent:

i) I is Zariski dense in S
ii) For the homogeneous manifold $X = S/I$ the set of analytic

hypersurfaces $\mathcal{H}(X)$ is empty.

These theorems yield partial solutions of the following

<u>Conjecture (Ahiezer)</u>. Let G be a complex Lie group and $H \subset G$ an arbitrary subgroup. Assume that every holomorphic function on G, which is invariant under the right action on H is constant, i.e., $\mathcal{O}(G)^H \cong \mathbb{C}$ and that H is not contained in any proper parabolic subgroup of G. Then a hypersurface F which is invariant under the right H-action on G is also invariant under the action of the commutator group G' of G, i.e., $\mathcal{H}(G)^H = \mathcal{H}(G)^{H \cdot G'}$.

<u>Remark</u>. If $\hat{H}$ denotes the smallest closed complex subgroup of G, which contains H, then by a theorem of Remmert-van de Ven [RV] we have $\mathcal{H}(G)^H = \mathcal{H}(G)^{\hat{H}}$ and $\hat{H}$ satisfies all conditions of the conjecture. Hence it is enough to consider the homogeneous manifold $X = G/\hat{H}$ and study the hypersurfaces there:

Let $X = G/\hat{H} \xrightarrow{p} G/J = Z$ be the hypersurface reduction of X (see section 1). If the conjecture is true, then $G' \subset J$ and the base Z is a complex abelian Lie group C with $\mathcal{O}(C) \cong \mathbb{C}$, which we refer to as a <u>Cousin group</u> (see section 1). In particular, it follows that the field of meromorphic functions $\mathcal{M}(X)$ is isomorphic (via pullback) to the field $\mathcal{M}(C)$. It is known that the conjecture is true in several special cases:

1) G is semisimple. In this case H is Zariski dense and the conjecture follows from Theorem 2 [HuM].
2) $G/\hat{H}$ is compact. Here the conjecture is an immediate consequence of Theorem 1, case ii) [GR].
3) G is nilpotent. For a nilpotent group the conjecture was proved independently in [A] and [Oe].

One of the results of this paper is the proof of the Ahiezer conjecture in two cases:

i) G is solvable

ii) $G = S \times R$, where S is semisimple and R is the radical of G, i.e., G is a group theoretic product of S and R.

A homogeneous complex manifold X is called a <u>solvmanifold</u> if there is a solvable complex Lie group acting holomorphically and transitively on X. The following theorem characterizes the structure of certain solvmanifolds.

<u>Main Theorem</u>. Let $X = G/H$ be a solvmanifold. Assume that X is i)

Kähler or ii) hypersurface separable. Then the holomorphic reduction $\pi : X = G/H \longrightarrow G/J = Y$ realizes X as a holomorphic, G-equivariant fiber bundle over a Stein solvmanifold Y. The fiber J/H is a connected abelian Lie group with $\mathcal{O}(J/H) \cong \mathbb{C}$, i.e., a Cousin group. Furthermore, $\pi_1(X)$ contains a nilpotent, normal subgroup of finite index.

This theorem will be proved in sections 2-7.

<u>Corollary a</u>. The Ahiezer conjecture is true, if G is solvable.

<u>Proof</u>: We only have to see that the base of the hypersurface reduction $z = G/I$ of $G/\hat{H}$ is a Cousin group. But $\mathcal{O}(Z) \cong \mathbb{C}$ and Z is separable by hypersurfaces. Hence Z is the fiber of the holomorphic reduction of Z and therefore a Cousin group.

<u>Corollary b</u>. Let $\tau : X = G/H \longrightarrow G/L$ be a holomorphic, G-equivariant fibration of the solvmanifold X. Assume that the fiber L/H and the base G/L are Stein manifolds. Assume furthermore that X is Kähler or hypersurface separable. Then X is Stein, i.e., the Serre problem has a positive answer in this setting.

<u>Proof</u>. Let $\pi : G/H \longrightarrow G/J$ be the holomorphic reduction of X. Then, by the main theorem, J/H is a Cousin group, and there is a diagram

$$\begin{array}{ccc} G/H & \longrightarrow & G/J \\ & \searrow \quad \swarrow & \\ & G/L & \end{array}$$

Since L/H is Stein and J/H is an orbit in L/H it follows that $J = H$. Hence X is Stein. □

The paper is organized as follows: In section 1, we give the necessary techniques, methods and known results, which will be used in the following. Sections 2-7 are devoted to the proof of the Main Theorem. In Section 8, we prove the special cases of the Ahiezer conjecture mentioned above and give a new proof of Theorem 2.

1. <u>Preliminaries</u>.

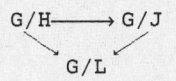

Throughout this paper we are concerned with complex solvmanifolds, i.e., with homogeneous spaces $X = G/H$, where G is a complex, solvable Lie group and H a closed complex subgroup of G. In this section we gather some basic definitions and known results about the complex analytic structure of $X = G/H$. An important differential

geometric concept for our approach is given in

<u>Definition 1</u>. A complex manifold X is weakly Kähler if there is a closed, real $(1,1)$-form ω which is positive semidefinite on X and positive definite in at least one point of X.

Note that by continuity a form given as in Definition 1 is positive definite on some open set $U \subset X$. Note also that if X satisfies Definition 1 then every convering space of X does as well. If moreover X is homogeneous then also every (immersed) complex submanifold of X satisfies Definition 1.

A basic technique to approach the structure of the algebra of holomorphic functions $\mathcal{O}(X)$ of $X = G/H$ is the holomorphic reduction $X = G/H \xrightarrow{\pi} G/J = Y$, which is given by the following equivalence relation: two points, $x,y \in X$ are equivalent if and only if $f(x) = f(y)$ for every $f \in \mathcal{O}(X)$.

This equivalence relation is $\mathrm{Aut}_{\mathcal{O}}(X)$-equivariant and the stabilizer $J \subset G$ of the equivalence class containing $x = H \in G/H$ is a closed complex subgroup of G (see [GH], Theorem 1.1). The base of the holomorphic reduction is holomorphically separable and $\mathcal{O}(X) = \pi^*\mathcal{O}(Y)$. Moreover Y is characterized by the universal property that every holomorphic map F from X into a holomorphically separable complex space factorizes through Y. In general there are no non-trivial conditions known that force the fiber J/H of the holomorphic reduction to have only constant holomorphic functions.

<u>Definition 2</u>. A complex abelian Lie group G with $\mathcal{O}(G) \cong \mathbb{C}$ is called a Cousin group.

It is clear that a Cousin group is abelian and of the form $\mathbb{C}^n/\Gamma$, where Γ is a discrete additive subgroup of $\mathbb{C}^n$. P. Cousin [Co] was the first who considered such groups. The following theorem is due to B. Gilligan and A.T. Huckleberry ([GH]).

<u>Theorem 3</u>. Let G be a nilpotent complex Lie group and H a closed complex subgroup. Then there is a closed complex subgroup $J \subset G$ containing H, such that

1) if $\pi : G/H \longrightarrow G/J$ denotes the bundle projection, then $\mathcal{O}(G/H) = \pi^*\mathcal{O}(G/J)$,
2) G/J is Stein,
3) the fiber J/H is connected
4) $\mathcal{O}(J/H) = \mathbb{C}$
5) J/H is a principal abelian Lie group tower.

Remarks.

a) In [Oe] it was shown that J/H in Theorem 3 is a principal Cousin group tower.

b) We will see later (see section 2) that Therorem 3 is false in the general solvable case, since the fiber of the holomorphic reduction may be even a Stein manifold.

We now give some facts about solvmanifolds. First, if G is a (not necessarily closed) complex Lie subgroup of $GL_{n+1}(\mathbb{C})$, then every orbit of G in $\mathbb{P}_n(\mathbb{C})$ is isomorphic to a product $(\mathbb{C}^*)^k \times \mathbb{C}^\ell$. Note that given $X = G/H$ and the normalizer $N = N_G(H^0)$ of the identity component H^0 of H, we have an equivariant embedding of G/N into some $\mathbb{P}_n(\mathbb{C})$. We refer to $G/H \longrightarrow G/N$ as the Tits fibration of G/H (see [HO1]). From the remark above and the universality of the holomorphic reduction $G/H \longrightarrow G/J$, it follows that we always have $J \subset N$.

Proposition 4. Let G be a simply connected solvable Lie group and H a connected complex subgroup. Then $G/H \cong \mathbb{C}^n$.

The proof of Proposition 4 follows easily by Grauert's Oka principle and induction on dimensions, using [HO1], I.2.5.7.

Grauert's Oka principle states the following ([G]):

A holomorphic fiber bundle over a Stein manifold with complex structure group is holomorphically trivial, if and only if it is topologically trivial.

The most recent result about solvmanifolds G/H gives in particular a characterization of the base G/J of the holomorphic reduction, which satisfies the maximal rank condition. By definition, a complex n-dimensional manifold X satisfies the maximal rank condition if there are $f_1, \ldots, f_n \in \mathcal{O}(X)$ such that $df_1 \wedge \ldots \wedge df_n \not\equiv 0$.

Theorem 5. Let $X = G/H$ be a complex solvmanifold satisfying the maximal rank condition. Then X is Stein. Moreover $\pi_1(X)$ contains a nilpotent normal subgroup of finite index.

This theorem is due to E. Oeljeklaus and A.T. Huckleberry (see [HO2]). It will be proved in section 3.

To prepare for this we need some terminology and facts about algebraic groups, and in particular, about embeddings of solvable groups into algebraic ones.

A subgroup of $GL_n(\mathbb{C})$ is called algebraic, if (as a set) it is defined by algebraic equations. If $G \subset GL_n(\mathbb{C})$ is algebraic, then a maximal torus of G is defined to be maximal abelian subgroup of

elements $g \in G$, such that $\mathrm{Ad}g$ is diagonizable, e.g. if $G =$ the upper triangular matrices, then the diagonal subgroup is a maximal torus of G.

An element g of an arbitrary complex Lie group G is called regular if the Zariski closure of $\mathbb{Z}g := \{\mathrm{Ad}g^n : n \in \mathbb{Z}\}$ in $\mathfrak{gl}(\mathfrak{G})$ contains a maximal torus of the Zariski closure of $\mathrm{Ad}(G)$. Here $\mathfrak{G}$ denotes the Lie algebra of G. For details concerning this notion we refer to [Mo].

A sort of algebraization yields the following (see [HM], Theorem 3.1). Let G be a simply connected solvable complex Lie group. Then there exists a solvable linear algebraic group G_a, which has the form

$$G_a = (\mathbb{C}^*)^k \ltimes G$$

containing G as a Zariski dense and topologically closed normal subgroup, i.e., $G_a = (\mathbb{C}^*)^k \times G$ as a manifold and $(\mathbb{C}^*)^k$ acts by conjugation on the normal part of G.

Note also that as a complex manifold, any simply connected solvable complex Lie group is isomorphic to some $\mathbb{C}^n$.

We will call G_a an abstract algebraic closure of G. A Zariski-dense subgroup G of an algebraic group S always contains a regular element of S. For G and G_a we have that the commutator subgroups G_a' of G_a and G' of G coincide (see [Bo]).

For every closed complex subgroup $H \subset G$, the complex manifolds G_a/H and $(\mathbb{C}^*)^k \times G/H$ are biholomorphically equivalent. In particular G_a/H is weakly Kähler if and only if G/H is weakly Kähler. If $G_a/H \longrightarrow G_a/J$ denotes the holomorphic reduction, then $J \subset G$ and $G_a/J = (\mathbb{C}^*)^k \times G/J$. It is also clear that $\pi_1(G/H)$ is nilpotent if and only if $\pi_1(G_a/H)$ is nilpotent.

Certain subgroups of an algebraic group are always algebraic, e.g. let G be any complex Lie subgroup of an algebraic group, then its commutator group G' and the normalizer of G are algebraic.

We close our short description of the algebraic setting with two remarks. First, any algebraic group has finitely many connected components. Second, any quotient G/H of solvable algebraic groups is Stein.

Later in sections 3 and 6 we will combine this fact with the following

Theorem (Matsushima and Morimoto [MM]). Let $E \longrightarrow B$ be a holomorphic

fiber bundle with fiber F and a complex structure group S. If S has finitely many connected components and F and B are Stein, then E is Stein.

One of the goals of this paper is to study the hypersurfaces (divisors with multiplicity one) of a solvmanifold X.

To study the set $\mathcal{H}(X)$ of hypersurfaces in $X = G/H$ it is convenient to have a reduction analogous to the holomorphic reduction. For this we define the following equivalence relation: Two points $x,y \in X$ are equivalent if and only if for every $H \in \mathcal{H}(X)$ with $x \in H$ it follows $y \in H$ and vice versa. This equivalence relation yields a holomorphic equivariant fibration

$$p : X = G/H \longrightarrow G/I = Z,$$

such that $\mathcal{H}(X) = \pi^*\mathcal{H}(Z)$, i.e., every hypersurface in X is the preimage of one in Z. Moreover, Z is separable by hypersurfaces, i.e., given $x,y \in Z$, $x \neq y$, there exists an $H \in \mathcal{H}(Z)$ such that $x \in H$, $y \notin H$. For details see ([Oe], Kap. I). A manifold Z is said to be locally separable by global hypersurfaces, if for every point $z \in Z$ there are hypersurfaces $F_1,\ldots,F_k$ such that $z \in F_i$, $i = 1,\ldots,k$ and $\bigcap_{i=i}^{k} F_i$ contains z as an isolated point. For a homogeneous manifold $X = G/\Gamma$, where Γ is discrete in G, this condition implies that X is weakly Kähler (see section 7).

2. <u>On a special class of complex solvmanifolds--Results of J.J. Loeb</u>.

The aim of this section is to consider complex solvmanifolds fulfilling certain additional conditions. The results are due to J.J. Loeb [Lo]. Let G be a solvable, simply connected, complex Lie group. Then G is isomorphic (as a complex manifold) to some $\mathbb{C}^n$. Let $\Gamma \subset G$ be a discrete subgroup satisfying the following properties:

i) There is a connected, real Lie group $G_0 \subset G$ such that $\Gamma \subset G_0$ and the quotient G_0/Γ is compact.

ii) The group G_0 is a <u>real form</u> of G, i.e., for the associated Lie algebras $\mathfrak{G}_0$ and $\mathfrak{G}$ one has $\mathfrak{G} = \mathfrak{G}_0 \oplus i\mathfrak{G}_0$.

<u>Remark</u>. The group G_0 is automatically closed in G and is isomormorphic (as a manifold) to $\mathbb{R}^n$.

We refer to triples (G,G_0,Γ) with the above properties as <u>totally real CR-solvmanifolds</u> (or totally real CRS). The main problem is to understand function theoretic and differential geometric properties of the solvmanifold $X = G/\Gamma$.

Before doing this we consider pairs (G,G_0), there G_0 is a <u>real form</u> of the simply connected, solvable, complex Lie group G. We refer to (G,G_0) as a <u>solvable pair</u>.

<u>Definition 1</u>. A solvable pair (G,G_0) is called <u>pseudoconvex</u>, if there is a C^∞, strictly plurisubharmonic function ψ on G which is invariant under the right G_0-action on G (right-G_0- invariant) and is an exhaustion function on G/G_0.

<u>Remark</u>. The quotient G/G_0 is isomorphic to $\mathbb{R}^n$.

<u>Definition 2</u>. A real Lie algebra $\mathfrak{G}_0$ has an imaginary spectrum, if for all $x \in \mathfrak{G}_0$ the operator $\operatorname{ad} x : \mathfrak{G} \longrightarrow \mathfrak{G}$ $(\mathfrak{G} = \mathfrak{G}_0 \oplus i\mathfrak{G}_0)$, $y \longmapsto [x,y]$ has only imaginary eigenvalues.

Now we are in a position to state the first main result of this section.

<u>Theorem 3</u>. A solvable pair (G,G_0) is pseudoconvex if and only if the Lie algebra $\mathfrak{G}_0$ of G has an imaginary spectrum.

We will only give the proof of the direction "(G,G_0) is pseudoconvex $\Rightarrow \mathfrak{G}_0$ has an imaginary spectrum." The proof of the converse is not essential for our further considerations and can be found in [Lo].

Before proving the theorem we are going to explain the condition on the imaginary spectrum in detail. We start by giving two important counterexamples.

<u>Example I</u>. Let A be the (unique) non-abelian, simply connected, two-dimensional, complex Lie group, i.e.

$$A = \left\{ \begin{pmatrix} e^z & 0 & 0 \\ 0 & 1 & 0 \\ b & z & 1 \end{pmatrix} \in GL_3(\mathbb{C}) \;\middle|\; z,b \in \mathbb{C} \right\}$$

and G_0 the real form of G given by

$$A_0 = \left\{ \begin{pmatrix} e^z & 0 & 0 \\ 0 & 1 & 0 \\ b & z & 1 \end{pmatrix} \in GL_3(\mathbb{C}) \;\middle|\; z,b \in \mathbb{R} \right\}.$$

Then in the Lie algebra $\mathfrak{A}_0$ of A_0 there are two elements u,v such that $[u,v] = v$, i.e., $\mathfrak{A}_0$ has not an imaginary spectrum. The group multiplication of A can be written in the form $(z,b)(z_0,b_0) = (z+z_0,e^{z_0}\cdot b+b_0)$.

<u>Example II</u>. Let $I = \begin{pmatrix} 1 & 0 \\ 0 & 1 \end{pmatrix}$ and $J = \begin{pmatrix} 0 & 1 \\ -1 & 0 \end{pmatrix}$. Fur $\mu \in \mathbb{R}$ define the matrix $A^\mu = I + \mu J = \begin{pmatrix} 1 & \mu \\ -\mu & 1 \end{pmatrix}$. Then A^μ is diagonizable over the complex numbers and the eigenvalues are $1 \pm i\mu$. Now define a group G^μ structure on $\mathbb{C}\times\mathbb{C}^2$ by

$$(z,b)\cdot(z_0,b_0) = (z+z_0,e^{A_\mu z_0}\cdot b+b_0) \ , \ z,z_0 \in \mathbb{C} \ , \ b,b_0 \in \mathbb{C}^2.$$

Let G_0^μ be the subgroup of G^μ for which $z \in \mathbb{R}$ and $b \in \mathbb{R}^2$. Clearly G_0^μ is a real form of G^μ. In the Lie algebra, $\mathfrak{G}_0^\mu \subset \mathfrak{G}^\mu$ there are three elements, x,u,v such that $[u,v] = 0$, $[x,u] = (1+i|\mu|)\cdot u \in \mathfrak{G}^\mu$ and $[x,v] = (1-i|\mu|)\cdot v \in \mathfrak{G}^\mu$. We see that the Lie algebra $\mathfrak{G}^\mu$ has not an imaginary spectrum.

The next step of our considerations is to study the plurisubharmonic functions in the two previous examples on the complex Lie group, which are right invariant under teh action of the real form. In order to do this for Example I we need

<u>Lemma 4</u>. There is a family $(f_\varepsilon)_{\varepsilon\in\mathbb{R}_+^*}$ of holomorphic functions on the closed unit disc $\bar{\Delta}$ satisfying the following conditions:

i) $|\mathrm{Im}\, f_\varepsilon| \le \pi$ for all $\varepsilon > o$.

ii) $f_\varepsilon(0) = c = c_1+ic_2$ const. for all $\varepsilon > 0$, $c_1,c_2 \in \mathbb{R}$.

iii) The function $b_\varepsilon = b(\varepsilon) := \int_{\partial\Delta} e^{\mathrm{Re} f_\varepsilon} d\theta$ is continuous and $\lim_{\varepsilon\to\infty} b(\varepsilon) = +\infty$. (Here $d\theta$ denotes the normalized Haar measure on $\partial\Delta \cong S^1$.)

<u>Proof</u>. Let log denote the principal part of the logarithm on the upper half plane and let

$$f_\varepsilon(z) = \log i \,\frac{1+\varepsilon+z}{1+\varepsilon-z} \ , \ z \in \bar{\Delta}.$$

It is easy to see that i) and ii) are satisfied and that b_ε is continuous. By the lemma of Fatou we obtain

$$\lim_{\varepsilon\to\infty} \int_{\partial\Delta} e^{\mathrm{Re}\, f_\varepsilon} d\theta \ge \int_{\partial\Delta} \lim_{\varepsilon\to\infty} e^{\mathrm{Re}\, f_\varepsilon} d\theta = \int_{\partial\Delta} \left|\frac{1+e^{i\theta}}{1-e^{i\theta}}\right| d\theta = +\infty$$

and the lemma is proved. □

<u>Proposition 5</u>. A plurisubharmonic function $\psi(z,b)$ on the complex Lie group A, which is right-A_0-invariant depends only on the variable $\operatorname{Im} z$.

<u>Proof</u>. We define a family of functions $(g_\varepsilon)_{\varepsilon>0}$ on $\bar{\Delta}$, which are continuous and holomorphic on Δ, by the properties

$$g_\varepsilon(0) = 0 \quad \text{and} \quad \operatorname{Im} g_\varepsilon(z) = e^{\operatorname{Re} f_\varepsilon} - b_\varepsilon \quad \text{for} \quad z \in \partial\Delta .$$

The functions g_ε are well defined since $e^{\operatorname{Re} f_\varepsilon}$ is C^∞ on $\partial\Delta$ and $\int_{\partial\Delta} e^{\operatorname{Re} f_\varepsilon} d\Theta = b_\varepsilon$.

Let ψ be a p.s.h. (plurisubharmonic) function on A, which is right-A_0-invariant. Then the function

$$h_\varepsilon(z) = \psi(f_\varepsilon(z), g_\varepsilon(z)+ib_\varepsilon)$$

is semicontinuous on $\bar{\Delta}$ and p.s.h on Δ.

By the right-A_0-invariance of ψ it follows that

$$\psi(f_\varepsilon(z), g_\varepsilon(z)+ib) = \psi(i \operatorname{Im} f_\varepsilon(z), ie^{-\operatorname{Re} f_\varepsilon(z)}(\operatorname{Im} g_\varepsilon(z)+b_\varepsilon)).$$

In particular for $z \in \partial\Delta$ we have

$$h_\varepsilon(z) = \psi(i \operatorname{Im} f_\varepsilon(z), i)$$

in view of the choice of g_ε. Hence $|h_\varepsilon(z)| \le K = \text{const.}$ for all $z \in \partial\Delta$ and all $\varepsilon > 0$ by Lemma 1.4 i).

Furthermore $h_\varepsilon(0) = (ic_2, ie^{-c_1} b_\varepsilon)$. Since h_ε is p.s.h. on Δ we conclude

$$h_\varepsilon(0) = \psi(ic_2, ie^{-c_1} b_\varepsilon) \le \int_{\partial\Delta} h_\varepsilon d\Theta \le K \quad \text{for all} \quad \varepsilon > 0.$$

Using Lemma 1.4 iii) it turns out that

$$\psi(ic_2, ib) \le K' = \text{const. for all} \quad b \in \mathbb{R} \ , \ b \ge 0.$$

Making the same argument for the function $\varphi(z,b) := \psi(z,-b)$ the final estimation is

$$\psi(ic_2, ib) \le K'' = \text{const. for all} \quad b \in \mathbb{R},$$

and ψ being right-A_0-invariants gives

$$\psi(ic_2, b) \le K'' \quad \text{for all} \quad b \in \mathbb{C}.$$

Hence by the theorem of Liouville $\psi(ic_2, b)$ does not depend on the

variable b. After replacing $\psi(z,b)$ by $\psi(a+z,b)$, where $a \in \mathbb{C}$ is a constant and using again the right-A_0-invariance the statement of the proposition follows. □

Proposition 6. Let $\mu \in \mathbb{R}$ and $\psi(z,b)$, $(z \in \mathbb{C}, b \in \mathbb{C}^2)$ be a p.s.h. function on G^μ being right-G_0^μ-invariant. Then ψ depends only on Im z.

Proof. For $\Theta \in \mathbb{R}$ define $\psi_\Theta(z,b) = \psi(z,e^{\Theta \cdot J}b)$, which is also right-$G_0^\mu$-invariant and p.s.h.

For all $(z,b) \in G^\mu$, $(z_0,b_0) \in G^\mu$ we have

$$\psi_\Theta(z,b) = \psi(z,e^{\Theta \cdot J}\cdot b) = \psi(z+z_0,e^{z_0A^\mu}\cdot e^{\Theta \cdot J}\cdot b+e^{\Theta \cdot J}\cdot b_0)$$

$$= \psi(z+z_0,e^{\Theta \cdot J}(e^{z_0A^\mu}b+b_0)) = \psi_\Theta(z+z_0,e^{z_0A^\mu}+b_0).$$

Since $(b \longmapsto e^{\Theta \cdot J}\cdot b)$ is holomorphic the function $\psi^A(z,b) = \int_0^{2\pi} \psi(z,e^{\Theta \cdot J}\cdot b)d\Theta$ is also right-G_0^μ-invariant and plurisubharmonic. Therefore,

$$\psi^A(z,b) = \psi^A(i\ \mathrm{Im}\ z,e^{-\mathrm{Re}\ z\cdot A^\mu}\cdot \mathrm{Im}\ b)$$

$$= \psi^A(i\ \mathrm{Im}\ z,e^{-\mathrm{Re}\ z\cdot \mu\cdot J}\cdot e^{-\mathrm{Re}\ z}\cdot \mathrm{Im}\ b)$$

$$= \psi^A(i\ \mathrm{Im}\ z,e^{-\mathrm{Re}\ z}\cdot \mathrm{Im}\ b),$$

since ψ^A is invariant under the Θ-action. Let $b \in \mathbb{R}^2$, then for all $z \in \mathbb{C}$, $\lambda \in \mathbb{C}$

$$\psi^A(z,\lambda b) = \psi^A(i\ \mathrm{Im}\ z,e^{-\mathrm{Re}\ z}\mathrm{Im}\ \lambda\cdot b).$$

By Proposition 1.5 it follows that

$$\psi^A(z,ib) = \psi^A(z,0),$$

i.e., ψ^A only depends on the variable Im z.

Now define the function

$$\psi^S(z,b) = \sup_{\Theta\in[0,2\pi]} \psi(z,e^{\Theta \cdot J}b).$$

Since ψ^S is invariant under the Θ-action the same argument as for ψ^A shows that ψ^S depends only on Im z. Hence $\psi^S(z,b) = \psi^S(z,0) = \psi(z,0) = \psi^A(z,0) = \psi^A(z,b)$ and therefore $\psi(z,e^{\Theta \cdot J}b) = \psi(z,0)$ for all $\theta \in [0,2\pi]$, which proves the proposition. □

Corollary 7. For the solvable pairs (A,A_0) resp. (G^μ,G_0^μ), any

plurisubharmonic function ψ on A, resp. G^{μ}, which is right-A_0, resp. G_0^{μ}-invariant is not strictly plurisubharmonic at any point of A, resp. G^{μ}.

Proof. This follows immediately from Proposition 1.5 and Proposition 1.6. □

Now we give a characterization of a real Lie algebra $\mathfrak{G}_0$, which does not have an imaginary spectrum.

Proposition 8. Let $\mathfrak{G}_0$ be a real Lie algebra not admitting an imaginary spectrum. Then $\mathfrak{G}_0$ contains a subalgebra of the form $\mathfrak{A}_0$ or $\mathfrak{G}_0^{\mu}$.

Proof. Let $\mathfrak{G} = \mathfrak{G}_0 \oplus i\mathfrak{G}_0$ be the complexification of $\mathfrak{G}_0$. In view of the assumption there are elements $x \in \mathfrak{G}_0$ and $z = u+iv \in \mathfrak{G}$, $(u,v \in \mathfrak{G}_0)$ such that

$$\begin{aligned}[x,z] &= [x,u+iv] = [x,u] + i[x,v] = (1+i\beta)\cdot z \\ &= (1+i\beta)(u+iv) = (u-\beta v) + i(v+\beta u)\end{aligned} \qquad (*)$$

1) Case $\beta = 0$. Then $[x,u] = u$ and $\mathfrak{G}_0$ contains $\mathfrak{A}_0$.

2) Case $\beta \neq 0$. From $(*)$ we conclude

$$[x,u] = u-\beta v$$
$$[x,v] = v+\beta u$$

Moreover: $[x,[u,v]] = [[x,u],v] + [u,[x,v]] = 2[u,v]$.

We may assume that $\mathfrak{G}_0$ does not contain $\mathfrak{A}_0$. Therefore $[u,v] = 0$. Hence the vectors $x,u,v \in \mathfrak{G}_0$ span a real Lie algebra, which is isomorphic to $\mathfrak{G}_0^{\beta}$. □

Proof of Theorem 1.3 ("⇒"). Let (G,G_0) be a solvable pseudo-convex pair. Then there is a strictly plurisubharmonic function ψ on G, which is right-G_0-invariant and an exhaustion on G/G_0. Assume that $\mathfrak{G}_0$ does not have an imaginary spectrum. By Proposition 1.8 the Lie algebra $\mathfrak{G}_0$ contains a subalgebra $\mathfrak{h}_0$ of type $\mathfrak{A}_0$ or $\mathfrak{G}_0^{\mu}$. Let $\mathfrak{h} = \mathfrak{h}_0 \oplus i\mathfrak{h}_0$ and $H \subset G$ the associated complex Lie group. The restriction of ψ to H is strictly plurisub- harmonic and right-H_0-invariant. But this is impossible in view of Corollary 1.7. Hence the theorem follows. □

Before stating the next main result, we need the following

Definition 9. A solvable pair (G,G_0) is called weakly Kähler (resp. Kähler) if there is a real, closed, (1,1)-differential form ω on G

which is right-G_0-invariant, positive semidefinite and positive definite in at least one point of G (resp. in all points of G).

Theorem 10. Let (G,G_0) be a solvable pair. Then the following conditions are equivalent:

i) the Lie algebra $\mathfrak{G}_0$ has an imaginary spectrum

ii) the pair (G,G_0) is pseudoconvex

iii) the pair (G,G_0) is weakly Kähler

iv) The pair (G,G_0) is Kähler.

Remark. The equivalence i) $\Leftrightarrow$ ii) is the statement of Theorem 1.3. It is clear that iv) $\Rightarrow$ iii). Furthermore ii) implies iv) by defining $\omega = i\partial\bar{\partial}\psi$ for a strictly plurisubharmonic, right-G_0-invariant function on G.

Hence the only direction we have to prove is iii) $\Rightarrow$ i): If (G,G_0) is weakly Kähler $\Rightarrow$ $\mathfrak{G}_0$ has an imaginary spectrum. In order to prove this we need some more preparatory techniques and notations.

For a real C^∞-manifold X let $\Lambda^p X$ denote the space of p-forms on X and ΛX the graded differential algebra with exterior differentiation d. For $x \in X$ the tangent space to X at x is denoted by $T_x X$.

For a complex manifold Y we write $\Lambda^{p,q}(Y)$ for the space of (p,q)-forms and $\Lambda^0(Y)$ for the graded differential algebra, which is the direct sum of the $\Lambda^{p,0}$, with exterior differentiation ∂.

Let $R : \Lambda Y \longrightarrow \Lambda^0 Y$ be the homomorphism of graded differential algebras sending a p-form to its (p,0)-component. For $y \in Y$ the vector space of $(\Lambda,0)$-tangent vectors to Y at y is denoted by $T_Y^{1,0}(Y)$.

Let (G,G_0) be a solvable pair and let $\pi : G \longrightarrow G/G_0$ denote the projection map. Then the pullback map

$$\pi^* : \Lambda(G/G_0) \longrightarrow \Lambda(G)$$

is a homomorphism of graded differential algebras. We define $\Lambda_I^{p,q}(G)$ to be the space of (p,q)-forms on G, which are right-G_0-invariant. It is clear that

$$\partial(\Lambda_I^{p,q}(G)) \subseteq \Lambda_I^{p+1,q}(G)$$

In the same way the algebra $\Lambda_I^0(G)$ is defined.

Proposition 11. For every pair (p,q), $0 \le p,q \le n-\dim_{\mathbb{R}} G_0$ there is a commutative diagram:

$$\begin{array}{ccc} \Lambda_I^{p,q}(G) & \xrightarrow{\sim} & (\Lambda^p(G/G_0))^{\binom{n}{q}} \\ \downarrow \partial & & \downarrow d \\ \Lambda^{p+1,q}(G) & \xrightarrow{\sim} & (\Lambda^{p+1}(G/G_0))^{\binom{n}{q}} \end{array}$$

In particular the map $R\circ\pi^*$ is an isomorphism of the graded differential algebras $\Lambda(G/G_0)$ and $\Lambda_I^0(G)$.

<u>Remark</u>: If we define for $p \geq 1$:

$$H_I^{p,q}(G) = \frac{\operatorname{Ker}\{\partial : \Lambda_I^{p,q}(G) \longrightarrow \Lambda_I^{p+1,q}(G)\}}{\operatorname{Im}\{\partial : \Lambda^{p-1,q}(G) \longrightarrow \Lambda_I^{p,q}(G)\}}$$

then Proposition 11 yields an isomorphism between $H_I^{p,q}(G)$ and $(H^p(G,\mathbb{C}))^{\binom{n}{q}}$, i.e., $H_I^{p,q}(G) = 0$, since $H^p(G,\mathbb{C}) = 0$ $(G \cong \mathbb{C}^n$ as a manifold).

<u>Proof of Proposition 11</u>. First we prove the statement about the map $R\circ\pi^*$. It is clear that $R\circ\pi^*$ is a homomorphism between graded differential algebras. For a $g \in G$ the tangential map π_* is an isomorphism between $T_g^{1,0}(G)$ and $T_{\pi(g)}(G/G_0)\otimes_{\mathbb{R}}\mathbb{C}$. From this it is easy to see that $R\circ\pi^*$ is injective. The above remark implies also the existence of an isomorphism π_* between the complex vector fields on G/G_0 and the right-G_0-invariant vector fields of type $(2,0)$ on G.

For $\omega \in \Lambda_I^{p,0}(G)$ we define a form α on G/G_0 by

$$\alpha(x_1,\ldots,X_p) \; : = \omega(\pi_*^{-1}(X_1),\ldots,\pi_*^{-1}(X_p)),$$

where $X_1,\ldots,X_p$ are vector fields on G/G_0.

We obtain

$$R\circ\pi^*\alpha(\pi^{-1}_*(X_1),\ldots,\pi^{-1}_*(X_p)) = \pi^*\alpha(\pi^{-1}_*(X_1),\ldots,\pi^{-1}_*(X_p)),$$

since the vector fields $\pi_*^{-1}(X_1)$ are of type $(1,0)$. Hence the forms $R\circ\pi^*\alpha$ and ω coincide on invariant vector field of type $(1,0)$ and therefore on all vector fields of type $(1,0)$. but $R\circ\pi^*\alpha$ and ω are forms of type $(p,0)$, i.e., $R\circ\pi^*\alpha = \omega$, which implies the surjectivity of $R\circ\pi^*$.

Finishing the proof, let $(\bar\omega_j)_{j\in J}$ be a base of the right-invariant $(0,q)$-forms on G. Observe that $|J| = \binom{m}{q}$. Each form $\bar\omega_j$ is antiholomorphic. Let $\beta \in \Lambda_I^{p,q}(G)$. Then β can uniquely be written as

$$\beta = \sum_{j\in I} \beta_j \wedge \bar\omega_j \;,\; \beta_j \in \Lambda_I^{p,0}(G).$$

Since the ω_j are antiholomorphic,

$$\partial\beta = \sum_{j\in I} \partial\beta_j \wedge \bar{\omega}_j. \quad (*)$$

The isomorphism between $\Lambda_I^{p,q}(G)$ and $(\Lambda^p(G/G_0))^{\binom{n}{q}}$ is now given by

$$\beta \longmapsto ((R\circ\pi^*)^{-1}\beta_j)_{j\in J}$$

and the diagram is commutative in view of (*).

<u>Proof of Theorem 10 (iii) ⇒ i))</u>. Indeed we prove (~i) ⇒ ~iii)), i.e., if $\mathfrak{G}_0$ does not have an imaginary spectrum, then the pair (G,G_0) is not weakly Kähler.

Assume that $\mathfrak{G}_0$ does not have an imaginary spectrum. By Proposition 8, there is a subalgebra $\bar{\mathfrak{G}}_0 \subset \mathfrak{G}_0$ of type $\mathfrak{A}_0$ or $\mathfrak{G}_0^{\mu}$, hence a solvable pair $(\bar{G},\bar{G}_0)$. If the pair (G,G_0) is weakly Kähler then clearly so is the pair $(\bar{G},\bar{G}_0)$. Therefore we have to show that if $\mathfrak{G}_0$ is a Lie algebra of type $\mathfrak{A}_0$ or $\mathfrak{G}_0^{\mu}$ then the pair $(\bar{G},\bar{G}_0)$ is not weakly Kähler.

In view of Corollary 7, it is enough to show that for a closed, real, right-$\bar{G}_0$-invariant form ω, there is a C^{∞}, right-$\bar{G}_0$-invariant function ψ on $\bar{G}$ such that $\partial\bar{\partial}\psi = \omega$.

Since $\bar{G}/\bar{G}_0$ is isomorphic to $\mathbb{R}^n$ the cohomology group $H^p(\bar{G}/\bar{G}_0,\mathbb{C}) = 0$ for all $p \geq 1$. The remark after Proposition 11 implies that there is a form $\alpha \in \Lambda_I^{0,1}(\bar{G})$ such that $\partial\alpha = \omega$. Since $\bar{\partial}\omega = 0$ we obtain $\partial(\bar{\partial}\alpha) = 0$. The form $\bar{\partial}\alpha$ is contained in $\Lambda_I^{0,2}(\bar{G})$, is $\bar{\partial}$-closed and antiholomorphic.

Now we consider the case where $\bar{G}_0 = G_0^{\mu}$. (The case $\bar{G}_0 = A_0$ can be treated in an analogous way.) First we look for a holomorphic (2,0)-form ν which is ∂-closed and right-G_0^{μ}- invariant. In general a holomorphic (2,0)-form on G^{μ} can be written as

$$\nu = dz \wedge H\cdot db + g\, db_1 \wedge db_2,$$

where $z \in \mathbb{C}$, $b = \begin{pmatrix} b_1 \\ b_2 \end{pmatrix} \in \mathbb{C}^2$ and $H = (H_1,H_2)$, g are holomorphic on G^{μ}.

Using the fact that ν is invariant under the right-G_0^{μ}-action, in particular under the $\{0\}\times\mathbb{R}^2$ action it is easy to see that H and g do not depend on the b variables. Furthermore, for all $(z,z_0) \in \mathbb{C}\times\mathbb{R}$ we have

$$H(z) = H(z+z_0)\cdot e^{z_0 A^{\mu}}$$

and

$$g(z) = g(z+z_0)\cdot e^{z_0 \mathrm{tr} A^\mu},$$

$\mathrm{tr}A^\mu$ denotes the trace of A^μ. After extending holomorphically, these equations are valid for all $z_0 \in \mathbb{C}$. We conclude that

$$H(z) = K_1 \cdot e^{-zA^\mu}$$

and

$$g(z) = K_2 \cdot e^{-z \mathrm{tr} A^\mu}$$

for fixed $K_1 \in \mathbb{C}^2$, $K \in \mathbb{C}$.

Since the form ν is closed, the function g is identically zero. On the other hand

$$H(z)dz = -\partial(K_1(A^\mu)^{-1}e^{-zA^\mu}),$$

hence

$$\nu = -\partial(K_1(A^\mu)^{-1}e^{-zA^\mu}db).$$

One can immediately check that the form $K_1(A^\mu)^{-1}e^{-zA^\mu}db$ is right-G_0^μ-invariant.

Going back to the original question, we see that $\partial\bar{\alpha} = \partial\delta$ where $\delta \in \Lambda_I^{1,0}(G^\mu)$ is holomorphic. Since $\partial\alpha = \omega = \bar{\omega} = \bar{\partial}\bar{\alpha}$, it follows that $\omega = \partial\alpha_1$ with $\alpha_1 = \bar{\alpha}-\delta \in \Lambda_I^{1,0}(G^\mu)$ and $\partial\alpha_1 = 0$. Applying again the remark following Proposition 11, α_1 can be written as $\alpha_1 = \partial\psi$, where $\psi \in \Lambda_I^{0,0}(G^\mu)$. Hence $\omega = \bar{\partial}\partial\psi$, with an invariant function ψ.p

<u>Definition 12</u>. A complex manifold X is called <u>weakly Kähler</u> if there is a closed, real, (1,1)-form on X, which is positive semi-definite for all points of X and positive for at least one point. Now we state the last main result of this section

<u>Theorem 13</u>. Let (G,G_0,Γ) be a totally real CRS (see Definition 6 of section 3). Then the following conditions are equivalent:

i) G/Γ is weakly Kähler,

ii) G/Γ is Kähler,

iii) G/Γ is Stein,

iv) $\mathfrak{G}_0$ has an imaginary spectrum.

<u>Proof</u>. Using Theorem 3 we see that iv) implies that the pair (G,G_0) is pseudoconvex. Therefore, since G_0/Γ is compact the manifold G/Γ admits a strictly plurisubharmonic exhaustion function, i.e., G/Γ is Stein. We proved iv) $\Rightarrow$ iii). The implications iii) $\Rightarrow$ ii) and ii) $\Rightarrow$

i) are obvious. So we only have to prove i) ⇒ iv).

Let ω be a weakly Kähler form on G/Γ. Since G_0 is solvable and G_0/Γ is compact, there is a normalized, left-G_0-invariant measure $\mu(g_0)$ on G_0/Γ. Now define a new (1,1)-form $\tilde{\omega}$ by

$$\tilde{\omega}_p(v_p,w_p) = \int_{G_0/\Gamma} \omega_{p\cdot g_0}(dv_{g_0}v_p, dr_{g_0}\omega_p)d\mu(g_0)$$

where $p \in G$, $v_p, w_p \in T_pG$ and the right Γ-invariant function $G_0 \longrightarrow \mathbb{R}$ such that

$$g_0 \longmapsto \omega_{pg_0}(dv_{g_0}v_p, dr_{g_0}w_p)$$

is considered as a function on G_0/Γ.

Pulling back $\tilde{\omega}$ to G we conclude that $\tilde{\omega}$ is a closed, real, (1,1)-form, which is positive semidefinite and positive definite in at least one point of G (since ω had this property). Furthermore $\tilde{\omega}$ is right-G_0-invariant, i.e., (G,G_0) is a weakly Kähler pair. By Theorem 10 the Lie algebra $\mathfrak{G}_0$ has an imaginary spectrum and Theorem 13 is proved. □

Finishing this section we give an example of a totally real CRS (G,G_0,Γ), where the Lie algebra $\mathfrak{G}_0$ does not have an imaginary spectrum.

For $K = \mathbb{Z},\mathbb{R},\mathbb{C}$ define a group structure G_K on $K\times K^2$ by

$$(z,b)(z_0,b_0) = (z+z_0, e^{z_0A}\cdot b+b_0),$$

for $z,z_0 \in K$, $b,b_0 \in K^2$ and the matrix $A = \log\begin{pmatrix}2 & 1\\ 1 & 1\end{pmatrix}$.

Since A has two distinct real eigenvalues, the Lie algebra $\mathfrak{G}_{\mathbb{R}}$ does not have an imaginary spectrum, i.e., the quotient $X = G_{\mathbb{C}}/G_{\mathbb{Z}}$ is not Stein. Therefore the fiber of the holomorphic reduction is positive dimensional (see section 3). In fact the holomorphic reduction is given by the fibration

$$X = G_{\mathbb{C}}/G_{\mathbb{Z}} \xrightarrow[\mathbb{C}^*\times\mathbb{C}^*]{\pi} G_{\mathbb{C}}/G'_{\mathbb{C}}\cdot G_{\mathbb{Z}} = \mathbb{C}^*$$

where $G'_{\mathbb{C}}$ denotes the commutator group of $G_{\mathbb{C}}$. The fiber is isomorphic to $\mathbb{C}^*\times\mathbb{C}^*$. As far as we know this manifold is the first counterexample to the Serre problem in the homogeneous setting. In this case the map π coincides also with the hypersurface reduction of X.

3. The holomorphically separable case.

As pointed out in the introduction, it is our goal to analyze the holomorphic reduction $\pi : X \longrightarrow Y$ of certain classes of solvmanifolds. One part of this analysis is to describe the base Y, i.e., to say as much about holomorphically separable solvmanifoldsas possible in general. Since there are no group theoretic algorithms for solvable groups as there are for instance for semisimple ones, we have to restrict the description to certain properties of the manifold, e.g., complex analytic or topological properties. However, it is one of the most pleasant aspects of the theory of solvmanifolds to see the connection between abstract group settings and abstract analytical properties of the associated manifolds.

The explanation of the following result, due to A.T. Huckleberry and E. Oeljeklaus will be the main content of this section. It is very far from being true for arbitrary homogeneous complex manifolds.

To state this result we first have to give a definition:

A complex n-dimensional manifold X satisfies the maximal rank condition if there are $f_1,\ldots,f_n \in \mathcal{O}(X)$ such that $df_1 \wedge \ldots \wedge df_n$ does not vanish identically.

Theorem 1. Let X be a complex solvmanifold satisfying the maximal rank condition. Then X is a Stein manifold. Moreover, the fundamental group $\pi_1(X)$ contains a nilpotent normal subgroup of finite index.

Remark. The condition on X to satisfy the maximal rank conditions is slightly weaker than separability by holomorphic functions. Also in the case of homogeneous manifolds these conditions are generally not the same, e.g. the universal covering of $\mathbb{C}^2\backslash\mathbb{R}^2$ satisfies the maximal rank condition but is not holomorphically separable.

Before we start with the proof of Theorem 1, we point out some aspects of the general philosophy behind this problem and some reasons why it is so delicate. First assume X is not a solvmanifold but acted on transitively by a complex reductive Lie group G. Let K denote a maximal compact subgroup of G, i.e., $G = K^{\mathbb{C}}$. Then a famous result of Matsushima [Ma] can be stated as follows:

The manifold X is Stein if and only if the maximal K-orbit in X is a totally real submanifold.

The idea to consider K-orbits is also decisive in the case where X itself is an arbitrary complex Lie group. In this case X is

$$B := \{g \in G : g(z) \in A\}$$

is a nowhere dense analytic set in G, and we take $g_0 \in G\backslash B$. Then the singular set of $F\times F\circ g_0 : X \longrightarrow \mathbb{C}^m\times\mathbb{C}^m$ is contained in A and of lower dimension than A. After finitely many steps we get a regular map as desired. □

Remark. Later we will give a proof of Lemma 3 which is substantially different from that above. It uses integration over currents and can also be used for the case of meromorphic separability.

Recall the basic theorem of Matsushima and Morimoto (see section 1). We now use this to reduce our problem to the case of discrete isotropy. If $X = G/H$ and H is discrete, we refer to X as being "complex parallelizable."

Proposition 4. Assume that Theorem 1 is true for every complex parallelizable homogeneous solvmanifold with maximal rank. Then it is true in general.

Proof. Let $X = G/H$. Consider the abstract algebraic closure $\bar{G}$ of G. Then

$$\bar{G}/H = (\mathbb{C}^*)^k\times G/H$$

and we have the complex normalizer fibration

$$\bar{G}/H \longrightarrow \bar{G}/N_{\bar{G}}(H^0).$$

If $\bar{G} = \bar{G}'\cdot N_{\bar{G}}(H^0)$, then $\bar{G}/N_{\bar{G}}(H^0)$ is an orbit of the nilpotent group $\bar{G}'$ and biholomorphically equivalent to $\mathbb{C}^m$.

The bundle $\bar{G}/H \longrightarrow \bar{G}/N_{\bar{G}}(H^0) = \mathbb{C}^m$ then splits analytically (Gauert's Oka principle [Gr])

$$\bar{G}/H = \mathbb{C}^m\times N_{\bar{G}}(H^0)/H,$$

and since $N_{\bar{G}}(H^0)/H$ has maximal rank, our assumption yields that $N_{\bar{G}}(H^0)/H$ is Stein. Then $\bar{G}/H$ and G/H are also Stein manifolds. There is also a normal subgroup $\tilde{H}$ of H of finite index such that $\pi_1(N_{\bar{G}}(H^0)/\tilde{H})$ is nilpotent. The relation

$$\pi_1(N_{\bar{G}}(H^0)/\tilde{H}) = \pi_1(\bar{G}/\tilde{H}) = \mathbb{Z}^k\times\pi_1(G/\tilde{H})$$

forces $\pi_1(G/\tilde{H})$ to be nilpotent.

Now we assume that $\bar{G} \neq \bar{G}'\cdot N_{\bar{G}}(H^0)$. Denote by $N^0_{\bar{G}}$ the identity

Stein if and only if K is a totally real subgroup.

Hence in the general setting it is certainly a reasonable strategy to attempt to find the compact real submanifolds M in X and relate the induced structure of M to that of X. For example, this works very well when X is a nilmanifold (see [Ma1]).

If $X = G/H$ is a nilmanifold, i.e., G is nilpotent, then it is easy to reduce function-theoretic considerations to the case of discrete isotropy, i.e., to the case where H is a discrete subgroup of G. In this case there is a uniquely determined connected real subgroup $G_{\mathbb{R}}(H)$ of G containing H such that $M := G_{\mathbb{R}}(H)/H$ is compact. It follows that X is Stein if and only if M is totally real in X. As a result, if X is a nilmanifold, then X is Stein if and only if it is holomorphically separable.

Now one might think that solvable is only a slightly more general structure than nilpotent. However, there are substantial difficulties:

1) While nilpotent (as reductive) groups are essentially algebraic, solvable groups are not and discrete subgroups may have a transcendental nature.
2) Given a discrete group H in G solvable it is possible that $G_{\mathbb{R}}(H)$ does not exist as in the nilpotent case.
3) Even when $G_{\mathbb{R}}(H)$ exists and $M = G_{\mathbb{R}}(H)/H$ is totally real in $X = G/H$ it is possible that X is not Stein or not even Kähler (see section 2).

To circumvent the above problems, A.T. Huckleberry and E. Oeljeklaus used the results of J.J. Loeb described in the preceding section and the relationship between G,H and their abstract algebraic closures described by Hochschild and Mostow [HM].

For the language and the facts used in the proof of Theorem 1 we refer to section 1. We start by remarking that if X is homogeneous and satisfies the maximal rank condition then X is Kähler.

<u>Lemma 3</u>. If $X = G/H$ and $\mathcal{O}(X)$ has maximal rank then there exists a regular holomorphic map $F : X \longrightarrow \mathbb{C}^N$ with discrete fibers. In particular X is a Kähler manifold.

<u>Proof</u>. Take $x \in X$ and a holomorphic map $F : X \longrightarrow \mathbb{C}^m$ with $\text{rank}_x F = \dim X$. Let $A_i, i \in I \subset \mathbb{N}$ be the irreducible components of the singular set A of F. Given $z \in A$, the set

component of $N_{\bar{G}}(H^0)$ and note that $\bar{G}/\bar{G}'\cdot N^0_{\bar{G}} = (\mathbb{C}^*)^r\times\mathbb{C}^s$ is a non-trivial linear algebraic abelian group which contains $A := G/G'\cdot(G \cap N^0_{\bar{G}})^0$ as a complex subgroup. Since G is Zariski-dense in $\bar{G}$ we see that $G \neq G'\cdot N_{\bar{G}}(H^0)$ and $A \neq \{0\}$. Thus there is a connected closed complex subgroup I of G, with $G'\cdot(G \cap N^0_{\bar{G}})^0 \subset I$, such that $G/I = \mathbb{C}$ or $\mathbb{C}^*$. The group $\hat{H} := H \cap N^0_{\bar{G}}$ has finite index in H and the holomorphic fibration

$$G/\hat{H} \longrightarrow G/I$$

splits analytically. Since I/H has maximal rank, we may assume by induction on $\dim G/H$ that $I/\hat{H}$ is a Stein manifold and that there exists a normal subgroup $\tilde{H}$ of $\hat{H}$ of finite index with $\pi_1(I/\tilde{H})$ nilpotent. Then $G/\hat{H} = G/I\times I/\hat{H}$ and G/H are Stein and $\pi_1(G/\tilde{H})$ is nilpotent. □

In view of Proposition 4 we will asume from now on that H is a discrete subgroup of G.

The next step in the paper of A. Huckleberry and E. Oeljeklaus is to prove

<u>Lemma 5</u>. Let L be a simply connected complex solvable Lie group and $\Gamma \subset L$ a discrete subgroup with $\mathcal{O}(L/\Gamma)$ having maximal rank. If Γ contains a regular element of L, then there exists a nilpotent subgroup $\tilde{\Gamma} \subset \Gamma$ with $\Gamma' \subset \tilde{\Gamma}$ and $\Gamma/\tilde{\Gamma}$ a finite abelian group.

This lemma is indeed the basic connection between the results of Loeb, the consideration of the compact orbits and the abstract methods given by the introduction of the abstract algebraic closure of a solvable Lie group. However, the proof given in [HO2] also works in a slightly more general context. Since we actually need this more general version later we will now also formally introduce the concept of compact orbits.

<u>Definition 6</u>. A triple (G,G_0,Γ) is called a CRS if G is a simply connected solvable complex Lie group, G_0 is a generic real Lie subgroup and $\Gamma \subset G_0$ is a discrete subgroup of G such that $M = G_0/\Gamma$ is a compact submanifold of $X = G/\Gamma$.

Here generic means that $\mathfrak{G}_0+i\mathfrak{G}_0 = \mathfrak{G}$ (the sum not necessarily being direct, i.e., it may be that $\mathfrak{G}_0 \cap i\mathfrak{G}_0 \neq \{0\}$).

Note that if (G,G_0,Γ) is a CRS and G_0 is a real form of G, i.e., $\mathfrak{G}_0 \cap i\mathfrak{G}_0 = \{0\}$ then the result of section 2 together with

Lemma 3 of this section say that the following are equivalent:

1) G/Γ is Kähler.

2) G/Γ satisfies the maximal rank condition.

3) G/Γ is Stein.

4) $\mathfrak{G}_0$ has an imaginary spectrum.

Note that in general 1) and 2) are conditions which can be introduced in the category of complex manifolds and which are compatible with (not necessarily closed) submanifolds and with coverings, i.e., if a complex manifold X satisfies 1) resp. 2) then every immersed submanifold and any covering $\tilde{X} \longrightarrow X$ satisfies 1) resp. 2). Applied to a CRS (G,G_0,Γ) with G_0 a real form of G the conditions 1) or 2) on G/Γ force 4) to be valid.

We will show later that for a CRS (G,G_0,Γ) one has 1) $\Rightarrow$ 4) in general (even for weakly Kähler) but the converse is not true in general, i.e., if $\mathfrak{G}_0 \cap i\mathfrak{G}_0 \neq \{0\}$ (see section 4).

Now it turns out that Lemma 5 can be generalized in the following sense. Assume $(*)$ is a condition which makes sense for complex manifolds. Moreover assume $(*)$ is compatible with submanifolds and coverings and for any CRS (G,G_0,Γ) one has the implication: If G/Γ satisfies $(*)$ then $\mathfrak{G}_0$ has an imaginary spectrum. Then one has

Lemma 5'. Let L be a simply connected solvable complex Lie group and $\Gamma \subset L$ a discrete subgroup such that L/Γ satisfies $(*)$. If Γ contains a regular element of L, then there exists a nilpotent subgroup $\tilde{\Gamma} \subset \Gamma$ containing Γ' such that $\Gamma/\tilde{\Gamma}$ is a finite (abelian) group.

Proof. Denote by $S_{\mathbb{R}}$ (resp. S) the minimal real (resp. complex) Lie subgroup of L' containing Γ'. Take a regular element $\gamma \in \Gamma$ of L, which is of course automatically a regular element of the abstract algebraic clusure $\bar{L}$ of L. There is exactly one 1-parameter real (resp. complex) subgroup $A_{\mathbb{R}}$ (resp. A) of L containing γ. Hence the algebraic closure of $\mathbb{Z}_\gamma := \{\gamma^n : n \in \mathbb{Z}\}$ in $\bar{L}$ contains $A_{\mathbb{R}}$ and since it normalizes the real algebraic closure $S_{\mathbb{R}}$ of $S \cap \Gamma$, we obtain the connected and simply connected subgroup $M_{\mathbb{R}} := A_{\mathbb{R}} \cdot S_{\mathbb{R}}$ of L. Of course this construction only works if $\mathbb{Z}_\gamma \simeq \mathbb{Z}$. But otherwise $\mathrm{Ad}\,\mathbb{Z}_\gamma \subset GL(\ell)$ would be finite and ℓ would already be nilpotent.

Now $(M = A \cdot S,\ M,\ \Gamma \cap M_{\mathbb{R}})$ is a CRS and $M/\Gamma \cap M$ is a covering of a submanifold of L/Γ and hence satisfies $(*)$. Thus $m_{\mathbb{R}}$ has an imaginary spectrum. It follows that for $g \in M_{\mathbb{R}}$ the restriction of

Adg to $m_{\mathbb{R}}$ has only eigenvalues in S^1. Since $\exp^{-1}(\Gamma \cap S_{\mathbb{R}}) \subset s_{\mathbb{R}}$ contains a uniform discrete lattice of $s_{\mathbb{R}}$ which is stabilized by $\mathrm{Ad}\gamma|s_{\mathbb{R}}$, we may assume that $\mathrm{Ad}_\gamma|s_{\mathbb{R}} \in SL(n,\mathbb{Z})$, $n = \dim_{\mathbb{R}} S_{\mathbb{R}}$. But if $g \in SL(n,\mathbb{Z})$ has only eigenvalues in S^1 then g^m is unipotent for some $m \in \mathbb{N}^+$. To see this note g having its eigenvalues in S^1 it follows that g^k has the same property for all $k \in \mathbb{N}$. Now that the set of normalized polynomials of one variable with integer coefficients having zeros only in S^1 is finite. (For this note that the coefficients of a polynomial can be written as symmetric functions of its roots.) Hence there are finitely many possibilities for the characteristic polynomials $X(g^k)$, $k \in \mathbb{N}$. This shows that $Y : \bigcup_{k\in\mathbb{N}} \{\text{eigenvalues of } g^k\}$ is finite and there exists $m \in N^+$ such that $\alpha^m = 1$ for all $\alpha \in Y$. Applying this to our situation we see that $\mathrm{Ad}_\gamma{}^m|s_{\mathbb{R}}$ is unipotent for some $m \in \mathbb{N}$ and then since $A_{\mathbb{R}}$ is abelian it follows that $A_\gamma{}^m|m_{\mathbb{R}}$ is unipotent. By our definition of regularity, the element γ^m is also a regular element of L; therefore we may take $m = 1$.

Now $\mathrm{Ad}\, \mathbb{Z}_\gamma \subset \mathrm{Ad}\, \bar{L}$ also operates on m as a group of unipotent transformations. For this note that $m = m_{\mathbb{R}} + i m_{\mathbb{R}}$ (not necessarily direct) and that for every $g \in \mathbb{Z}_\gamma$ the transformation $(\mathrm{Adg}|m_{\mathbb{R}} - \mathrm{id}_{m_{\mathbb{R}}})$ is nilpotent. Hence the algebraic closure $(\mathrm{Ad}\, \mathbb{Z}_\gamma)_a$ of $\mathrm{Ad}\, \mathbb{Z}_\gamma$ in $\mathrm{Ad}\, \bar{L}$ acts unipotently on m. But $(\mathrm{Ad}\, \mathbb{Z}_\gamma)_a$ contains a maximal torus T of $\mathrm{Ad}\, \bar{L}$ and it follows that the stabilizer of m in $\mathrm{Ad}\, \bar{L}$ acts unipotently on m. The latter obviously contains $\mathrm{Ad}\, L_a(\Gamma)$, where $L_a(\Gamma)$ is the identity component of the algebraic closure of Γ. Now let $\tilde{\Gamma} = \Gamma \cap L_a(\Gamma)$. The fact that $L_a(\Gamma)' = L_a(\Gamma') \subset S \subset M$ implies that $\mathrm{AD}\, L_a(\Gamma)$ consists of unipotent transformations of the Lie algebra of $L_a(\Gamma)'$. As a consequence $L_a(\Gamma)$ is nilpotent. Note that $\Gamma' \subset \tilde{\Gamma}$ and $\Gamma/\tilde{}\Gamma$ is finite. □

As an immediate consequence of this lemma we obtain the following result.

Lemma 6. Let L be a connected solvable linear algebraic group and $\Gamma \subset L$ a Zariski-dense discrete subgroup such that L/Γ satisfies (*). Then L is nilpotent.

Proof. We consider the universal covering $\pi : L_0 \longrightarrow L$ of L and denote by $\bar{L}_0$ an abstract algebraic closure of L_0. Then for $\Gamma_0 = \pi^{-1}(\Gamma)$ we have $L/\Gamma = L_0/\Gamma_0$. Since Γ is Zariski-dense in L, it contains a regular element γ of L and every element of $\pi^{-1}(\gamma) \subset$

Γ_0 is regular with respect to L_0. By Lemma 5′ we obtain a nilpotent subgroup $\tilde{\Gamma}_0 \subset \Gamma_0$ of finite index and $\tilde{\Gamma} = \pi(\tilde{\Gamma}_0)$ is a nilpotent subgroup of finite index in Γ. Obviously $\tilde{\Gamma}$ is Zariski-dense in L and as a consequence L is nilpotent. □

As already mentioned, we will need the lemmas 5′ and 6 later, where (*) will be a local Kähler condition. But now we come back to the original problem of this section. So from now on let (*) be the maximal rank condition again.

Then as an immediate consequence of Lemma 6 and a result of [GH] we have

<u>Corollary 7</u>. Let L be a connected solvable linear algebraic group and $\Gamma \subset L$ a Zariski-dense subgroup such that L/Γ satisfies the maximal rank condition. Then L is nilpotent and L/Γ is a Stein manifold.

Finally we complete the proof of Theorem 1. By Proposition 4 we may assume that $\Gamma := H$ is discrete. Moreover it is sufficient to prove the theorem for the solvmanifold $\bar{G}/\Gamma$. Let L_a denote the identity component of the algebraic closure of Γ in $\bar{G}$ and consider the holomorphic fibration

$$\bar{G}/\Gamma \longrightarrow \bar{G}/L_a\Gamma$$

with connected fiber $L_a/\Gamma \cap L_a$. Since $\bar{G}/L_a\Gamma$ is Stein (see section 1) and since $L_a/\Gamma \cap L_a$ is Stein (Lemma 6), it follows from the theorem of Matsushima-Morimoto that $\bar{G}/\Gamma$ is Stein. Since $\tilde{\Gamma} := \Gamma \cap L_a$ is nilpotent, of finite index in Γ and contains Γ', the proof of the theorem is complete. □

4. <u>CR-Solvmanifolds</u>.

As we saw in the previous section, it si reasonable to study those solvmanifolds which have big compact orbits of subgroups. In particular, nilmanifolds are of this type and Lemma 5′ in section 3 tells how to use them in a general setting. In general it is also clear that the case of discrete isotropy causes the most difficulties. Hence up to standard methods from algebraic groups and complex homogeneous spaces, the objects which have to be studied are the CRS = CR-Solvmanifolds. These already have been defined in section 3 an are triples (G, G_0, Γ), where G is a complex simply connected Lie group G, a real Lie subgroup G_0 and a discrete subgroup $\Gamma \subset G_0$ of G

such that

i) $\mathfrak{G}_0 + i\mathfrak{G}_0 = \mathfrak{G}$

ii) G_0/Γ is compact.

As said before, given a simply connected complex nilpotent Lie group G and a discrete subgroup Γ in G then there exists a unique real Lie subgroup G_0 in G such that $\Gamma \subset G_0$ and G_0/Γ is compact (see [Ma1]). Of course in this case $\mathfrak{G}_0$ does not in general satisfy i), but the complexification $\hat{G}_0$ of G_0 in G, i.e., $\langle \exp(\mathfrak{G}_0 + i\mathfrak{G}_0)\rangle = \hat{G}_0$ is closed in G and it is easy $G/\hat{G}_0 \simeq_{bihol} \mathbb{C}^n$. Thus one has $G/\Gamma = \hat{G}_0/\Gamma \times \mathbb{C}^n$, which is a reduction to a CRS again.

A complex manifold X is called weakly Kähler if it admits a real closed positive semidefinite (1,1)-form ω, which is positive definite in at least one point of X.

If (G,G_0,Γ) is a CRS and G/Γ is weakly Kähler, we call (G,G_0,Γ) a weakly Kähler CR-Solvmanifold (KCRS).

The goal of this section is to prove that the condition weakly Kähler is of the type (*) described in section 3 (of course for homogeneous manifolds).

Theorem 1. Let (G,G_0,Γ) be a KCRS. Then $\mathfrak{G}_0$ has an imaginary spectrum.

Note that this theorem together with the ideas of section 3 reduces our problem to the case of nilmanifolds. This case will be studied in th next section.

In a more general setting compact orbits in Kähler manifolds have been studied in [Ri]. let (G,G_0,Γ) be a CRS. Then the intersection $\mathfrak{G}_0 \cap i\mathfrak{G}_0$ is a complex ideal in $\mathfrak{G}$. In the following we will denote it by $\mathfrak{m}$.

Now suppose (G,G_0,Γ) is a KCRS, i.e., on G/Γ we have a real positive semidefinite closed (1,1)-form ω, which is positive definite for some $p \in G_0/\Gamma$. Moreover since ω is continuous it is positive definite on some neighborhood of p in G/Γ.

Identify ω with its right-Γ-invariant pullback to G. Since G_0/Γ is compact, it carries a left-G_0-invariant, finite, normalized measure μ. Now define $\tilde{\omega}$ on G as follows

$$\omega_q(v_q,w_q) \ := \int_{G_0/\Gamma} \omega_{q\cdot g}(dr_g v_q, dr_g w_q)\, d\mu(g),$$

where $q \in G$, $v_q, w_q \in TG_q$ and the right-Γ-invariant function

$G_0 \longrightarrow \mathbb{R}$, $g \longmapsto \omega_{q\cdot g}(dr_g v_q, dr_g w_q)$ is considered as a function on G_0/Γ. Let $\pi : G \longrightarrow G/G_0$ denote the quotient map given by G_0. Furthermore let $V = \pi(U)$, where U is an open right-Γ-invariant subset of G, such that $\omega|U$ is positive definite and $U \cap G_0 \neq \phi$. Since π is an open map V is open in G/G_0 and $\tilde{U} = \pi^{-1}(U)$ is open and right-G_0-invariant in G. Let $p \in \tilde{U}$ and let J denote the complex structure tensor on G. Then for $V \in TG_p$ we have

$$\tilde{\omega}(v_p, Jv_p) = \int_{G_0/\Gamma} w_{p\cdot y}(dr_g v_p, Jdr_g v_p)d\mu(g) > 0,$$

since $\omega|\tilde{U} \geq 0$ and $\tilde{U} \cap U \neq \emptyset$. Moreover $\tilde{\omega}$ is a closed real positive semidefinite (1,1)-form which is right-G_0-invariant.

This consideration shows that (G, G_0, Γ) is a KCRS if and only if there is a closed real positive semidefinite right-G_0-invariant (1,1)-form ω on G and a right-G_0-invariant neighborhood of G_0 in G, where ω is positive definite.

In the following we always assume a KCRS to be given in such a way. The main ingredient for Theorem 1 is section 2 and

<u>Proposition 2</u>. Let G be a complex Lie group. Suppose G admits a Kähler metric which is invariant with respect to the right (resp. left) multiplication of G. Then G is abelian and the given Kähler metric is both left- and right-invariant.

<u>Proof</u>. Let h denote a Kähler metric on G which is as we may assume right-G-invariant. Consider the Lie algebra $\mathfrak{G}$ of right invariant vector fields on G. If J denotes the complex structure tensor on G, then the Kähler form ω associated to h is given by

$$(X,Y) = h(X,JY).$$

Since ω is closed and right-invariant, we have

$$(1) \qquad 0 = \omega(X[Y,Z]) + \omega(Z,[X,Y]) + \omega(Y,[Z,X]),$$

for all $X,Y,Z \in \mathfrak{G}$.

Assume there is a non abelian, two-dimensional complex subalgebra $\mathfrak{A} \subset \mathfrak{G}$, i.e., we have two vector fields $X,Y \in \mathfrak{G}$ such that $[X,Y] = Y$. By (1) we have

$$0 = \omega(X[Y,JY]) + \omega(JY,[X,Y]) + \omega(Y,[JY,X]),$$
$$= 2\omega(JY,Y) = 2h(JY,JY) > 0,$$

which is a contradiction. Thus every two-dimensional complex

subalgebra of $\mathfrak{G}$ is abelian.

Now let B denote a maximal solvable subalgebra of $\mathfrak{G}$. Let C denote the center of B. By (1) we have

$$0 = \omega(X,[Y,Z])$$

for all $X \in C$ and $Y,Z \in B$. Thus $\omega(X,X') = 0$ for all $X' \in B'$. Thus we have $C \cap B' = 0$. On the other hand Lie's flag theorem yields a one-dimensional ideal in B'. Since every two-dimensional subalgebra of B is abelian, this ideal has to be contained in C. This shows that $B = C$ is abelian. Since it is well known that a complex Lie algebra is abelian if and only if its maximal solvable subalgebras are abelian the proposition is proved. □

<u>Corollary 3</u>. Let (G,G_0,Γ) be a KCRS. Then $\mathfrak{m} = \mathfrak{G}_0 \cap i\mathfrak{G}_0$ is abelian.

<u>Proof</u>. Let $G_{\mathfrak{m}} = \langle \exp \mathfrak{m} \rangle$ denote the Lie subgroup of G_0 generated by $\mathfrak{m}$. Take a right-G_0-invariant neighborhood U of G_0 in G and a closed right-G_0-invariant (1,1)-form ω which is strictly positive definite on U. Then $\omega | G_{\mathfrak{m}}$ is a right-$G_{\mathfrak{m}}$-invariant Kähler form on $G_{\mathfrak{m}}$. Since $G_{\mathfrak{m}}$ is a complex Lie group, the corollary now follows from Proposition 2. □

After this preparation we now come to the

<u>Proof of Theorem 1</u>. Assume that $\mathfrak{G}_0$ does not have an imaginary spectrum. Then G_0 contains a subgroup of the form $A_{\mathbb{R}}$ on G_μ (see section 2). At first assume it contains $A_{\mathbb{R}}$. Let A denote the complexification of $A_{\mathbb{R}}$ in G, i.e., $A = \langle \exp (\mathfrak{A} + i\mathfrak{A}) \rangle$. Now $A_{\mathbb{R}}$ is real two-dimensional and not abelian, hence A cannot be abelian and in particular it follows that $\dim_{\mathbb{C}} A = 2$, i.e., $A_{\mathbb{R}}$ has to be a real form of A. This, however, contradicts Theorem 1.10.

Now assume that G_0 contains a group of type G_μ. Again consider the complexification $G_\mu^{\mathbb{C}}$ of G_μ in G. Also in this case Theorem 1.10 excludes the possibility that G_μ is a real form of $G_\mu^{\mathbb{C}}$. Hence $G_\mu^{\mathbb{C}}$ has to be a two-dimensional complex Lie subgroup of G, and it cannot be abelian since it contains G_μ. Thus the Lie algebra $\mathfrak{G}_\mu^{\mathbb{C}}$ of $G_\mu^{\mathbb{C}}$ contains exactly one non-zero proper ideal, namely its commutator algebra $(\mathfrak{G}_\mu^{\mathbb{C}})'$. This shows that $(\mathfrak{G}_\mu^{\mathbb{C}})' = \mathfrak{G}_\mu \cap i\mathfrak{G}_\mu$. Also since $\mathfrak{G}_\mu^{\mathbb{C}}$ is two-dimensional, every eigenvector belonging to a non-zero eigenvalue of an element $\operatorname{ad} X$, $X \in \mathfrak{G}_\mu^{\mathbb{C}}$ has to be contained in $(\mathfrak{G}_\mu^{\mathbb{C}})'$. Now take $X \in \mathfrak{G}_\mu$ and let $\lambda \in \mathbb{C}$ be a non-zero eigenvalue of

ad X. Then there is a $Y \in (\mathfrak{G}_\mu^{\mathbb{C}})'$, such that

$$[X,Y] = \lambda \cdot Y.$$

Let $\lambda = \alpha + i\beta$, $\alpha, \beta \in \mathbb{R}$. Then we have

$$0 = \omega(x,[Y,iY]) + \omega(iY,[X,Y]) + \omega(Y,[iY,X])$$

because $Y \in \mathfrak{G}_\mu \cap i\mathfrak{G}_\mu \subset \mathfrak{G}_0$. Hence it follows that

$$0 = 2\omega(iY,[X,Y]) = 2\omega(iY,\alpha Y + i\beta Y) = 2\alpha\omega(iY,Y).$$

Since $\omega(iY,Y) > 0$ this shows that $\alpha = 0$. Hence ad X has only imaginary eigenvalues. Thus ad X has only imaginary eigenvalues, which contradicts the fact that $\mathfrak{G}_\mu$ has a non-imaginary spectrum. Therefore, by Theorem 1.10 it follows that $\mathfrak{G}_0$ has an imaginary spectrum. □

Before we come to analyze the holomorphic reduction of weakly Kählerian nilmanifolds in the next section, let us give one further technical remark on CR-solvmanifolds, which is easy but quite useful.

Lemma 4. Let (G, G_0, Γ) be a CRS and let H be a closed complex sib-group of G containing Γ. Let $H_0 := H \cap G_0$. Then G_0/H_0 is totally real in G/H if and only if the following conditions are satisfied:

i) $\mathfrak{h}_0 + i\mathfrak{h}_0 = \mathfrak{h}$

ii) $\mathfrak{m} \subset \mathfrak{h}$.

Proof. The G_0-orbit of $p \simeq H$ in G/H is totally real if and only if

$$0 = \mathfrak{m}_p := T(G_0/H_0)_p \cap iT(G_0/H_0)_p.$$

Since $T(G_0/H_0)_p + iT(G_0/H_0)_p = T(G/H)_p$, we have

$$\dim_{\mathbb{R}}\mathfrak{m}_p = 2\dim_{\mathbb{R}} T(G_0/H_0)_p - \dim_{\mathbb{R}} T(G/H)_p$$
$$= 2\dim_{\mathbb{R}}\mathfrak{G}_0 - 2\dim_{\mathbb{R}}\mathfrak{h}_0 - \dim_{\mathbb{R}}\mathfrak{G} + \dim_{\mathbb{R}}\mathfrak{h}.$$

On the other hand we have

$$\dim_{\mathbb{R}}\mathfrak{m} = 2\dim_{\mathbb{R}}\mathfrak{G}_0 - \dim_{\mathbb{R}}\mathfrak{G},$$

since $\mathfrak{G}_0 + i\mathfrak{G}_0 = \mathfrak{G}$. Thus we obtain

$$\dim_{\mathbb{R}}\mathfrak{m}_p = \dim_{\mathbb{R}}\mathfrak{m} - 2\dim_{\mathbb{R}}\mathfrak{h}_0 + \dim_{\mathbb{R}}\mathfrak{h}$$

and hence $\dim_{\mathbb{R}}\mathfrak{m}_p = 0$ if and only if

$$0 = \dim_{\mathbb{R}}\mathfrak{m} - 2\dim_{\mathbb{R}}\mathfrak{h}_0 + \dim_{\mathbb{R}}\mathfrak{h}.$$

Note that $\dim_{\mathbb{R}}(\mathfrak{h}_0 + i\mathfrak{h}_0) = 2\dim_{\mathbb{R}}\mathfrak{h}_0 - \dim_{\mathbb{R}}(\mathfrak{h}_0 \cap i\mathfrak{h}_0)$, i.e., $\mathfrak{m}_p = 0$

if and only if

$$0 = \dim_{\mathbb{R}} \mathfrak{m} - \dim_{\mathbb{R}}(\mathfrak{h}_0 + i\mathfrak{h}_0) - \dim_{\mathbb{R}}(\mathfrak{h}_0 \cap \mathfrak{h}_0) + \dim_{\mathbb{R}} \mathfrak{h}.$$

Since $\mathfrak{h}_0 + i\mathfrak{h}_0 \subset \mathfrak{h}$ and $\mathfrak{h}_0 \cap i\mathfrak{h}_0 \subset \mathfrak{m}$, the last equation holds if and only if

i) $\mathfrak{h}_0 + i\mathfrak{h}_0 = \mathfrak{h}$,

ii) $\mathfrak{h}_0 \cap i\mathfrak{h}_0 = \mathfrak{m}$. □

<u>Remark</u>. Let (G, G_0, Γ) be a CRS and H, H_0 be given as in Lemma 4. If moreover $\mathfrak{h}_0$ is an ideal in $\mathfrak{G}_0$, then $(G/H^0, G_0/H_0^0, H_0/H_0^0)$, H^0 (resp. H_0^0) = identity component of H (resp. H_0), is a CRS and G_0/H_0^0 is a real form of G/H^0 if and only if i) and ii) in Lemma 4 are satisfied.

5. <u>Nilmanifolds</u>.

In this section we consider weakly Kähler nilmanifolds. A nilmanifold X is a quotient space $X = G/H$, where G is a nilpotent complex Lie group and $H \subset G$ a closed complex subgroup. (Of course, we may assume that G is simply connected.) Let $\pi : G/H \longrightarrow G/J$ denote the holomorphic reduction of G/H. Then the following is known (see [GH]): $\mathcal{O}(J/H) = \mathbb{C}$, G/J is a Stein manifold and J/H is connected. Moreover J/H is a principal abelian Lie group tower.

If G/H is weakly Kähler then clearly the same is true for J/H. Our aim, and this is another cornerstone of the general solvable case, is to prove

<u>Theorem 1</u>. Let X be a weakly Kähler nilmanifold and let $\pi : X \xrightarrow{F} Y$ denote the holomorphic reduction of X. Then Y is Stein and F is a Cousin group.

<u>Theorem 2</u>. Let G be a nilpotent complex Lie group and $\Gamma \subset G$ a discrete subgroup such that

1) $\mathcal{O}(G/\Gamma) = \mathbb{C}$,

2) G/Γ is weakly Kähler.

Then G is abelian.

Before we begin to prove these theorems we will gather some remarks about weakly Kähler nilmanifolds, which are not contained in them, but come out of the proof and are very useful.

<u>Remark 3</u>. Let $X = G/H$ be a complex nilmanifold. Then there exists

a uniquely determined nilpotent complex Lie gruop N such that

$$G/H = \mathbb{C}^n \times N/\Gamma,$$

and Γ is a "maximal" discrete subgroup of N. (Here "maximal" means that Γ is not contained in any proper connected complex subgroup of N (see [LOR]).) Thus G/H is Kähler if and only if N/Γ is Kähler.

Now let N_0 denote the smallest connected real subgroup of N containing Γ (see [Mal]). Then we shall see that the weak Kähler assumption on G/H yields a splitting

$$\mathfrak{n}_0 = \mathfrak{U} \oplus \mathfrak{m},$$

where $\mathfrak{m}$ is a complex subspace of the center of $\mathfrak{n}$ and $\mathfrak{U}$ is a subgroup of $\mathfrak{n}_0$ such that $\mathfrak{U} \cap i\mathfrak{U} = \{0\}$.

Conversely assume that there is such a splitting of $\mathfrak{n}_0$. Let $\mathfrak{U}^{\mathbb{C}} = \mathfrak{U} + i\mathfrak{U}$ and A (resp. $A^{\mathbb{C}}$) = $\langle \exp \mathfrak{U} \rangle$ (resp. $\langle \exp \mathfrak{U}^{\mathbb{C}} \rangle$). Then we have

$$N_0 = A \times G_{\mathfrak{m}}, \quad G_{\mathfrak{m}} = \langle \exp \mathfrak{m} \rangle,$$

and

$$N = A^{\mathbb{C}} \times G_{\mathfrak{m}}.$$

Moreover there exists a right-A-invariant Kähler form on $A^{\mathbb{C}}$ (see [Lo]). Since $G_{\mathfrak{m}}$ is abelian and central in N, this yields a right-N_0-invariant Kähler form on N and hence a Kähler metric on N/Γ.

Remark 4. For practical use, i.e., to check whether a nilmanifold $G/H = N/\Gamma \times \mathbb{C}^n$ is Kahler or not, it is convenient to note that a Kähler assumption on N/Γ implies that $\mathfrak{m}$ is contained in the center of $\mathfrak{G}$ and

$$\mathfrak{n}_0' \cap \mathfrak{m} = 0.$$

Remark 5. By remark 3 it follows that a nilmanifold is weakly Kähler if and only if it is Kähler

In order to prove Theorem 2, we first rephrase it, using the following Theorem of Malcev [Mal].

Theorem. Let G be a simply connected nilpotent Lie group and $H \subset G$ a closed subgroup. Then there exists a unique connected closed subgroup $G_0 \subset G$ containing H, such that

1) G_0/H is compact,
2) $G/G_0 \cong \mathbb{R}^n$,
3) the bundle $G/H \longrightarrow G/G_0$ is real analytically trivial.

Taking G_0 in the situation of Theorem 2, implies that $G_0^{\mathbb{C}}$ is a closed complex subgroup of G containing Γ. Here we use the fact that any connected subgroup of a simply connected solvable group is closed. Thus we obtain $G/G_0^{\mathbb{C}} = \mathbb{C}^k$ and the condition $\mathcal{O}(G/\Gamma) = \mathbb{C}$ forces $G_0^{\mathbb{C}} = G$, i.e., G_0 is a generic subgroup of G. Hence it is enough to prove an equivalent version of Theorem 2.

<u>Theorem 2'</u>. Let (G,G_0,Γ) be a KCRS. Suppose G is nilpotent and $\mathcal{O}(G/\Gamma) = \mathbb{C}$. Then G is abelian.

<u>Proof</u>. Let ω denote a right-G_0-invariant closed real positive (1,1)-form on G, which is strictly positive on a right-G_0-invariant open neighborhood U of G_0 in G. For $p \in G$ and $v_p, w_p \in TG_p$ define

$$h(v_p,w_p) = \omega(Jv_p,w_p),$$

where J denotes the complex structure tensor on G. Then $h|U$ is a right-G_0-invariant Kähler metric on U. Since G_0 is a generic subgroup of G, the center C_0 of $\mathfrak{G}_0$ is given by

$$\mathfrak{C}_0 = \mathfrak{C} \cap \mathfrak{G}_0,$$

where $\mathfrak{C}$ denotes the center of $\mathfrak{G}$.

At first we show that

$$\mathfrak{C} \cap \mathfrak{m} = \{X \in \mathfrak{m} : \omega(X,[\mathfrak{G},\mathfrak{m}]) = 0\}. \tag{1}$$

For this let

$$\mathfrak{U} = \{X \in \mathfrak{G}_0 : \omega(X,\mathfrak{m}) = 0\}.$$

Then $\mathfrak{U}$ is a real subalgebra of $\mathfrak{G}_0$ and we have

$$\mathfrak{G}_0 = \mathfrak{U} \oplus \mathfrak{m}.$$

To see this consider h and let $(v_1,\ldots,v_{2k}, w_1,\ldots,w_\ell)$ be an orthogonal basis of $\mathfrak{G}_0$, where $(v_1,\ldots,v_{2k})$ is an orthogonal basis of $\mathfrak{m}$, such that for $j \le k$

$$v_{j+k} = Jv_j.$$

Using the relation between ω and h, it is easy to see that $\mathfrak{U}$ is just the real span of $(w_1,\ldots,w_\ell)$ we have

$$\mathfrak{G}_0 = \mathfrak{U} \oplus \mathfrak{m}$$

as a vector space. To verify that $\mathfrak{U}$ is an algebra note that for $X,Y,Z \in \mathfrak{G}_0$ we have

(2) $$0 = \omega(X,[Y,Z]) + \omega(Z,[X,Y]) + \omega(Y,[Z,X]),$$

and use the fact that $\mathfrak{m}$ is an ideal.

Now $\mathfrak{m}$ is abelian (see Corollary 4.3) and it is easy to check that $[\mathfrak{G},\mathfrak{m}] = [\mathfrak{G}_0,\mathfrak{m}]$. Hence we obtain

$$[\mathfrak{G},\mathfrak{m}] = [\mathfrak{U},\mathfrak{m}].$$

Using (2) it follows that

$$\mathfrak{C} \cap \mathfrak{m} \subset \{X \in \mathfrak{m} : \omega(X,[\mathfrak{G},\mathfrak{m}]) = 0\}.$$

To verify the opposite inclusion it is enough to show that for

$$X_0 \in \{X \in \mathfrak{m} : \omega(X,\mathfrak{m}) = 0\}$$

one has

$$[X_0,\mathfrak{U}] = 0.$$

For this let $Y \in \mathfrak{U}$ and $Z \in \mathfrak{m}$. Then by (2) we have

$$0 = \omega(X_0,[Y,Z]) + \omega(Z,[X_0,X]) + \omega(Y,[Z,X_0]).$$

Since $[Y,Z] \in [\mathfrak{G},\mathfrak{m}]$ and $[Z,X_0] \in \mathfrak{m}$, it follows that

$$\omega(Z,[X_0,Y]) = 0$$

for all $Z \in \mathfrak{m}$, $Y \in \mathfrak{U}$.

But $X_0 \in \mathfrak{m}$ implies that $[X_0,Y] \in \mathfrak{m}$. Since $\mathfrak{m}$ is a complex subspace of $\mathfrak{G}$, we know that $\omega|\mathfrak{m}$ is non-degenerate. Therefore we have $[X_0,Y] = 0$ and (1) is proved.

Now since $\mathfrak{G}$ is nilpotent the center $\mathfrak{C}$ of $\mathfrak{G}$ is non-zero and non-trivially intersects every non-zero ideal of $\mathfrak{G}$. The equation (1) shows that the ideal $[\mathfrak{G},\mathfrak{m}]$ which is contained in $\mathfrak{m}$ does not intersect $\mathfrak{C}$. Hence we have $[\mathfrak{G},\mathfrak{m}] = \{0\}$ or equivalently $\mathfrak{m} \subset \mathfrak{C}$. Now we claim that

$$\mathfrak{C} = \mathfrak{C}_0 + i\mathfrak{C}_0.$$

For this note that

$$\mathfrak{G}_0' = \mathfrak{U}' + [\mathfrak{U},\mathfrak{m}] = \mathfrak{U}', \text{ i.e., } \mathfrak{G}_0' \cap i\mathfrak{G}_0' = \mathfrak{U}' \cap i\mathfrak{U}' = 0.$$

Now take $\mathfrak{C} \ni Z = X + iY$, $X,Y \in \mathfrak{G}_0$. then for any $X' \in \mathfrak{G}_0$ we have

$$0 = [Z,X'] = [X,X'] + i[Y,X\ ], \text{ i.e.}$$
$$[X,X'],\ [Y,X'] \in \mathfrak{G}_0 \cap i\mathfrak{G}_0' = 0,$$

which proves our claim.

Note that $\mathcal{O}(G/\Gamma) = \mathbb{C}$ implies that the orbits of the center Z of G are closed in G/Γ ([BO]). Thus we may consider the fibrations

$$\begin{array}{ccc} G_0/\Gamma & \hookrightarrow & G/\Gamma \\ \downarrow & & \downarrow \\ G_0/Z_0\Gamma & \hookrightarrow & G/Z\Gamma \end{array}$$

Unluckily we cannot apply induction, since we do not know if ω is Z-invariant. However, we can use Lemma 4, which shows that $G_0/Z_0\Gamma$ is a totally real submanifold of $G/Z\Gamma$. Thus $G/Z\Gamma$ is a Stein manifold ([GH] p. 47). Now $\mathcal{O}(G/\Gamma) = \mathbb{C}$ forces $G = Z$, i.e., G is abelian. □

<u>Proof of Theorem 1</u> Using [GH] again it reduces to the remark that for a nilmanifold $X = G/H$ (let us say G simply connected) the holomorphic reduction $X = G/H \xrightarrow{F} G/J = Y$, $F = J/H$ has a Stein base and moreover $\mathcal{O}(F) = \mathbb{C}$. The latter implies $H^0 \cap J^0$ and of course J^0/H^0 is simply connected. Hence we can apply Theorem 2 to $F = \tilde{G}/\tilde{\Gamma}$, where $\tilde{G} = J^0/H^0$ and $\tilde{\Gamma} = J^0 \cap H/H^0$. □

Now we give two examples, the first demonstrating that the holomorphic reduction $X \xrightarrow{F} Y$ of a Kähler manifold need not be a holomorphically trivial bundle. The second example shows how useful Remark 4 is to decide whether a given nilmanifold is Kähler or not.

<u>Example 6</u>. Consider the groups of matrices

$$G_K = \left\{ \begin{pmatrix} 1 & x & z & w \\ 0 & 1 & y & 0 \\ 0 & 0 & 1 & 0 \\ 0 & 0 & 0 & 1 \end{pmatrix} : x,y,z,w \in K \right\},$$

where $K = \mathbb{C}, \mathbb{R}, \mathbb{Z}$.

It is clear that $G_{\mathbb{R}}$ is a real form of $G_{\mathbb{C}}$. Since $G_{\mathbb{R}}$ is nilpotent, it has an imaginary spectrum. Therefore $G_{\mathbb{C}}$ admits a right-$G_{\mathbb{R}}$-invariant Kähler form ω (see [Lo]). In view of the product structure of $G_{\mathbb{C}}$, we may assume that ω restricted to the w-direction is the standard Kähler form $dw \wedge d\bar{w}$. Now take the discrete subgroup Γ of $G_{\mathbb{C}}$ generated by $G_{\mathbb{Z}}$ and the element

$$\begin{pmatrix} 1 & 0 & \sqrt{2} & i\sqrt{3} \\ 0 & 1 & 0 & 0 \\ 0 & 0 & 1 & 0 \\ 0 & 0 & 0 & 1 \end{pmatrix} \in G_{\mathbb{C}}.$$

Then ω is right-Γ-invariant, i.e., $G_{\mathbb{C}}/\Gamma$ is a Kähler manifold.

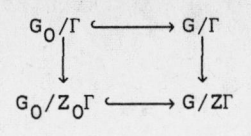

Let $Z = \left\{ \begin{pmatrix} 1 & 0 & z & w \\ 0 & 1 & 0 & 0 \\ 0 & 0 & 1 & 0 \\ 0 & 0 & 0 & 1 \end{pmatrix} : z,w \in \mathbb{C} \right\} \subset G_{\mathbb{C}}$, i.e., Z is the center of $G_{\mathbb{C}}$. The base of the fibration

$$G_{\mathbb{C}}/\Gamma \longrightarrow G_{\mathbb{C}}/Z\Gamma$$

is $\mathbb{C}^* \times \mathbb{C}^*$ and the fiber is isomorphic to the Cousin group $\mathbb{C}^2/\tilde{\Gamma}$, where $\tilde{\Gamma} = \left\langle \begin{pmatrix} 1 \\ 0 \end{pmatrix}, \begin{pmatrix} 0 \\ 1 \end{pmatrix}, \begin{pmatrix} \sqrt{2} \\ i\sqrt{3} \end{pmatrix} \right\rangle_{\mathbb{Z}} \subset \mathbb{C}^2$.

Since Γ is not abelian this bundle cannot be (even topologically) trivial.

<u>Example 7</u>. Let $G = \left\{ \begin{pmatrix} 1 & x & z \\ 0 & 1 & y \\ 0 & 0 & 1 \end{pmatrix} : x,y,z \in \mathbb{C} \right\}$, i.e., G is the three-dimensional Heisenberg group. Consider the discrete subgroup

$$\Gamma = \left\{ \begin{pmatrix} 1 & m & k+i\ell \\ 0 & 1 & n \\ 0 & 0 & 1 \end{pmatrix} : n,m,k,\ell \in \mathbb{Z} \right\}$$

and let

$$Z = \left\{ \begin{pmatrix} 1 & 0 & z \\ 0 & 1 & 0 \\ 0 & 0 & 1 \end{pmatrix} : z \in \mathbb{C} \right\},$$

which is the center and at the same time the commutator group of G. The fibration $G/\Gamma \longrightarrow G/Z\Gamma$ realizes G/Γ as a torus bundle over $\mathbb{C}^* \times \mathbb{C}^*$. In this example the group G_0 given by

$$G_0 = \left\{ \begin{pmatrix} 1 & x & y \\ 0 & 1 & z \\ 0 & 0 & 1 \end{pmatrix} : x,y \in \mathbb{R}\ ,\ z \in \mathbb{C} \right\},$$

i.e., (G,G_0,Γ) is a CRS. Since $\mathfrak{m} = \mathfrak{G}_0' \cap i\mathfrak{G}_0'$, we have $\mathfrak{G}_0' \cap \mathfrak{m} \neq \{0\}$. Hence by Remark 4 it follows G/Γ is not Kähler.

Finishing this section let us note that the holomorphic reduction $G/\Gamma \longrightarrow G/J$, where Γ is maximal in G, is holomorphically trivial if and only if it is topologically trivial (Grauert's Oka principle, see section 1).

6. The holomorphic reduction of Kähler-solvmanifolds.

We are now ready to prove our main theorem, concerning the holomorphic reduction of a weakly Kähler-solvmanifold. Using an embedding into an algebraic setting and the theory of the KCRS the problem will be reduced to the nilpotent case. What we prove is

Theorem 1. Let X be a weakly Kähler solvmanifold. Then the holomorphic reduction $X \longrightarrow Y$ realizes X as a Cousin group bundle over a Stein manifold. Moreover $\pi_1(X)$ contains a nilpotent normal subgroup of finite index.

Since X is a solvmanifold we may write $X = G/H$, where G is a simply connected solvable complex Lie group, which acts almost effectively on X, i.e., $\bigcap_{g \in G} gHg^{-1}$ is a discrete subgroup of G.

An abstract algebraic closure of G will be denoted by G_a. This is a linear solvable linear algebraic complex Lie group of the form

$$G_a = (\mathbb{C}^*)^k \ltimes G,$$

containing G as a Zariski dense, but topologically closed, normal subgroup. For details see section 1.

By Theorem 4.1 we know that the Lemmas 3.5′ and 3.6 are valid for a KCRS.

Proof of Theorem 1. First note that in the course of the proof we always may replace H by normal complex subgroup $\tilde{H} \subset H$ of finite index. For any subgroup $L \subset G$ let $G_a(L)$ denote the Zariski closure of L in G_a. The group $H_0 := (G_a(H))^0 \cap H$ is a normal subgroup of H and has finite index. Thus by the remark above we may assume that $H_0 = H$ and $G_a(H)$ is connected. Let us now consider the case where H is a discrete subgroup of G. It is clear that $G_a(H)/H$ is weakly Kähler. Thus by Lemma 3.6 it follows that $G_a(H)$ is nilpotent. Let $G_a(H)/H \longrightarrow G_a(H)/J$ denote the holomorphic reduction of the $G_a(H)$-orbit of H in $G_a(H)$. Then by Theorem 5.1 we know that $G_a(H)/J$ is Stein and J/H is a Cousin group. Since quotients of solvable, algebraic groups are Stein and $G_a(H)$ is an algebraic subgroup of G_a, we also know that $G_a/G_a(H)$ is Stein (see [HO1]). Thus we have the holomorphic fibrations

$$G_a/H \longrightarrow G_a/J \longrightarrow G_a/G_a(H).$$

By a theorem of Matsushima and Morimoto (see section 1) it follows that G_a/J is Stein. Thus, since $\mathcal{O}(J/H) = \mathbb{C}$, it follows that

$G_a/H \longrightarrow G_a/J$ is the holomorphic reduction of G_a/H. It is also clear that $J \subset G$ (see section 1). Hence G/J is Stein, J/H is a Cousin group and H is nilpotent $(H \subset G_a(H)!)$.

If H is not discrete in G, let $N = N_{G_a}(H^0)$. This is an algebraic subgroup of G_a (see [HO2]) and H/H^0 is a discrete subgroup of N/H^0. Since also N^0 is algebraic and $H \subset G_a(H)$, where the latter is assumed to be connected, it follows that $H \subset N^0$. Thus by the discrete case we know that the holomorphic reduction $N^0/H \longrightarrow N^0/J$ has a Stein base and a Cousin group as fiber. Since G_a/N^0 is Stein it follows (again by the theorem of Matsushima-Morimoto) that G_a/J is Stein and $G_a/H \longrightarrow G_a/J$ is the holomorphic reduction. Thus again we have $J \subset G$ and G/J is Stein and $G/H \longrightarrow G/J$ is the holomorphic reduction of G/H. Moreover, again by the discrete case, H/H^0 contains a nilpotent normal subgroup of finite index. Thus $\pi_1(G/H) \cong H/H^0$ contains a nilpotent normal subgroup of finite index and the theorem is proved. □

7. Complex hypersurfaces of solvmanifolds.

Let X be a solvmanifold. In order to study complex hypersurfaces in X it is convenient to consider the hypersurface reduction $\pi : X \longrightarrow Z$ (see section 1). Its base Z is separable by hypersurfaces and every hypersurface in X is the inverse image of one in Z, i.e.

$$\mathcal{H}(X) = \pi^* \mathcal{H}(Z).$$

Thus in order to describe $\mathcal{H}(X)$ it is sufficient to understand Z. It turns out that a complex solvmanifold X, which is locally separable by global hypersurfaces (see section 1), inherits the same structure as a weakly Kähler solvmanifold.

Theorem 1. Let X be a complex solvmanifold which is locally separable by global hypersurfaces. Then the holomorphic reduction $X \xrightarrow{F} Y$ has a Stein base Y and the fiber G is a Cousin group. Moreover $\pi_1(X)$ contains a normal nilpotent subgroup of finite index.

The basic technical ingredient for the proof of Theorem 1 is

Theorem 2. Let G be an arbitrary complex Lie group and $\Gamma \subset G$ a discrete subgroup. If $X = G/\Gamma$ is locally separable by global hypersurfaces, then X admits an exhaustion $(U_m)_{m \in \mathbb{N}}$ and a sequence

$(\omega_m)_{m\in\mathbb{N}}$ of real closed positive semidefinite (1,1)- forms, such that $\omega_m|U_m$ is positive definite. If X is locally separable by meromorphic functions, then X is Kähler.

Corollary 3. Let X be given as in Theorem 2, then X is weakly Kähler.

In order not to be technical we will first show how Theorem 2 implies Theorem 1 and then give a brief sketch of the proof of Theorem 2.

Proof of Theorem 1. The proof is almost the same as that of Theorem 6.1. If $X = G/H$ and H is discrete then Theorem 1 immediately follows from Corollary 3 an Theorem 6.1.

If H is not discrete, let G_a denote an abstract algebraic closure of G (see section 1). Then G_a/H is also locally separable by global hypersurfaces. Let $N = N_{G_a}(H^0)$. This is an algebraic subgroup of G_a and H/H^0 is a discrete subgroup of N/H^0. Since N^0 is also algebraic and since we may assume $H \subset G_a(H)$ (see the proof of 6.1), we also may assume that $H \subset N_0$. Thus by Theorem 6.1 and Corollary 3 we know that the holomorphic reduction $N^0/H \longrightarrow N^0/J$ has a Stein base and a Cousin group as fiber. The rest of the proof goes exactly like the proof of Theorem 6.1. □

Sketch of the proof of Theorem 2. Assume X is locally separable by global hypersurfaces. Let $p = \Gamma \in G/\Gamma$. By assumption we have a hypersurface H in G/Γ, such that every irreducible component H_i of H is smooth in p and $\bigcap_{i=1} H_i$ contains p as an isolated point $(n = \dim_{\mathbb{C}} X)$. Now H defines a real closed positive (1,1) current

$$T_H(\varphi) := \int_{H^*} \varphi,$$

where $\varphi \in A_c^{n-1,n-1}(X)$ is an (n-1,n-1)-form with compact support and H^* is the set of smooth points of H.

It is almost classical to show the following

Claim: For every $\eta \in A_c^{n-1,0}(X)$ such that $p \in \operatorname{supp}(\eta)$ we have

$$(\tfrac{i}{2})^{n-1}(-1)^{(n-1)(n-2)/2}\, T_H(\eta \wedge \bar{\eta}) > 0.$$

Note that the discreteness of Γ in G makes it possible to lift T_H to a right-Γ-invariant closed, real, positive (1,1) current on G, which is strictly positive on forms $\eta \wedge \bar{\eta}$, $\eta \in A_c^{(n-1,0)}(G)$ such

that $\operatorname{supp}(\eta) \cap \Gamma \neq \emptyset$. Let T denote this current.

Now we take an exhaustion $(K_m)_{m\in\mathbb{N}}$ of G, $K_m \subset\subset K_{m+1}$ and a subordinate sequence of cut-off functions $(\chi_m)_{m\in\mathbb{N}}$.

Then we smooth T by convolution with χ_m, using a right invariant volume form μ on G, i.e., we let

$$T_m := \chi_m * T = \int_G \chi_m(g) 1_g^* T \, d\mu(g).$$

Now T_m is given by integration against a $(1,1)$-form ω_m, which can be chosen right-Γ-invariant and has the properties required in Theorem 2, if U_m is taken to be the image of K_m in G/Γ.

If G/Γ is locally separable by meromorphic functions, then using a meromorphic function for the construction of H, it is easy to see that T is strictly positive in G and hence yields a Kähler form on G/Γ. □

8. Applications.

In this section we apply some of the methods developed in the preceding sections to related problems in the theory of homogenous complex manifolds.

First we give a new proof of Theorem 0.2 (Huckleberry-Margulis [HuM]).

Theorem. Let S be a semisimple complex Lie group and $I \subset S$ an arbitrary subgroup. Then I is Zariski dense if and only if $\mathcal{H}(S)^I = \emptyset$.

Proof. Let $\hat{I}$ be the smallest closed complex Lie group of S containing I. We may replace I by $\hat{I}$ in the theorem. Assume that $\hat{I}$ is not Zariski dense. Then $\mathcal{H}(S)^{\hat{I}} \neq \emptyset$, since as a quotient of algebraic groups $S/\bar{I}$ ($\bar{I}$ = Zariski closure of $\hat{I}$ in S) has analytic hypersurfaces. This proves the "⇐"-direction.

Now let $\hat{I}$ be Zariski dense in S. Then $\hat{I}^0$ (the identity component of $\hat{I}$) is normal n S and we may assume that $\hat{I}$ is discrete.

Assuming the existence of analytic hypersurfaces on $S/\hat{I}$ the hypersurface reduction $p : S/\hat{I} \longrightarrow S/L$ has a non-trivial base. As before we have $L^0 \triangleleft S$ and we are reduced to the case that $S/\hat{I}$ is hypersurface separable and $\hat{I}$ is discrete.

By Theorem 7.2 the manifold $X = S/\hat{I}$ is weakly Kähler. The group $\hat{I}$ being Zariski dense in S implies the existence of a semisimple element $\alpha \in \hat{I}$ with $\Lambda^\alpha = \{\alpha^n | n \in \mathbb{Z}\} \cong \mathbb{Z}$ so that the Zariski closure Z^α of Λ^α is a connected diagonizable algebraic group, i.e., $Z^\alpha = (\mathbb{C}^*)^n$ see e.g. [Ti]).

In particular we have that Z^α/Λ^α is a Cousin group, i.e., $\mathcal{O}(Z^\alpha/\Lambda^\alpha) \cong \mathbb{C}$. Now we consider the covering space $S/\Lambda^\alpha \longrightarrow S/\hat{I}$. The manifold S/Λ^α is also weakly Kähler with a weak Kähler form ω.

Since S is Stein and $\Lambda^\alpha \cong \mathbb{Z}$ the isomorphism between group cohomology and Čech cohomology yields $H^2(S/\Lambda^\alpha,\mathbb{Z}) = 0$, hence $H^2(S/\Lambda^\alpha,\mathbb{C}) = 0$. Therefore $\omega = d\eta$ for a 1-form η. Since S/Λ^α is homogeneous, we may assume that the restriction $\tilde{\omega}$ of ω on the Cousin group Z^α/Λ^α is also an exact weak Kähler form.

After integration over the maximal compact subgroup K of Z^α/Λ^α we obtain an exact, K-invariant weak Kähler form $\tilde{\tilde{\omega}}$ on Z^α/Λ^α, i.e., $\tilde{\tilde{\omega}} = d\beta$ with the property that $\tilde{\tilde{\omega}}$ is positive definite on an open neighborhood U of K.

Let $\emptyset \neq X \in \mathfrak{k} \cap i\mathfrak{k}$ ($\mathfrak{k}$ = Lie algebra of K) be an invariant vector field in the maximal complex subalgebra of K. In the open set U we have

$$0 < \tilde{\tilde{\omega}}(x,Jx) = d\beta(X,JX) = \tfrac{1}{2}(X\beta(JX) - JX\beta(X) - \beta([X,JX]) = 0,$$

since JX is also an invariant vector field in $\mathfrak{k} \cap i\mathfrak{k}$.

Contradiction! this implies that S/Γ has no analytic hypersurfaces and the theorem is proved. □

In the introduction we proved the Ahiezer conjecture in the solvable setting. Now we handle another special case.

<u>Ahiezer Conjecture in the product case</u>.

Let $G = R \times S$ be a simply connected complex Lie group, which is a group theoretic product of its radical R and a maximal semisimple subgroup S. For this kind of complex Lie groups the Ahiezer conjecture is true.

<u>Proof</u>. Let $H \subset G$ be an arbitrary subgroup of G such that $\mathcal{O}(G)^H \cong \mathbb{C}$ and $H \subset P$ for any proper parabolic subgroup P. Let $\hat{H}$ be the smallest closed complex subgroup of G containing H. As we already have seen it is enough to consider the honogeneous manifold $G/\hat{H}$ and assume that $G/\hat{H}$ is hypersurface separable.

Let $N = N_G(\hat{H}^0)$ be the normalizer of $\hat{H}^0$ in G. It is clear

that $\hat{H} \subset N$. Look at the holomorphic fibration $G/\hat{H} \longrightarrow G/N$. The base is equivariantly imbedded in some $\mathbb{P}_N(\mathbb{C})$ and $\mathcal{O}(G/N) \cong \mathbb{C}$. If $N \neq G$, by a remark of [HO1] (Proposition 5, p. 48) the group N is contained in a proper parabolic subgroup. This is excluded by the assumption of the conjecture. Therefore $N = G$, i.e., $\hat{H} \triangleleft G$ and we may assume that $\hat{H}$ is a discrete subgroup of G.

Now Proposition 9, p. 53 of [HO1] implies that the R-orbits in $G/\hat{H}$ are closed, i.e., there is a fibration $G/\hat{H} \longrightarrow G/\hat{H}\cdot R = S/\Lambda$, where $\Lambda = S \cap (\hat{H}\cdot R)$ is a discrete group of S. Since $\mathcal{O}(S/\Lambda) = \mathbb{C}$ and Λ is not contained in any parabolic subgroup of S, it follows that $\Lambda \subset S$ is Zariski dense in S.

We consider the discrete group $S \cap \hat{H}$. We claim that if $S \neq e$ then $S \cap \hat{H}$ has infinitely many elements. Assume that $S \cap \hat{H}$ is finite. The the projection $\rho : G \longrightarrow R$ has a finite kernel. Therefore the Zariski closure of Λ would be solvable. This is not true. Therefore $S \cap \hat{H}$ has infinitely many elements.

Since Λ normalizes $S \cap \hat{H}$ in S, we see that the Zariski closure $\tilde{S} := \overline{S \cap \hat{H}}$ is normalized by S, i.e., $\tilde{S}$ is a semisimple normal subgroup of S containing $S \cap \hat{H}$ as a Zariski dense subgroup, i.e., $\mathcal{H}(S/\tilde{S} \cap \hat{H}) = \emptyset$. But this contradicts the fact that the $\tilde{S}$-orbit in $G/\hat{H}$, which is $\tilde{S}/S \cap \hat{H}$, is separable by hypersurfaces. (See Theorem 0.2, which is proved at the beginning of this section.) Therefore $S = \{e\}$ and the proof is finished by using the solvable case (see Introduction). □

References

[A] Ahiezer, D., Invariant analytic hypersurfaces in complex nilpotent Lie groups, Ann. Global. Anal. and Geom., Vol. 2, No. 2 (1984), 129-140.

[Bo] Borel, A. Linear algebraic groups, New York, Benjamin (1969).

[BO] Barth, W., Olte, M., Über fast uniforme Untergruppen komplexe Lie gruppen und anflösbare komplexe Mannigfaltikgeiten, Comm. Math. Helv. (1969), 269-287.

[BR] Borel, A., Remmert, R., Über kompakte homogene Kählersche Mannigfaltikgeiten, Math. Ann. 145 (1962), 429-439.

[Co] Cousin, P., Sur les fonctions triplement périodiques de deux variables, Acta Math. 10, (1910), 105-232.

[GH] Gilligan, B., Huckleberry, A.T., On non-compact complex nilmanifolds, Math. Ann. 238 (1978), 39-49.

[Gr] Grauert, H., Analytische Faserungen über holomorph völlstandigen Räumen, Math. Ann. 135 (1958), 263-273.

[GR] Grauert, H., Remmert R., Über kompakte homogene komplexe Mannigfaltikgeiten, Arch. Math. 13 (1962), 498-507.

[HM] Hochschild, G., Mostow, G.D., On the algebra of representative functions of an analytic group, II, Am. J. Math. 86 (1964), 869-887.

[HuM] Huckleberry, A.T., Margulis, G.A., Invariant analytic hypersurfaces, Invent. Math. 71 (1983), 235-240.

[HO1] Huckleberry, A.T., Oeljeklaus, K., Classification theorems for almost homogeneous spaces, Revue de l'Institut Elie Cartan 9 (Nancy), Janvier 1984.

[HO2] Huckleberry, A.T., Oeljeklaus, K., On holomorphically separable complex solvmanifolds, (to appear in Ann. de l'Institut Fourier).

[Lo] Loeb, J.J., Actions d'une forme de Lie reelle d'un groupe de Lie complexe sur les fonctions plurisubharmoniques, Annales de l'Institut Fourier, 35.4 (1985), 59-97.

[LOR] Loeb, J.J., Oeljeklaus, K., Richthofer, W.R., A decomposition theorem for nilmanifolds (to appear).

[Mal] Malcev, A.J., On a class of homogeneous spaces. AMS Transl. 39 (1951).

[Ma] Matsushima, Y., Espaces honogenès de Stein des groupes de Lie complexes I, Nagoya Math. J. 16 (1960), 205-278.

[MM] Matsushima, Y., Morimoto, A., Sur certain espaces fibrés holomorphes sur un variéte de Stein, Bull. Soc. Math. France 88 (1960), 137-155.

[Mo] Mostow, G.D., Factor spaces of solvable groups, Ann. of Math. 60 (1954), 1-27.

[Oe] Oeljeklaus, K., Hyperflachen und Geradenbündel auf homogenen komplexen Mannigfaltigkeiten, Schriftenreihe der Univ. Münster, Ser. 2, Heft 36, 1985.

[RV] Remmert, R., van de Veen, T., Zur Functionentheorie homogenen komplexen Mannigfaltigkeiten, Topology 2 (1963), 137-157.

[Ri] Richthofer, W., Homogene CR-Mannigfaltigkeiten, Dissertation Bochum, 1985.

[Ti] Tits, J. Free subgroups in linear groups, J. Algebra 20, (1974), 250-270.

Rhur-Universität Bochum
Institut für Mathematik
4630 Bochum 1
Federal Republic of Germany

KÄHLER-EINSTEIN METRICS FOR THE CASE OF POSITIVE FIRST CHERN CLASS[1]

Yum-Tong Siu

The existence of Kähler-Einstein metrics for compact Kähler manifolds with trivial or negative anticanonical line bundle has been known for some time ([1,2,14]) For compact Kähler manifolds with positive anticanonical line bundle in general there are obstructions [4,5,6,7,8,9,10,11] to the existence of Kähler-Einstein metrics. All the known obstructions are related to the presence of nonzero holomorphic vector fields. There is a conjecture that any compact Kähler manifold of positive anticanonical line bundle without any nonzero holomorphic vector field admits a Kähler-Einstein metric. The conjecture is still open. The only known examples of Kähler-Einstein metrics of Kähler manifolds of positive anticanonical line bundle are those of Hermitian symmetric manifolds or homogeneous manifolds or certain noncompact manifolds [3]. So far there is no known way of proving the existence of Kähler-Einstein metrics of compact Kähler manifolds of positive anticanonical line bundle by using the continuity method with reasonable additional assumptions such as the nonexistence of nonzero holomorphic vector fields. In this talk we discuss a method to prove the existence of Kähler-Einstein metrics for compact Kähler manifolds of positive anticanonical line bundle under the additional assumption of the existence of a suitable finite or compact group of symmetry. As examples this method is applied to prove the existence of a Kähler-Einstein metric on the Fermat cubic surface and the surface obtained by blowing up three points of the complex projective surface $\mathbb{P}_2$. The method is also applicable to higher-dimensional Fermat hypersurfaces. We sketch in this talk the main ideas and the key steps of this method. Details will later appear in a full length paper [13].

Before we discuss our method, we would like first to make some observations. In the case of positive anticanonical line bundle the only difficulty in getting a Kähler-Einstein metric is the lack of a zeroth order *a priori* estimate for the

---

[1]Presented at the Conference on Geometric Theory of Several Complex Variables at University of Maryland on April 18, 1986. Research partially supported by a National Science Foundation grant and a Guggenheim Fellowship.

solution of the Monge-Ampère equation by the continuity method. Such a zeroth order *a priori* estimate would follow if one has a suitable positive lower bound for the first nonzero eigenvalue of the Laplacian. Eigenvalues are determined by the minimax principle. If one imposes suitable symmetry requirements, the number of functions eligible for consideration in the minimax principle may be so reduced that a suitable positive lower bound would exist for the first nonzero eigenvalue of the Laplacian. This idea of using a finite symmetry group to help solve problems involving estimates was first used by Moser [12] to prove a theorem on conformal metric deformation on the 2-sphere with antipodal symmetry. Though the motivation for using a finite group of symmetry is to increase the first nonzero eigenvalue of the Laplacian, we do not know how to quantitatively relate properties of a finite group of symmetry to the lower bound of the first nonzero eigenvalue of the Laplacian. We will instead directly use some properties of the finite group of symmetry to get a zeroth order *a priori* estimate of the solution of the Monge-Ampère equation without going through the consideration of the first nonzero eigenvalue of the Laplacian. The technical key step is to apply the simple inequality $uv \leq u\log u + e^{v-1}$ to the Green's formula for the restriction, to a complex curve, of the solution of the Monge-Ampère equation so that one can transform the product of the Green's kernel and the Laplacian on the curve of the solution of the Monge-Ampère equation into a sum. Since the Laplacian of the solution of the Monge-Ampère equation is bounded by the exponential of a constant times the difference of its supremum and infimum, one would get a zeroth order *a priori* estimate if the curve passes through a supremum point and an infimum point and the area of the curve is small relative to the constant in the exponent of the estimate of the Laplacian of the solution. The use of symmetry has the same effect as reducing the area of the curve by taking the quotient with respect to the finite group of symmetry. The constant in the exponent of the estimate of the Laplacian of the solution of the Monge-Ampère equation is linked to the lower bound of the bisectional curvature of the manifold. As a consequence the conditions on the finite group of symmetry is related to the lower bound of the bisectional curvature of the manifold for two orthonormal directions, the area of a (possibly reducible) curve joining two arbitrary points, and the number of points in a branch of the curve which are congruent under the group. Because in general the computation of a good explicit lower bound of the bisectional curvature for two orthonormal directions is rather difficult, in our applications we have to modify the argument so that the more easily computable bisectional curvature with a conformal factor is used instead of the usual bisectional curvature.

Table of Contents

§1. *The Monge-Ampere equation.*

Let M be a compact Kähler manifold of complex dimension m. Let $g_{i\bar{j}}$ be a Kähler metric of M whose Kähler form is in the anticanonical class of M. The Ricci curvature $R_{i\bar{j}}$ of $g_{i\bar{j}}$ is given by

$$R_{i\bar{j}} = -\partial_i \partial_{\bar{j}} \log \det(g_{i\bar{j}}).$$

There exists a real-valued smooth function F on M such that

$$R_{i\bar{j}} - g_{i\bar{j}} = \partial_i \partial_{\bar{j}} F.$$

Consider the Monge-Ampère equation

$$(*) \qquad \frac{\det(g_{i\bar{j}} + \partial_i \partial_{\bar{j}} \varphi)}{\det(g_{i\bar{j}})} = \exp(-t\varphi + F)$$

for the function $\varphi = \varphi_t$ on M $(0 \le t \le 1)$. Let $g'_{i\bar{j}} = g_{i\bar{j}} + \partial_i \partial_{\bar{j}} \varphi$ and $R'_{i\bar{j}}$ be the Ricci curvature of $g'_{i\bar{j}}$. One has

$$R'_{i\bar{j}} = t g'_{i\bar{j}} + (1-t) g_{i\bar{j}}$$

so that the metric $g'_{i\bar{j}}$ for $t = 1$ is a Kähler-Einstein metric.

We try to produce a Kähler-Einstein metric on M by solving the Monge-Ampère equation (*) by the continuity method. The method of Calabi and Yau used for the case of trivial and negative anticanonical line bundle works partially for this case of positive anticanonical line bunlde. There is no trouble with the openness in the continuity method when $t < 1$, because the derivative of the funciton $\log \frac{\det(g_{i\bar{j}} + \partial_i \partial_{\bar{j}} \varphi)}{\det(g_{i\bar{j}})} + t\varphi$ along the vector ψ is $\Delta'\psi + t\psi$ and the operator $\Delta' + t$ is invertible as a result of the positivity of the matrix $R'_{i\bar{j}} - t g'_{i\bar{j}}$. Here

Δ and Δ' denote respectively the (negative) Laplacian for the Kähler metrics $g_{i\bar{j}}$ and $g'_{i\bar{j}}$. One gets also in the same way the first, second, and third order *a priori* estimates for the function φ, provided that one has the zeroth order *a priori* estimate for φ. So the difficult part is the zeroth order *a priori* estimate for φ. Because of known obstructions such as the reductivity of the automorphism group [9,10,11] and the vanishing of the Futaki invariant [6,7], we know that in general one cannot have the the zeroth order *a priori* estimate for φ in the case of positive anticanonical line bundle. However, it is possible to estimate the supremum of $-\varphi$ (respectively φ) from above in terms of the supremum of φ (respectively $-\varphi$). The first of these two estimates is needed for our method. The precise statement of these two estimates are as follows.

§2. *Relation Between* $\sup_M \varphi$ *and* $\inf_M \varphi$.

(2.1) Proposition. Given any positive number ε and any $0 < t_0 \le 1$ there exists a positive constant C such that if φ is a solution of (*) on M for $0 \le t < t_0$, then $\sup_M(-\varphi) \le (m+\varepsilon)\sup_M\varphi + C$ and $\sup_M\varphi \le (m+\varepsilon)\sup_M(-\varphi) + C$ on M for $0 \le t < t_0$.

We sketch the proof of the first inequality in the conclusion of Proposition (2.1). The proof of the second inequality is analogous. By using $R'_{i\bar{j}} = tg'_{i\bar{j}} + (1-t)g_{i\bar{j}}$ and a Bochner type formula to get lower eigenvalue estimates, we obtain for $t+s > 0$ the Poincaré type inequality

$$\int_M |f|^2 e^{-s\varphi} dV' \le \frac{1}{t+s}\int_M \langle\bar\partial f,\bar\partial f\rangle' e^{-s\varphi} dV' + \frac{1}{t+s}\,\frac{\left(\int_M f e^{-s\varphi} dV'\right)^2}{\int_M e^{-s\varphi} dV'}$$

for any smooth function f on M, where dV' is the volume form of $g'_{i\bar{j}}$ and $\langle\cdot,\cdot\rangle'$ is the inner product with respect to $g'_{i\bar{j}}$. Using $\Delta'\varphi \le m$ and $f = e^{s\varphi}$ and Hölder's inequality, we get

$$\int_M e^{s\varphi - t\varphi + F} dV \le \frac{1}{1 + \frac{ms}{t+s}}\int_M e^{(t+s)\varphi - F} dV.$$

Choosing $s = -(\frac{1}{m+1} - \epsilon)t$ for some small positive number ϵ and using $\Delta'\varphi \leq m$ and the fact that the Green's function for Δ' is bounded from below by an *a priori* constant, we obtain $\sup_M(-\varphi) \leq (m+\epsilon)\sup_M\varphi + C$.

§3. *Estimate of* $m+\Delta\varphi$.

We need a second-order *a priori* estimate of φ which is a slightly modified form of that given in [14] because we would like to use the more easily computable bisectional curvature *with a conformal factor* instead of the usual bisectional curvature. The method used in [14] to establish the second-order *a priori* estimate of φ can easily be modified to give the form we want.

(3.1) Definition. Let $g_{i\bar{j}}$ be a Kähler metric of a complex manifold M and $R_{i\bar{j}k\bar{\ell}} = -\partial_i\partial_{\bar{j}}g_{k\bar{\ell}} + g^{\lambda\bar{\mu}}\partial_i g_{k\bar{\mu}}\partial_{\bar{j}}g_{\lambda\bar{\ell}}$ be its curvature tensor. Let ψ be a smooth real-valued function on M. We say that *with the conformal factor* $e^{-\psi}$ *the bisectional curvature of* $g_{i\bar{j}}$ *for two orthonormal vectors is bounded from below by* A if $(\psi_{i\bar{j}} + R_{i\bar{j}k\bar{\ell}})\xi^i\overline{\xi^j}\eta^k\overline{\eta^\ell} \geq A$ for any (ξ^i) and (η^i) that satisfy $g_{i\bar{j}}\xi^i\overline{\xi^j} = g_{i\bar{j}}\eta^i\overline{\eta^j} = 1$ and $g_{i\bar{j}}\xi^i\overline{\eta^j} = 0$.

(3.2) Proposition. Suppose $h_{i\bar{j}}$ is a Kähler metric in the anticanonical class of M and ψ is a smooth real-valued function on M. Assume that with the conformal factor $e^{-\psi}$ the bisectional curvature of $h_{i\bar{j}}$ for two orthonormal vectors is bounded from below by some real number $-\kappa$. Let γ be a nonnegative number $> \kappa$. Then there exist positive constants C and C' such that if φ is a solution of the Monge-Ampère equation (*) on M for $0 \leq t < t_0$, then $m+\Delta\varphi \leq C\exp(\gamma\varphi - (\gamma+1)\inf_M\varphi) + C'$ on M for $0 \leq t < t_0$. Here Δ means the (negative) Laplacian with respect to the Kähler metric $g_{i\bar{j}}$.

In our application the following conformal factor $e^{-\psi}$ is used. Suppose the bisectional curvature of $h_{i\bar{j}}$ for two orthonormal vectors is bounded from above by some real number κ_1 and the holomorphic sectional curvature of $h_{i\bar{j}}$ for unit vectors is bounded from above by some real number κ_2. Suppose μ is a Hermitian metric along the fibers of the anticanonical line bundle K_M^{-1} whose curvature form dominates σ times $h_{i\bar{j}}$ for some real number σ. Choose a real-valued smooth function ψ on M so that $\partial_i\partial_{\bar{j}}\psi + R^h_{i\bar{j}}$ equals the curvature form of the

Hermitian metric μ of K_M^{-1}, where $R^h_{i\bar{j}}$ is the Ricci curvature form of $h_{i\bar{j}}$. Then with the conformal factor $e^{-\psi}$ the bisectional curvature of $h_{i\bar{j}}$ for two orthonormal vectors is bounded from below by $\sigma - \kappa_2 - (m-2)\kappa_1$.

§4. *The use of a finite group of symmetry.*

First let us make some simple remarks. On the complex line $\mathbb{C}$ with coordinate z our Laplacian $\Delta = g^{i\bar{j}}\partial_i\partial_{\bar{j}}$ becomes $\frac{\partial^2}{\partial z\partial\bar{z}}$. For any smooth function f on $\mathbb{C}$ with compact support the Cauchy integral formula for smooth functions gives

$$f(0) = \frac{1}{2\pi\sqrt{-1}}\int_{\mathbb{C}} \frac{\frac{\partial f}{\partial\bar{z}}(z)dz\wedge d\bar{z}}{z} = \frac{1}{2\pi\sqrt{-1}}\int_{\mathbb{C}}(\frac{\partial}{\partial z}\log|z|^2)\frac{\partial f}{\partial\bar{z}}(z)dz\wedge d\bar{z}$$

$$= \int_{\mathbb{C}}(\frac{1}{2\pi}\log|z|^2)\sqrt{-1}\partial\bar{\partial}f(z).$$

So the Green's function is $\frac{-1}{2\pi}\log|z|^2$. For the general case of a nonsingular complex curve Γ with a Kähler metric, the dominant term of the Green's function $G_\Gamma(x,y)$ for Γ is $\frac{-1}{2\pi}\log \operatorname{dist}(x,y)^2$ near $x = y$; and for any smooth function f on Γ we have

$$(\#) \qquad f(x) = \frac{1}{\operatorname{Vol}(\Gamma)}\int_\Gamma f + \int_{y\in\Gamma} G_\Gamma(x,y)(-\sqrt{-1}\partial\bar{\partial}f)(y).$$

Consider now our compact Kähler manifold M of complex dimension m with a Kähler metric $g_{i\bar{j}}$ on which we would like to solve by the continuity method the Monge-Ampère equation (*). Assume that G is a finite subgroup of automorphisms of M. We require that the Kähler metric $g_{i\bar{j}}$ and the function F both be invariant under G. We consider only solutions φ that are invariant under G. We are going to discuss how the finite group of symmetry G would help us to get our zeroth-order *a priori* estimate of φ. To illustrate the idea, let us consider only the simplest situation. Take a nonsingular complex curve Γ in M which contains both a point P where $\sup_M\varphi$ is achieved and a point Q where $\inf_M\varphi$ is achieved. Let $\omega = \sqrt{-1}g_{i\bar{j}}dz^i\wedge\overline{dz^j}$ and $\omega' = \sqrt{-1}g'_{i\bar{j}}dz^i\wedge\overline{dz^j}$ be the Kähler forms of the two Kähler metrics $g_{i\bar{j}}$ and $g'_{i\bar{j}} = g_{i\bar{j}} + \partial_i\partial_{\bar{j}}\varphi$ on M. Let $G_\Gamma(x,y)$ be the Green's function for the Laplacian of the restriction of the Kähler metric

$g_{i\bar{j}}$ to Γ. Then for any smooth function f on Γ, we have the Green's formula (#). Let K be a positive number such that $G_\Gamma(x,y) + K \geq 0$. Then

$$f(x) = \frac{1}{\mathrm{Vol}(\Gamma)}\int_\Gamma f\,\omega + \int_{y\in\Gamma}(G_\Gamma(x,y) + K)(-\sqrt{-1}\partial\bar{\partial}f)(y).$$

Let P_μ and Q_μ $(1 \leq \mu \leq N)$ be points of Γ so that Q_μ $(1 \leq \mu \leq N)$ are *distinct*. Assume that all the P_μ's are congruent to P under G and all the Q_μ's are congruent to Q under G. Since $\omega' = \omega + \sqrt{-1}\partial\bar{\partial}\varphi$ is positive definite, it follows that $-\sqrt{-1}\partial\bar{\partial}\varphi < \omega$ and $\sqrt{-1}\partial\bar{\partial}\varphi < \omega'$. Applying the above formula to φ and $-\varphi$ at the points P_μ and Q_μ respectively, we get

$$\begin{aligned}\varphi(P_\mu) &= \frac{1}{\mathrm{Vol}(\Gamma)}\int_\Gamma \varphi\,\omega + \int_{y\in\Gamma}(G_\Gamma(P_\mu,y) + K)(-\sqrt{-1}\partial\bar{\partial}\varphi)(y).\\ &\leq \frac{1}{\mathrm{Vol}(\Gamma)}\int_\Gamma \varphi\,\omega + \int_{y\in\Gamma}(G_\Gamma(P_\mu,y) + K)\omega.\\ &= \frac{1}{\mathrm{Vol}(\Gamma)}\int_\Gamma \varphi\,\omega + K\mathrm{Vol}(\Gamma).\end{aligned}$$

$$\begin{aligned}\varphi(Q_\mu) &= \frac{-1}{\mathrm{Vol}(\Gamma)}\int_\Gamma \varphi\,\omega + \int_{y\in\Gamma}(G_\Gamma(Q_\mu,y) + K)(\sqrt{-1}\partial\bar{\partial}\varphi)(y).\\ &\leq \frac{-1}{\mathrm{Vol}(\Gamma)}\int_\Gamma \varphi\,\omega + \int_{y\in\Gamma}(G_\Gamma(Q_\mu,y) + K)\omega'(y).\end{aligned}$$

We use the inequality $uv \leq u\log u + e^{v-1}$ with u equal to the quotient of $\frac{1}{2\pi-\epsilon}\,\omega'|\Gamma$ by $\omega|\Gamma$ and v equal to $(2\pi-\epsilon)(\Sigma_{\mu=1}^N G_\Gamma(Q_\mu,y) + NK)$. By using also the fact that the quotient of $\omega'|\Gamma$ by $\omega|\Gamma$ is $\leq m + \Delta\varphi$, we get

$$\begin{aligned}\Sigma_{\mu=1}^N -\varphi(Q_\mu) &\leq \Sigma_{\mu=1}^N \frac{-1}{\mathrm{Vol}(\Gamma)}\int_\Gamma \varphi\,\omega + \int_{y\in\Gamma}(\Sigma_{\mu=1}^N G_\Gamma(Q_\mu,y) + NK)\omega'(y)\\ &\leq \Sigma_{\mu=1}^N \frac{-1}{\mathrm{Vol}(\Gamma)}\int_\Gamma \varphi\,\omega + \frac{1}{2\pi-\epsilon}\left[\sup_\Gamma \log(m + \Delta\varphi) + \log\frac{1}{2\pi-\epsilon}\right]\mathrm{Vol}(\Gamma)\\ &\quad + \int_{y\in\Gamma}\exp\left[(2\pi-\epsilon)(\Sigma_{\mu=1}^N G_\Gamma(Q_\mu,y) + NK)\right]\omega\end{aligned}$$

Here ϵ is any small positive number. By adding together the inequalities for $\Sigma_{\mu=1}^N \varphi(P_\mu)$ and $\Sigma_{\mu=1}^N -\varphi(Q_\mu)$, we get

$$\begin{aligned}\Sigma_{\mu=1}^N\left[\varphi(P_\mu)-\varphi(Q_\mu)\right] \leq\ & \frac{1}{2\pi-\epsilon}\left[\sup_\Gamma \log(m + \Delta\varphi)(y) + (2\pi-\epsilon)NK + \log\frac{1}{2\pi-\epsilon}\right]\mathrm{Vol}(\Gamma)\\ &+ \int_{y\in\Gamma}\exp\left[(2\pi-\epsilon)(\Sigma_{\mu=1}^N G_\Gamma(Q_\mu,y) + NK)\right]\omega.\end{aligned}$$

Finally we use $m+\Delta\varphi \le C\exp(\gamma\varphi - (\gamma+1)\inf_M\varphi) + C'$ and $\varphi(P_\mu) = \sup_M\varphi$ and $\varphi(Q_\mu) = \inf_M\varphi$ to conclude that

$$N(\sup_M\varphi - \inf_M\varphi) \le \frac{1}{2\pi-\epsilon}\left[\gamma\sup_M\varphi - (\gamma+1)\inf_M\varphi\right]\mathrm{Vol}(\Gamma) + C^{\#}$$

for some constant $C^{\#}$ independent of φ. So for example when N is greater than $\frac{\gamma+1}{2\pi}\mathrm{Vol}(\Gamma)$ we have a zeroth-order *a priori* estimate of φ. In actual applications the situation is more complicated and a sequence of possibly singular irreducible curves is required to join a supremum point of φ to an infimum point of φ in order to assure that there are enough congruent points on each irreducible curve with small volume to make the argument work. The relation between the supremum of φ and the infimum of φ has to be used in the situation when no supremum point of φ and no infimum point of φ lie on the same irreducible curve with enough congruent points and small enough volume. To precisely state our main result we need some definitions.

(4.1) <u>Definition</u>. Let S be a connected complex manifold. A holomorphic family $\mathcal{F}$ of (possibly singular) complex curves Γ_s with base point O_s $(s \in S)$ is said to be *smoothly simultaneously uniformizable* if there exist

(i) a differentiable manifold $\mathcal{G}$ and a smooth submersion $\theta: \mathcal{G} \to S$ whose fibers are compact of real dimension 2 and

(ii) two smooth maps $\sigma: S \to \mathcal{G}$ and $\tau: \mathcal{G} \to M$ such that for every $s \in S$

(a) $\theta^{-1}(s)$ can be given a complex structure so that $\theta^{-1}(s)$ is the normalization of Γ_s under the map τ and

(b) $\tau(\sigma(s)) = O_s$.

(4.2) <u>Definition</u>. By the *curve volume* of the family $\mathcal{F}$ of holomorphic curves we mean the volume of the curve Γ_s with respect to the anticanonical class of M (which is independent of the choice of s in S and independent of the choice of the Kähler metric in the anticanonical class).

(4.3) <u>Definition</u>. We say that the *orbit cardinality* of the family $\mathcal{F}$ of holomorphic curves is at least N if there exists some positive number δ such that for every $s \in S$ there are at least N points $Q_1, \cdots, Q_N$ in the orbit GO_s (under the group G) of the base point O_s of the curve Γ_s with the distance between Q_i and $Q_j \ge \delta$ for all $i \ne j$ with respect to some Kähler metric of M.

(4.4) <u>Definition</u>. Let S' be a subset of S. The holomorphic subfamily $\mathcal{F}' = \{\Gamma_s\}_{s \in S'}$ of $\mathcal{F}$ is said to a *strictly smaller open* subfamily if S' is a relatively compact connected open subset of S.

(4.5) <u>Definition</u>. Suppose we have a finite collection of smoothly simultaneously uniformizable families $\mathcal{F}_\mu$ of holomorphic curves in M with base points $(1 \le \mu \le p)$. Suppose each family $\mathcal{F}_\mu$ contains a strictly smaller open subfamily $\mathcal{F}'_\mu$. Let the symbols S_μ, S'_μ, Γ^μ_s, O^μ_s carry the meanings analogous to those of the corresponding symbols without the subscript or superscript μ. Let A_μ be the curve volume of $\mathcal{F}_\mu$ and assume that the orbit cardinality of $\mathcal{F}_\mu$ is at least N_μ. Let P and Q be two points of M. We say that Q *is linked to* P *via* the collection $\{\mathcal{F}'_\mu\}$ of families of curves if there exist $\mu_1, \cdots, \mu_k$ and there exists $s_\nu \in S_{\mu_\nu}$ for $1 \le \nu \le k$ such that $Q = O^{\mu_1}_{s_1}$ and $O^{\mu_{\nu+1}}_{s_{\nu+1}} \in \Gamma^{\mu_\nu}_{s_\nu}$ for $1 \le \nu < k$ and $P \in \Gamma^{\mu_k}_{s_k}$. Let γ be a nonnegative number. By the *γ-linking constant from* Q *to* P we mean the infimum of

$$-c_1 \cdots c_k + \textstyle\sum_{\lambda=1}^{k} d_\lambda (\prod_{\nu=\lambda+1}^{k+1} c_\nu) - \frac{1}{m}$$

over all such choices of $\mu_1, \cdots, \mu_k$ and $s_\nu \in S_{\mu_\nu}$ $(1 \le \nu \le k)$ where

$$c_\nu = \left(1 - \frac{A_{\mu_\nu}\gamma}{2\pi N_{\mu_\nu}}\right)^{-1} \text{ and } d_\nu = c_\nu \frac{A_{\mu_\nu}(\gamma+1)}{2\pi N_{\mu_\nu}} \quad (1 \le \nu \le k) \text{ and } c_{k+1} = 1$$

(when all the c_ν's are positive).

We are now ready to state our main result.

(4.6) <u>Main Theorem</u>. Suppose M is a compact complex manifold with a Kähler metric $h_{i\bar{j}}$ in its anticanonical class. Let G be a finite subgroup of the automorphism group of M. Suppose ψ is a smooth real-valued function on M such that with the conformal factor $e^{-\psi}$ the bisectional curvature of $h_{i\bar{j}}$ for two orthonormal vectors is bounded from below by some real number $-\kappa$. Let γ be a nonnegative number $> \kappa$. Suppose $\{\mathcal{F}_\mu\}$ is a finite collection of smoothly simultaneously uniformizable holomorphic family of complex curves in M with base points and each family $\mathcal{F}_\mu$ contains a strictly smaller open subfamily $\mathcal{F}'_\mu$. Let ρ and ϵ be two positive numbers. Suppose for every point Q of M there exists an open ball B_Q of radius ρ in M (with respect to $h_{i\bar{j}}$) such that Q is linked to every point in B_Q via $\{\mathcal{F}'_\mu\}$ with the γ-linking constant $< -\epsilon$. Then there exists a Kähler-Einstein metric on M. Moreover, the Kähler-Einstein

metric can be obtained by solving by the continuity method the Monge-Ampère equation (*) with both $g_{i\bar{j}}$ and F invariant under the action of G.

§5. *Applications.*

This method can be applied to the Fermat cubic surface to give us a Kähler-Einstein metric on it. It can also be applied to higher dimensional Fermat hypersurfaces. With a rather minor modification it can be applied to give us a Kähler-Einstein metric on the surface obtained from $\mathbb{P}_2$ by blowing up three points. The modification involves using the torus group action and an additional step involving the mean value property of harmonic functions on domains in $\mathbb{C}$.

References

1. E. Calabi, The variation of Kähler metrics, II: A minimum problem, Amer. Math. Soc. Bull. 60 (1954), Abstract No. 294, p.168.

2. E. Calabi, The space of Kähler metrics, Proc. Intern. Congress of Math., Amsterdam, 1954, Vol.II, 206-207.

3. E. Calabi, Metriques kählleriennes et fibres holomorphes, Ann. Sci. Ec. Norm. Sup. Paris 12 (1979), 269-294.

4. E. Calabi, Extremal Kähler metrics, Seminar on Differential Geometry, ed. S.-T. Yau, Ann. of Math. Studies 102, Princeton University Press, 1982, pp.259-290.

5. E. Calabi, Extremal Kähler metrics II. In: Differential Geometry and Complex Analysis (dedicated to H. E. Rauch) ed. I. Chavel & H. M. Farkas, Springer-Verlag 1985, pp.95-114.

6. A. Futaki, An obstruction to the existence of Einstein Kähler metics, Invent. Math. 73 (1983), 437-443.

7. A. Futaki and S. Morita, Invariant polynomials of the automorphism group of a compact complex manifold, J. Diff. Geom. 21 (1985), 135-142.

8. J. L. Kazdan and F. W. Warner, Curvature functions for compact 2-manifolds, Ann. of Math. 99 (1974), 14-47.

9. M. Levine, A remark on extremal Kähler metrics, J. Diff. Geom. 21 (1985), 73-77.

10. A. Lichnerowicz, Sur les transformations analytiques des variétés kähleriennes, C. R. Acad. Sci. Paris 244 (1957), 3011-3014.

11. Y. Matsushima, Sur la structure du groupe d'homeomorphismes analytiques d'une certaine variété kählerienne, Nagoya Math. J. 11 (1957), 145-150.

12. J. Moser, A sharp form of an inequality of N. Trudinger, Indiana Univ. Math. J. 20 (1971), 1077-1092.

13 Y.-T. Siu, The existence of Kähler-Einstein metrics on manifolds with positive anticanonical line bundle and a suitable finite symmetry group.
14. S.-T. Yau, On the Ricci curvature of a compact Kähler manifold and the complex Monge-Ampère equation I, Comm. Pure Appl. Math. 31 (1978), 339-411.

Author's address: Department of Mathematics, Harvard University, Cambridge, MA 01138, U.S.A.

ALGEBROID REDUCTION OF NEVANLINNA THEORY

Wilhelm Stoll*

Dedicated to Maurice Heins

1. INTRODUCTION.

Let M and N be connected, complex manifolds of dimension m. Let $\pi: M \longrightarrow N$ be a proper, surjective, holomorphic map of generic sheet number q. The field of meromorphic functions on M is a finite algebraic extension of the field of meromorphic functions on N. For each meromorphic function $f: M \longrightarrow \mathbb{P}_1$ there are unique meromorphic functions $g_0, \ldots, g_{q-1}$ on N such that

$$f^q + (g_{q-1} \circ \pi) f^{q-1} + \ldots + (g_1 \circ \pi) f + (g_0 \circ \pi) = 0 \tag{1.1}$$

on M. These coefficients define a meromorphic map

$$g = (g_0, g_1, \ldots, g_{q-1}): N \longrightarrow \mathbb{P}_q. \tag{1.2}$$

Beginning with Valiron [35], the relation between the value distribution theories of f and g has been studied extensively. A few contributions are listed under references, but the vast literature shall not be recapitulated here nor is this paper concerned with the specific problem of algebroid <u>functions</u>. Here, given any meromorphic <u>map</u> $f: M \longrightarrow \mathbb{P}_n$, an algebroid reduction map $g: N \longrightarrow \mathbb{P}_p$ will be associated where $p+1 = \binom{n+q}{q}$. The value distribution of f and g will

---

* This research was supported in part by NSF Grant DMS 84-04921.

be compared mainly in respect to the First Main Theorem. Now a rough outline of the results will be given without definitions. These are contained in the text.

Assume that τ is a parabolic exhaustion of N. Then $\tilde{\tau} = \tau \circ \pi$ is a parabolic exhaustion of M. Let $\varsigma > 0$ be the volume of the parabolic pseudospheres on N. Then $\tilde{\varsigma} = q\varsigma$ is the volume of the parabolic pseudospheres on M. If $S \subseteq N$, denote $\tilde{S} = \pi^{-1}(S)$. If $R \subseteq M$, denote $R' = \pi(R)$ and $\hat{R} = \pi^{-1}(R') = \tilde{R}'$. The blow up

$$E = \{x \in M \mid \dim_x \pi^{-1}(\pi(x)) > 0\} \tag{1.3}$$

is analytic in M with $E \neq M$ and $\dim E' \leq m-2$. The set E is called the <u>exceptional set</u>. The <u>branching set</u> B is the set of all points $x \in M$ such that π is not locally biholomorphic at x. Here B, B' and $\hat{B}$ are analytic. If $B \neq \phi$, then B is pure $(m-1)$-dimensional and $\dim B' \leq m-1$. Obviously, $E \subseteq B \neq M$. Also $\pi: M-\hat{B} \longrightarrow N-B'$ is a q-sheeted covering space. For $x \in M-E$, let $\nu_\pi(x)$ be the local mapping degree of π. The zero-multiplicity of the Jacobian of π defines a non-negative divisor β on M called the <u>branching divisor</u> whose support is B. Then

$$q = \sum_{x \in \tilde{z}} \nu_\pi(x) \qquad \text{for all } z \in N-E'. \tag{1.4}$$

There is an analytic subset S_0 of B' with $\dim S_0 \leq m-2$, such that

$$\beta(x) = \nu_\pi(x) - 1 \qquad \text{for all } x \in M-\hat{S}_0. \tag{1.5}$$

For each divisor $\nu: M \longrightarrow \mathbb{Z}$ a <u>direct image divisor</u> $\pi_*(\nu)$: $N \longrightarrow \mathbb{Z}$ is defined. Then there is an analytic subset S_1 of N with $E' \subseteq S_1$ and $\dim S_1 \leq m-2$ such that

$$\pi_*(\nu)(z) = \sum_{x \in \tilde{z}} \nu(x) \qquad \text{for all } z \in N-S_1. \tag{1.6}$$

For the divisor ν a counting function n_ν and a valence function N_ν are assigned and the same is true for $\pi_*(\nu)$. Then (1.6) implies

$$n_{\pi_*(\nu)}(r) = n_\nu(r) \qquad N_{\pi_*(\nu)}(r,s) = N_\nu(r,s) \tag{1.7}$$

for all $r > s > 0$.

The Ricci function Ric_τ measures the growth of the Ricci form of υ^m. The classical Riemann-Hurwitz formula is extended by the identity

$$\mathrm{Ric}_{\tilde{\tau}}(r,s) = q\,\mathrm{Ric}_\tau(r,s) + N_\beta(r,s). \tag{1.8}$$

Let V be a complex vector space of dimension $n+1 > 1$. Define $V_* = V - \{0\}$. Let $\mathbb{P}(V) = V_*/\mathbb{C}_*$ be the associated projective space. Let $\mathbb{P}: V_* \longrightarrow \mathbb{P}(V)$ be the residual map. Let $\amalg_q V$ be the qth symmetric tensor product of V. For every meromorphic map $f: M \longrightarrow \mathbb{P}(V)$ a meromorphic reduction map $g: N \longrightarrow \mathbb{P}(\amalg_q V)$ is defined. Let I_f be the indeterminacy of f and take $z \in N-(I_f' \cup E')$, then

$$g(z) = \amalg_{x \in \tilde{z}} f(x)^{\nu_\pi(x)}. \tag{1.9}$$

Let $\mathfrak{S}_g$ be the permutation group of $\mathbb{N}[1,q]$. The cartesian symmetric product $\mathbb{P}(V)^{(q)} = \mathbb{P}(V)^q/\mathfrak{S}_g$ is naturally imbedded into $\mathbb{P}(\amalg_q V)$ and g maps into this algebraic subvariety $\mathbb{P}(V)^{(q)}$. Given any irreducible non-degenerate meromorphic map $g: N \longrightarrow \mathbb{P}(V)^{(q)}$, there exists a connected, complex manifold M of dimension m, a proper, surjective holomorphic map $\pi: M \longrightarrow N$ of sheet number q and a holomorphic map $f: M \longrightarrow \mathbb{P}(V)$ such that g is the algebroid reduction of f for π.

Again let $f: M \longrightarrow \mathbb{P}(V)$ be a meromorphic map. Let T_f be its characteristic. Let $g: N \longrightarrow \mathbb{P}(\amalg_q V)$ be the algebroid reduction of f and let T_g be the characteristic of g. Then

(1.10) $$|T_f(r,s) - T_g(r,s)| \leq (1/2)\varsigma q \sum_{\lambda=1}^{n} \frac{1}{\lambda} .$$

Let V^* be the dual vector space of V. Each $a \in \mathbb{P}(V^*)$ represents a hyperplane $E[a]$ in $\mathbb{P}(V)$. Assume $f(M) \not\subseteq E[a]$. Then the intersection divisor $\mu_{f,a}$ of f with $E[a]$ is defined. Let $n_{f,a}$ and $N_{f,a}$ be the counting function and valence function of $\mu_{f,a}$ respectively. Let $m_{f,a}$ be the compensation function. The well-known First Main Theorem states

(1.11) $$T_f(r,s) = N_{f,a}(r,s) + m_{f,a}(r) - m_{f,a}(s).$$

If f is not constant, $T_f(r,s) \longrightarrow \infty$ for $r \longrightarrow \infty$ and the defect

(1.12) $$0 \leq \delta_f(a) = \lim_{r\to\infty} \frac{m_{f,a}(r)}{T_f(r,s)} = 1 - \overline{\lim_{r\to\infty}} \frac{N_{f,a}(r,s)}{T_f(r,s)} \leq 1$$

measures the degree of avoidance of f from $E[a]$. Let $\varphi\colon \mathbb{P}(V^*) \longrightarrow \mathbb{P}(\amalg_q V^*)$ be the Veronese map. Define $b = \varphi(a) = a^q$. Then $g(N) \not\subseteq E[b]$. Then we obtain

(1.13) $$n_{g,b}(r) = n_{f,a}(r) \qquad N_{g,b}(r,s) = N_{f,a}(r,s)$$

(1.14) $$m_{g,b}(r) \leq m_{f,a}(r) \leq m_{g,b}(r) + (1/2)\varsigma q \sum_{\lambda=1}^{n} \frac{1}{\lambda}$$

(1.15) $$\delta_f(a) = \delta_g(b)$$

In this sense value distribution is immune to algebroid reduction and algebroid lifting. However the Second Main Theorem and the Defect Relation do not transfer easily, since general position and linear non-degeneracy are not invariant.

The map $f\colon M \longrightarrow \mathbb{P}(V)$ is said to <u>separate</u> the fibers of π, if there exists at least one fiber over a point in $N-B'$ such that f is holomorphic and injective on this fiber. If so, then we introduce a meromorphic <u>discriminant map</u>

$$(1.16) \qquad d\colon N \longrightarrow \mathbb{P}(\amalg_{q(q-1)}(V\wedge V)),$$

a <u>discriminant</u> <u>divisor</u> $\delta \geq 0$ and a <u>discriminant</u> <u>compensation</u> function $m_\Delta \geq 0$ of f such that

$$(1.17) \qquad T_d(r,s) + N_\delta(r,s) + m_\Delta(r) - m_\Delta(s) = 2(q-1)\, T_f(r,s)$$

for $0 < s < r$. Also we prove $\beta \leq \delta$. Hence $N_\beta(r,s) \leq N_\delta(r,s)$, which implies a Theorem of J. Noguchi [12].

$$(1.18) \qquad N_\beta(r,s) \leq 2(q-1)\, T_f(r,s) + m_\Delta(s)$$

for all $0 < s < r$.

Although value distribution on the parabolic manifolds M and N can be studied independently, it seems to be worthwhile to correlate the two theories. Junjiro Noguchi encouraged me to study the reduction problem when I was at the Research Institute for Mathematical Science (RIMS) in Kyoto. I thank him for the many discussions on the topic and I thank RIMS and the NSF for supporting these investiga-

2. OUTLINE OF VALUE DISTRIBUTION THEORY

a) <u>Parabolic manifolds</u> Let M be a connected, complex manifold of dimension m. Let $\tau \geq 0$ be a non-negative, continuous function on M. For $A \subseteq M$ and $0 \leq r \in \mathbb{R}$ define

$$(2.1) \qquad A(r) = \{x \in A \mid \tau(x) < r^2\} \qquad A[r] = \{x \in A \mid \tau(x) \leq r^2\}$$

$$(2.2)\qquad A\langle r\rangle = \{x \in A \mid \tau(x) = r^2\} \qquad A_* = \{x \in A \mid \tau(x) > 0\}$$

The function τ is said to be an <u>exhaustion</u> of M, if $M[r]$ is compact for each $r \in \mathbb{R}$ with $r \geq 0$ and if τ is not bounded. If $\tau \not\equiv 0$ is of class C^∞ on M, define υ on M and ω and σ on M_* by

$$(2.3)\qquad \upsilon = dd^c\tau \qquad \omega = dd^c \log \tau \qquad \sigma = d^c \log \tau \wedge \omega^{m-1}.$$

The function τ is said to be <u>parabolic</u> on M if

$$(2.4)\qquad \omega \geq 0 \qquad \omega^m \equiv 0 \not\equiv \upsilon^m \quad \text{on } M_*$$

which implies $\upsilon \geq 0$ on M. If $\upsilon > 0$, then τ is said to be <u>strict.</u> If τ is a parabolic exhaustion, then (M,τ) is said to be a <u>parabolic</u> <u>manifold</u>, <u>strict</u> if τ is strict. On a parabolic manifold, $M[r]$ are called the <u>closed pseudoballs</u>, $M(r)$ the <u>open</u> <u>pseudoballs</u> and $M\langle r\rangle$ the <u>pseudosphere</u>.

There are many parabolic manifolds. A non-compact Riemann surface is parabolic if and only if every subharmonic function bounded above is constant. Let $(M_1,\tau_1),\ldots,(M_n,\tau_n)$ be parabolic manifolds. Define $M = M_1 x \ldots x\, M_n$ and $\tau: M \longrightarrow \mathbb{R}$ by

$$(2.5)\qquad \tau(x_1,\ldots,x_n) = \tau_1(x_1)+\ldots+\tau_n(x_n).$$

Then (M,τ) is a parabolic manifold. Hence $\mathbb{C}^{m-1}x(\mathbb{C}-\mathbb{Z})$ is a parabolic manifold. It is not affine algebraic. $(\mathbb{C}^m,\tau_0)$ is a strictly parabolic manifold with

$$(2.6)\qquad \tau_0(z_1,\ldots,z_m) = |z_1|^2 + \ldots + |z_m|^2.$$

Let M and N be connected, complex manifolds of dimension m. Let $\pi: M \longrightarrow N$ be a proper, surjective holomorphic map. Let τ be a parabolic exhaustion of N. Then $\tilde{\tau} = \tau\circ\pi$ is a parabolic exhaustion

of M. If M is a connected, affine algebraic manifold of dimension m, then there is a proper, surjective holomorphic map $\pi: M \longrightarrow \mathbb{C}^m$. Thus every affine algebraic manifold is parabolic.

Uniformization Theorem (Stoll [25]). Let (M,τ) be a strictly parabolic manifold of dimension m. Then there is a biholomorphic map $h: M \longrightarrow \mathbb{C}^m$ such that $\tau = \tau_0 \circ h$.

Let X be an irreducible, reduced complex space of dimension m. Let $\Sigma(X)$ be the singular set and $\Re(X) = X - \Sigma(X)$ be the regular set. Let $G \neq \phi$ be an open subset of X. Let dG be the largest boundary submanifold of class C^∞ of $\Re(G)$ in $\Re(X)$. If $dG \neq \phi$, orient dG to the exterior of G. Then G is called a Stokes domain, if $\bar{G}$ is compact, if ∂G has locally finite Hausdorff measure of dimension $2m-1$ and if $\partial G - dG$ has zero Hausdorff measure of dimension $2m-1$. If G is a Stokes domain in X and if χ is a form of class C^1 and degree $2m-1$ on X, then Stokes theorem holds

$$\int_{dG} \chi = \int_G d\chi. \tag{2.7}$$

If Y is an irreducible, reduced complex space of dimension m, if $\pi: Y \longrightarrow X$ is a proper, surjective holomorphic map, if G is a Stokes domain in X, then $H = \pi^{-1}(G)$ is a Stokes domain in Y. (See Tung [34]).

Let (M,τ) be a parabolic manifold of dimension m. Define $\mathbb{R}^+ = \{x \in \mathbb{R} \mid x > 0\}$ and

$$\mathfrak{C}_\tau = \{r \in \mathbb{R}^+ \mid M(r) \text{ is a Stokes domain}\} \tag{2.8}$$

$$\hat{\mathfrak{C}}_\tau = \{r \in \mathbb{R}^+ \mid d\tau(x) \neq 0 \text{ for all } x \in M\langle r\rangle\} \tag{2.9}$$

Clearly, $\hat{\mathfrak{C}}_\tau \subseteq \mathfrak{C}_\tau$. By Sard's Lemma, $\mathbb{R}^+ - \hat{\mathfrak{C}}_\tau$ has measure zero. For many purposes $\hat{\mathfrak{C}}_\tau$ suffices, but if a proper map π is given as above

$\pi^{-1}(M(r))$ is a Stokes domain, but may not have a smooth boundary. Thus $\mathfrak{C}_\tau$ is needed. If $r \in \hat{\mathfrak{C}}_\tau$, then $\partial M(r) = dM(r) = M\langle r\rangle$. If $r \in \mathfrak{C}_\tau$, this may not be the case, but we write $\int_{M\langle r\rangle} = \int_{dM(r)}$ in an abuse of notation. For $r \in \mathfrak{C}_\tau$, the volume

$$(2.10) \qquad \varsigma = \int_{M\langle r\rangle} \sigma > 0$$

is positive and constant. For $r \in \mathbb{R}_+ = \{x \in \mathbb{R} \mid x \geq 0\}$ we have

$$(2.11) \qquad \int_{M[r]} \upsilon^m = \int_{M(r)} \upsilon^m = \varsigma r^{2m}.$$

b) Divisors. Since M is a manifold, a divisor can be defined by its multiplicity function. Let $G \neq \emptyset$ be an open subset of M. Let $f: G \longrightarrow \mathbb{C}$ be a holomorphic function. Take $x \in G$. Let $\alpha: U \longrightarrow U'$ be a biholomorphic map of an open neighborhood U of x in G onto a ball U' in $\mathbb{C}^m$ with $\alpha(x) = 0$. Then there exist a unique sequence of homogeneous polynomials P_λ on $\mathbb{C}^m$ with $\deg P_\lambda = \lambda$ such that

$$(2.12) \qquad f|U = \sum_{\lambda=0}^{\infty} P_\lambda \circ \alpha.$$

Define $I = \{\lambda \in \mathbb{Z}_+ \mid P_\lambda \not\equiv 0\}$. Here $I = \emptyset$ if and only $f|U \equiv 0$. The zero multiplicity

$$(2.13) \qquad \mu_f^0(x) = \begin{cases} \text{Min } I & \text{if } I \neq \emptyset \\ +\infty & \text{if } I = \emptyset \end{cases}$$

of f at x does not depend on the choice of α. A function $\nu: M \longrightarrow \mathbb{Z}$ is said to be a divisor if for every point $x \in M$ there is an open, connected neighborhood U of x and if there are holomorphic functions $g \not\equiv 0 \not\equiv h$ on U such that $\nu|U = \mu_g^0 - \mu_h^0$. The set $\mathfrak{D} = \mathfrak{D}(M)$ of divisors on M is a module. If $\nu \equiv 0$, then ν is said to be the null divisor. If $0 \not\equiv \nu \in \mathfrak{D}$, the support $S = \operatorname{supp} \nu$ is an analytic subset of pure dimension $m-1$ of M, if $\nu \equiv 0$, then

supp $\nu = \phi$. The divisor ν is non-negative as a function, if and only if for each point $x \in M$, there is an open, connected neighborhood U of x and a holomorphic function $g \not\equiv 0$ on U with $\nu | U = \mu_g^0$. For each analytic subset S of pure dimension $m-1$ in M, there is one and only one divisor $\nu_S \in \mathfrak{D}$ with $\nu_S(x) = 1$ for all $x \in \mathfrak{R}(S)$. Here $\nu_S \geq 0$. Let $\nu \not\equiv 0$ be a divisor on M. Let $\mathfrak{B} \neq \phi$ be the set of all branches of supp ν. For each $B \in \mathfrak{B}$ there is an integer $q_B \neq 0$ such that $\nu(x) = q_B$ for all $x \in B \cap \mathfrak{R}(S)$. Then

$$\nu = \sum_{B \in \mathfrak{B}} q_B \, \nu_B \tag{2.14}$$

where the sum is locally finite. Define $\mathfrak{B}^+ = \{B \in \mathfrak{B} \mid q_B > 0\}$ and $\mathfrak{B}^- = \{B \in \mathfrak{B} \mid q_B < 0\}$.
Then

$$\nu^+ = \sum_{B \in \mathfrak{B}^+} q_B \, \nu_B \qquad \nu^- = - \sum_{B \in \mathfrak{B}^-} q_B \, \nu_B \tag{2.15}$$

are non-negative divisors with $\nu = \nu^+ - \nu^-$ and with

$$\dim(\operatorname{supp} \nu^+ \cap \operatorname{supp} \nu^-) \leq m-2. \tag{2.16}$$

If $\nu_1 \geq 0$ and $\nu_2 \geq 0$ are divisors with $\dim(\operatorname{supp} \nu_1 \cap \operatorname{supp} \nu_2) \leq m-2$ such that $\nu = \nu_1 - \nu_2$ then $\nu_1 = \nu^+$ and $\nu_2 = \nu^-$. Here $\nu \geq 0$ if and only if $\nu = \nu^+$, meaning $q_B > 0$ for all $B \in \mathfrak{B}$. Let $\{\nu_\lambda\}_{\lambda \in \Lambda}$ be a family of divisors such that $\{\operatorname{supp} \nu_\lambda\}_{\lambda \in \Lambda}$ is locally finite. Let $\{q_\lambda\}_{\lambda \in \Lambda}$ be a family of integers, then

$$\sum_{\lambda \in \Lambda} q_\lambda \, \nu_\lambda \tag{2.17}$$

is a divisor on M. If ν_1 and ν_2 are divisors on M, if S is analytic in M with $\dim S \leq m-2$ and if $\nu_1 | (M-S) = \nu_2 | (M-S)$, then $\nu_1 = \nu_2$, since $\nu_1 - \nu_2$ is a divisor whose support is contained in S.

<u>PROPOSITION 2.1</u> Let M be a connected complex manifold of dimension m. Let T be an analytic subset of M with $\dim T \leq m-2$.

Let ν be a divisor on $M-T$. Then one and only one divisor $\tilde{\nu}$ exists on M with $\tilde{\nu}|(M-T) = \nu$. If $\nu \geq 0$, then $\tilde{\nu} \geq 0$.

<u>Proof</u>. The case $\nu \equiv 0$ is trivial. Assume that $\nu \not\equiv 0$. Then $S = \operatorname{supp} \nu$ is a pure $(m-1)$-dimensional analytic subset of $M-T$. By Remmert and Stein [16], the closure $\bar{S}$ of S in M is an analytic subset of pure dimension $m-1$ of M. Let $\mathfrak{B}$ be the set of branches of S. Then $\tilde{\mathfrak{B}} = \{\bar{B} | B \in \mathfrak{B}\}$ is the set of branches of $\bar{S}$. Here $\nu_{\bar{B}}|(M-T) = \nu_B$. For each $B \in \mathfrak{B}$, a constant integer $q_B \neq 0$ exists such that $\nu(x) = q_B$ for all $x \in B \cap \mathfrak{R}(S)$. Then

$$\tilde{\nu} = \sum_{B\in\mathfrak{B}} q_B \, \nu_{\bar{B}} \tag{2.18}$$

is a divisor on M with $\tilde{\nu}|(M-T) = \nu$. The uniqueness of the extension is trivial. If $\nu \geq 0$, then $q_B > 0$ for all $B \in \mathfrak{B}$. Hence $\tilde{\nu} \geq 0$, q.e.d.

Let (M,τ) be a parabolic manifold of dimension m. Let $\nu: M \longrightarrow \mathbb{Z}$ be a divisor with $S = \operatorname{supp} \nu$. For all $t > 0$, the <u>counting function</u> n_ν of ν is defined by

$$n_\nu(t) = t^{2-2m} \int_{S[t]} \nu \, \upsilon^{m-1} \qquad \text{if } m > 1 \tag{2.19}$$

$$n_\nu(t) = \sum_{x\in S[t]} \nu(x) \qquad \text{if } m = 1 \tag{2.20}$$

Then the limit $n_\nu(t) \longrightarrow n_\nu(0)$ for $t \longrightarrow 0$ with $t > 0$ exists. If $m > 1$, then

$$n_\nu(t) = \int_{S_*[t]} \nu\omega^{m-1} + n_\nu(0) \qquad \text{for all } t > 0 \tag{2.21}$$

For $0 < s < r$, the <u>valence function</u> N_ν of ν is defined by

$$N_\nu(r,s) = \int_s^r n_\nu(t) \, \frac{dt}{t}. \tag{2.22}$$

If $\nu \geq 0$, then $n_\nu \geq 0$ and $N_\nu(\square, s) \geq 0$ increase and the limits

$$(2.23) \qquad \lim_{r\to\infty} \frac{N_\nu(r,s)}{\log r} = \lim_{t\to\infty} n_\nu(t) = n_\nu(\infty) \leq \infty$$

exist. If $n_\nu(\infty) < \infty$, then ν is said to have <u>affine growth</u>. If $(M,\tau) = (\mathbb{C}^m, \tau_0)$ a divisor $\nu \geq 0$ is affine algebraic if and only if $n_\nu(\infty) < \infty$.

Let $f \not\equiv 0$ be a meromorphic function on M. Then the zero divisor $\mu_f^0 \geq 0$ and the pole divisor $\mu_f^\infty \geq 0$ are defined. If $z \in M$, then there are holomorphic functions $g \not\equiv 0$ and $h \not\equiv 0$ on an open, connected neighborhood U of x such that $\dim g^{-1}(0) \cap h^{-1}(0) \leq m-2$ and such that $hf = g$ on U. Then $\mu_f^0|U = \mu_g^0$ and $\mu_f^\infty|U = \mu_h^0$. The divisor $\mu_f = \mu_f^0 - \mu_f^\infty$ is called the <u>divisor</u> of f. If $0 < s < r \in \mathfrak{C}_\tau$ with $s \in \mathfrak{C}_\tau$, <u>Jensen's Formula</u> holds

$$(2.24) \qquad N_{\mu_f}(r,s) = \int_{M\langle r\rangle} \log|f|\sigma - \int_{M\langle s\rangle} \log|f|\sigma.$$

Easily it implies, that a bounded holomorphic function on M is constant.

c) <u>Meromorphic maps</u> Let M and N be complex manifolds of pure dimensions m and n respectively. Let S be a thin analytic subset of M, where <u>thin</u> means that $A = M-S$ is dense in M. Let $f: A \longrightarrow N$ be a holomorphic map with graph $\Gamma = \{(x,f(x)) \mid x \in A\}$. Let $\overline{\Gamma}$ be the closure of Γ in $M\times N$ and let $\pi: \overline{\Gamma} \longrightarrow M$ be the projection. Then f is said to be a <u>meromorphic map on M</u> if $\overline{\Gamma}$ is analytic and π is proper. If so, there is a unique, smallest, closed subset I_f of S such that f continues to a holomorphic map $f: M-I_f \longrightarrow N$. Then I_f is called the <u>indeterminacy</u> of f. We have $\dim I_f \leq m-2$.

Let ν be a divisor on N with support S. Then f is said to be <u>free</u> for ν if $f^{-1}(S)$ is thin. Then there is one and only one

<u>pullback</u> <u>divisor</u> $f^*(\nu)$ defined on M as follows: Take $x \in M-I_f$. Then there are open, connected neighborhoods U of x in $M-I_f$ and W of $f(x)$ in N with $f(U) \subseteq W$ and holomorphic functions $g \not\equiv 0$ and $h \not\equiv 0$ on W with $\operatorname{supp} \mu_g^0 \subseteq S$ and $\operatorname{supp} \mu_h^0 \subseteq S$ such that $\nu|W = \mu_g^0 - \mu_h^0$. Then $g\circ f \not\equiv 0 \not\equiv h\circ f$ and

$$f^*(\nu)|U = \nu_{g\circ f}^0 - \mu_{h\circ f}^0. \tag{2.25}$$

Now $f^*(\nu)$ extends from $M-I_f$ to M by Proposition 2.1.

Let W be a holomorphic vector bundle over N. Let $\alpha \in \Gamma(N,N)$ be a holomorphic section in W over N. Let $Z(\alpha) = \{x \in N|\alpha(x) = 0\}$ be the zero set of α. Assume that $Z(\alpha)$ is thin. Then there is one and only one divisor μ_α called the <u>zero divisor</u> of α: Take any $x \in N$. Then there is a holomorphic function $g \not\equiv 0$ and a holomorphic section $\beta \in \Gamma(U,E)$ on an open, connected neighborhood U of x in N such that $\alpha|U = g\cdot\beta$ and $\dim Z(\beta) \le n-2$. Then $\mu_\alpha|U = \mu_g^0$. Obviously $\mu_\alpha \ge 0$ and $\operatorname{supp} \mu_\alpha \subseteq Z(\alpha)$. If W is a line bundle, $\operatorname{supp} \mu_\alpha = Z(\alpha)$.

The bundle W pulls back to a holomorphic vector bundle

$$f^*(W) = \{(x,w) \in (M-I_f) \times W \mid w \in W_{f(x)}\}. \tag{2.26}$$

Let $\pi: f^*(W) \longrightarrow M-I_f$ and $\tilde{f}: f^*(W) \longrightarrow W$ be the projection. Then π is the bundle projection. A holomorphic section $\alpha_f \in \Gamma(M-I_f, f^*(W))$ is defined by $\alpha_f(x) = (x,\alpha(f(x)))$ for all $x \in M-I_f$. Then $\tilde{f}\circ\alpha_f = \alpha\circ f$. The map f is said to be <u>free</u> for α if $Z(\alpha_f)$ is thin, which is the case if and only if $f(M_0-I_f) \subseteq Z(\alpha)$ for each component M_0 of M. If so, the non-negative <u>intersection</u> <u>divisor</u> $\mu_{f,\alpha} = \mu_{\alpha_f}$ is defined and extends to M. Here $\mu_{f,\alpha} \ge f^*(\mu_\alpha)$. If W is a line bundle, $\mu_{f,\alpha} = f^*(\mu_\alpha)$.

Let $\varkappa$ be a hermitian metric along the fibers of W. If $e \in W$, then the length of e by $\varkappa$ is denoted by $\|e\|_\varkappa$. Let (M,τ) be a

parabolic manifold of dimension m take $0 < s < r \in \mathfrak{C}_\tau$ with $s \in \mathfrak{C}_\tau$. Then the <u>Green Residue Theorem</u> (Stoll [30] Theorem 3.1) holds

$$\int_s^r \int_{M[t]} dd^c \log\|\alpha\circ f\|_\varkappa^2 \upsilon^m \frac{dt}{t^{2m-1}} + N_{\mu_{f,\alpha}}(r,s)$$

(2.27)

$$= \int_{M\langle r\rangle} \log\|\alpha\circ f\|_\varkappa \sigma - \int_{M\langle s\rangle} \log\|\alpha\circ f\|_\varkappa \sigma.$$

d) <u>The First Main Theorem for line bundles</u>. We retain the situation of c) but assume that $W = L$ is a holomorphic line bundle. Its <u>Chern form</u> for $\varkappa$ is denoted by $c(L,\varkappa)$. Then

$$c(L,\varkappa) = - dd^c \log \|\alpha\|_\varkappa^2 \qquad \text{on } N - Z(\alpha). \tag{2.28}$$

For $0 < t \in \mathbb{R}$ the <u>spherical image</u> A_f is defined by

$$A_f(t;L,\varkappa) = t^{2-2m} \int_{M[t]} f^*(c(L,\varkappa))\wedge\upsilon^{m-1}. \tag{2.29}$$

For $0 < s < r \in \mathbb{R}$ the <u>characteristic</u> T_f is defined by

$$T_f(r,s;L,\varkappa) = \int_s^r A_f(t;L,\varkappa) \frac{dt}{t}. \tag{2.30}$$

The <u>counting function</u> and the <u>valence function</u> for the divisor $\mu_{f,\alpha}$ are denoted by $n_f(t,\alpha,L) \geq 0$ and $N_f(r,s;\alpha,L) \geq 0$. For $r \in \mathfrak{C}_\tau$ the <u>compensation function</u> is defined by

$$m_f(r;\alpha,L,\varkappa) = \int_{M\langle r\rangle} \log \frac{1}{\|\alpha\circ f\|_\varkappa} \sigma. \tag{2.31}$$

If $0 < s < r \in \mathfrak{C}_\tau$ and $s \in \mathfrak{C}_\tau$, the Green-Residue Theorem yields the <u>FIRST MAIN THEOREM</u>.

$$T_f(r,s;L,\varkappa) = N_f(r,s;\alpha,L) + m_f(r;\alpha,L,\varkappa) - m_f(s;\alpha,L,\varkappa). \tag{2.32}$$

Since T_f and N_f are continuous in r and s, the function

$m_f(\square;\alpha,L,\varkappa)$ extends continuously to the positive axis $\mathbb{R}^+$ such that (2.32) holds for all $0 < s < r$.

If $c(L,\varkappa) \geq 0$, then $A_f \geq 0$ and $T_f \geq 0$ are increasing functions. The limit $A_f(t;L,\varkappa) \longrightarrow A_f(0;L,\varkappa)$ for $t \longrightarrow 0$ with $t > 0$ exists and we have

$$(2.33) \qquad A_f(t;L,\varkappa) = \int_{M[t]} f^*(c(L,\varkappa)) \wedge \omega^{m-1} + A_f(0;L,\varkappa)$$

$$(2.34) \qquad \lim_{r\to\infty} \frac{T_f(r,s,L,\varkappa)}{\log r} = \lim_{t\to\infty} A_f(t;L,\varkappa) = A_f(\infty;L,\varkappa) \leq \infty$$

If $\|\alpha\|_\varkappa \leq 1$, then $m_f(r;\alpha,L,\varkappa) \geq 0$, If N is compact, this can be achieved by multiplying with a constant. If $M = N$ and if f is the identity, drop the index "f".

e) <u>THE RICCI FUNCTION</u>. Now we consider the situation where $M = N$ and where $K = K(M)$ is the canonical bundle and f is the identity. Let Θ be a positive form of degree 2m and class C^∞ on M. One and only one hermitian metric $\varkappa_\Theta$ along the fibers of K is defined by Θ. For each $0 \leq a \in \mathbb{Z}$ put

$$(2.35) \qquad i_a = \left(\frac{i}{2\pi}\right)^a (-1)^{\frac{a(a-1)}{2}} a!$$

Take $x \in M$ and $\varphi \in K_x$ and $\psi \in K_x$. Then

$$(2.36) \qquad i_m\, \varphi\wedge\bar{\psi} = \varkappa_\upsilon(\varphi,\psi)\, \Theta(x).$$

The <u>Ricci form</u> of Θ is defined as the Chern form for $\varkappa_\Theta$

$$(2.37) \qquad \mathrm{Ric}\,\Theta = c(K,\varkappa_\Theta)$$

and we write

$$(2.38) \qquad \mathrm{ric}(t,\Theta) = A(t,K,\varkappa_\Theta) = t^{2-2m} \int_{M[t]} (\mathrm{Ric}\,\Theta) \wedge \upsilon^{m-1}$$

(2.39) $$\mathrm{Ric}(r,s,\Theta) = \int_s^r \mathrm{ric}(t,\Theta)\,\frac{dt}{t} = T(r,s;\,K,\varkappa_\Theta).$$

However $\mathrm{Ric}\,\upsilon^m$ can be computed on $M^+ = \{x \in M \mid \upsilon(x) > 0\}$ only.

Define a function $\nu \geq 0$ of class C^∞ on M by $\upsilon^m = \nu\Theta$. Then the set $\mathfrak{C}_\tau^0$ of all $r \in \mathfrak{C}_\tau$ such that $(\log \nu)\sigma$ is integrable over $\partial M(r)$ is independent of Θ and $\mathbb{R}^+ - \mathfrak{C}_\tau^0$ has measure zero. For $0 < s < r \in \mathfrak{C}_\tau^0$ with $s \in \mathfrak{C}_\tau^0$ the <u>Ricci function</u> Ric_τ is independently of Θ defined by

(2.40)
$$\mathrm{Ric}_\tau(r,s) = (1/2)\int_{M\langle r\rangle} (\log \nu)\sigma - (1/2)\int_{M\langle s\rangle} (\log \nu)\sigma + \mathrm{Ric}(r,s,\Theta).$$

If τ is strictly parabolic on M; i.e. if $(M,\tau) = (\mathbb{C}^m, \tau_0)$ then $\mathrm{Ric}\,\upsilon^m = 0$ and the choice $\upsilon^m = \Theta$ yields $\nu = 1$ and $\mathrm{Ric}_\tau(r,s) \equiv 0$. If M is Stein, the following unsolved questions remain:

1. <u>Question</u>: Is $\mathrm{Ric}\,\upsilon^m \geq 0$ on M^+?

2. <u>Question:</u> Is $\mathrm{Ric}_\tau(r,s) \geq 0$ for all $0 < s < r \in \mathfrak{C}_\tau^0$ with $s \in \mathfrak{C}_\tau^0$

3. <u>Question:</u> If $\mathrm{Ric}_\tau(r,s) = 0$ for all $0 < s < r \in \mathfrak{C}_\tau^0$, with $s \in \mathfrak{C}_\tau^0$, is τ strict?

f) <u>Meromorphic maps into projective space</u>. Let V be a complex vector space of dimension $n+1 > 1$. Put $V_* = V - \{0\}$. Let $\mathbb{P}(V) = V_*/\mathbb{C}_*$ be the projective space associated to V and let $\mathbb{P}: V_* \longrightarrow \mathbb{P}(V)$ be the residual map. For $A \subseteq V$, define $\mathbb{P}(A) = \{\mathbb{P}(\mathfrak{x}) \mid 0 \neq \mathfrak{x} \in A\}$. If $x \in \mathbb{P}(V)$ then $\mathfrak{x} \in V_*$ exists with $\mathbb{P}(\mathfrak{x}) = x$. The line $E(x) = \mathbb{C}\mathfrak{x}$ is well defined. Let V^* be the dual vector space of V. The <u>inner product</u> of $\mathfrak{x} \in V$ and $\mathfrak{y} \in V^*$ is denoted by $\langle \mathfrak{x},\mathfrak{y}\rangle$. Each of $y = \mathbb{P}(\mathfrak{y}) = \mathbb{P}(V^*)$ defines an n-dimensional linear subspace

$$E[y] = \operatorname{Ker} \mathfrak{y} = \{\mathfrak{x} \in V \mid \langle \mathfrak{x}, \mathfrak{y} \rangle = 0\} \tag{2.41}$$

and a hyperplane $\ddot{E}[y] = \mathbb{P}(E[y])$ in $\mathbb{P}(V)$.

The <u>tautological bundle</u> $\mathcal{O}(-1)$ with projection π over $\mathbb{P}(V)$ is defined by

$$\mathcal{O}(-1) = \{(x, \mathfrak{x}) \in \mathbb{P}(V) \times V \mid \mathfrak{x} \in E(x)\} \tag{2.42}$$

where $\pi\colon \mathcal{O}(-1) \longrightarrow \mathbb{P}(V)$ and $\sigma\colon \mathcal{O}(-1) \longrightarrow V$ are the projections and $\iota\colon \mathcal{O}(-1) \longrightarrow \mathbb{P}(V)\times V$ is the inclusion. Here $\pi\colon \mathcal{O}(-1) \longrightarrow \mathbb{P}(V)$ is a holomorphic line bundle where $\mathcal{O}(-1)_x = E(x)$ are identified. Also $\sigma\colon \mathcal{O}(-1) \longrightarrow V$ in the <u>Hopf</u> σ <u>process</u> which blows up $0 \in V$. The inclusion $\iota\colon \mathcal{O}(-1) \longrightarrow \mathbb{P}(V)\times V$ is a bundle map. The dual line bundle $H = \mathcal{O}(1) = \mathcal{O}(-1)^*$ is called the <u>hyperplane</u> <u>section</u> <u>bundle</u> over $\mathbb{P}(V)$ and the surjective bundle map $\varepsilon = \iota^*\colon \mathbb{P}(V)\times V^* \longrightarrow H$ is called the <u>evaluation map</u>. If $\mathfrak{y} \in V^*$, a section $\tilde{\mathfrak{y}}$ of $\mathbb{P}(V)\times V^*$ is defined by $\tilde{\mathfrak{y}}(x) = (x, \mathfrak{y})$. Then $\varepsilon\circ\tilde{\mathfrak{y}}$ is a holomorphic section of H and all sections are so obtained. If $\mathfrak{y} \in V^*$ and $\mathfrak{z} \in V^*$ then $\varepsilon\circ\tilde{\mathfrak{y}} = \varepsilon\circ\tilde{\mathfrak{z}}$ if an only if $\mathfrak{y} = \mathfrak{z}$. Hence $\Gamma(\mathbb{P}(V), H)$ can be identified with V^* by $\mathfrak{y} = \varepsilon\circ\tilde{\mathfrak{y}}$. If $x \in \mathbb{P}(V)$, then $\varepsilon\circ\tilde{\mathfrak{y}}(x) \in E[x]^*$ with $\langle \mathfrak{x}, \varepsilon\circ\tilde{\mathfrak{y}}(x)\rangle = \langle \mathfrak{x}, \tilde{\mathfrak{y}}\rangle$ for all $\mathfrak{x} \in E(x)$.

Let ℓ be a hermitian metric on V. We write $\ell(\mathfrak{x}, \mathfrak{y}) = (\mathfrak{x} \mid \mathfrak{y})$ and $\sqrt{\ell(\mathfrak{x},\mathfrak{x})} = \sqrt{(\mathfrak{x}\mid\mathfrak{x})} = \|\mathfrak{x}\|$ is the <u>norm</u> of $\mathfrak{x}$. Here ℓ induces a hermitian metric along the fibers of $\mathbb{P}(V)\times V$ and by restriction on $\mathcal{O}(-1)$. By duality ℓ induces a hermitian metric on V^* and along the fibers of $\mathbb{P}(V)\times V^*$ and H. The Chern form of H for ℓ is called the <u>Fubini-Study form</u> on $\mathbb{P}(V)$:

$$\Omega = c(H, \ell) > 0. \tag{2.43}$$

If $\tau_o\colon V \longrightarrow \mathbb{R}^+$ is defined by $\tau_o(\mathfrak{x}) = \|\mathfrak{x}\|^2$, then

$$\mathbb{P}^*(\Omega) = dd^c \log \tau_o \qquad \text{on } V_*. \tag{2.44}$$

If $\mathfrak{x} \in V$ and $\alpha \in V^*$ we have the <u>Schwarz inequality</u>

$$(2.45) \qquad |\langle \mathfrak{z}, \mathfrak{a}\rangle| \le \|\mathfrak{z}\| \, \|\mathfrak{a}\|.$$

If $x = \mathbb{P}(\mathfrak{z}) \in \mathbb{P}(V)$ and $a = \mathbb{P}(\mathfrak{a}) \in \mathbb{P}(V^*)$, the <u>projective distance</u> is defined as

$$(2.46) \qquad 0 \le \Box x,a\Box = \frac{|\langle \mathfrak{z}, \mathfrak{a}\rangle|}{\|\mathfrak{z}\| \, \|\mathfrak{a}\|} \le 1.$$

<u>LEMMA 2.2</u>. Take $\mathfrak{a} \in V^*$ with $\|\mathfrak{a}\| = 1$. Define $a = \mathbb{P}(\mathfrak{a}) \in \mathbb{P}(V^*)$. Take $x \in \mathbb{P}(V)$. Then $\|\varepsilon\circ\tilde{\mathfrak{a}}(x)\|_{\ell} = \Box x,a\Box$.

<u>Proof</u>. Take any $\mathfrak{z} \in V_*$ with $x = \mathbb{P}(\mathfrak{z})$ and $\|\mathfrak{z}\| = 1$. Then $\mathfrak{z}$ is an orthonormal base of $E(x)$. A vector $\mathfrak{b} \in V^*$ exists such that $\varepsilon\circ\tilde{\mathfrak{b}}(x)$ is an orthonormal base of $E(x)^* = H_x$ dual to $\mathfrak{z}$. Thus $1 = \langle \mathfrak{z}, \varepsilon\circ\tilde{\mathfrak{b}}(x)\rangle = \langle \mathfrak{z}, \mathfrak{b}\rangle$. Then $\mathfrak{z}\circ\tilde{\mathfrak{a}}(x) = \lambda\cdot\varepsilon\circ\tilde{\mathfrak{b}}(x)$ with $\lambda \in \mathbb{C}$. Then $\|\varepsilon\circ\tilde{\mathfrak{a}}(x)\|_{\ell} = |\lambda|$. Also

$$\langle \mathfrak{z}, \mathfrak{a}\rangle = \langle \mathfrak{z}, \varepsilon\circ\tilde{\mathfrak{a}}(x)\rangle = \lambda\langle \mathfrak{z}, \varepsilon\circ\tilde{\mathfrak{b}}(x)\rangle = \lambda.$$

Hence $\|\varepsilon\circ\tilde{\mathfrak{a}}(x)\|_{\ell} = |\lambda| = |\langle \mathfrak{z}, \mathfrak{a}\rangle| = \Box x,a\Box$; q.e.d.

Let M be a connected, complex manifold. Let S be an analytic subset of M with $A = M-S \ne \emptyset$. Let $f: A \longrightarrow \mathbb{P}(V)$ be a holomorphic map. Let $U \ne \emptyset$ be open in M. A holomorphic map $\mathfrak{v}: U \longrightarrow V$ is said to be a <u>representation</u> of f (at x if $x \in U$) if and only if the analytic set $\mathfrak{v}^{-1}(0)$ is thin in U and if $f = \mathbb{P}\circ\mathfrak{v}$ on $A\cap U-\mathfrak{v}^{-1}(0)$. If in addition $\dim \mathfrak{v}^{-1}(0) \le m-2$, the representation is said to be <u>reduced</u>. The map f is meromorphic on M if and only if there is a representation at every point of S. If so, then there is a reduced representation at every point of M. A representation $\mathfrak{v}: U \longrightarrow V$ is reduced if an only if $\mathfrak{v}^{-1}(0) = I_f\cap U$. If $\mathfrak{v}: U \longrightarrow V$ and $\mathfrak{w}: U \longrightarrow V$ are representations of f, then there is a meromorphic function W on U such that $\mathfrak{w} = W\mathfrak{v}$. If $\mathfrak{v}$ is reduced, then W is holomorphic. If also $\mathfrak{w}$ is reduced, then W has no zeroes. Let $\mathfrak{w}: U \longrightarrow V$ be a representation. Then $\mathfrak{w}$ defines a non-negative divisor $\mu_{\mathfrak{w}}$: Take

$x \in U$. Then there is a reduced representation $\mathfrak{v} : U_x \longrightarrow V$, where U_x is an open neighborhood of x in U, and a holomorphic function $W \not\equiv 0$ on U_x such that $\mathfrak{w}|U_x = W.\mathfrak{v}$. Then $\mu_{\mathfrak{w}}|U = \mu_W^0$. Obviously $\mathfrak{w}$ is reduced if and only if $\mu_{\mathfrak{w}} \equiv 0$. Also $\tilde{\mathfrak{w}}$ is a section in $U \times V$ where $\tilde{\mathfrak{w}}(z) = (z, \mathfrak{w}(z))$ for all $z \in U$. Then $\mu_{\mathfrak{w}} = \mu_{\tilde{\mathfrak{w}}}$.

Again consider $M \times V$ as the trivial bundle. By Stoll [30] Proposition 2.1 there is a holomorphic line bundle L_f, unique upto isomorphism, on M called the <u>hyperplane section bundle</u> of f and a holomorphic section F of $(M\times V) \otimes L_f$ over M called the <u>reduced representation section</u> of f such that the following property is satisfied: Let $\mathfrak{v} : U \longrightarrow V$ be a reduced representation of f, then there is a holomorphic frame $\mathfrak{v}^\Delta$ of f over U such that

$$F|U = \tilde{\mathfrak{v}} \otimes \mathfrak{v}^\Delta \qquad \text{with } \tilde{\mathfrak{v}}(x) = (x, \mathfrak{v}(x)) \tag{2.47}$$

Here $L_f|(M - I_f)$ is isomorphic to $f^*(H)$. A holomorphic section G of $(M\times V) \otimes L_f$ over M is called a <u>representation section</u> of f, if for every point $x \in M$ there is an open, connected neighborhood U with representations $\mathfrak{g} : U \longrightarrow V$ and $\mathfrak{v} : U \longrightarrow V$ where $\mathfrak{v}$ is reduced, and where $\tilde{\mathfrak{g}} : U \longrightarrow M\times V$ is defined by $\tilde{\mathfrak{g}}(z) = (z, \mathfrak{g}(z))$ for all $z \in U$ such that $G|U = \tilde{\mathfrak{g}} \otimes \mathfrak{v}^\Delta$.

Take $a \in \mathbb{P}(V^*)$. Then f is said to be <u>free</u> for a, if $f(M - I_f) \not\subseteq \ddot{E}[a]$. If so, the <u>intersection divisor</u> $\mu_{f,a} \geq 0$ is defined by equivalent definitions. Take $\alpha \in V^*$ with $\mathbb{P}(\alpha) = a$.

1. <u>Definition</u>: Let $\mathfrak{v} : U \longrightarrow V$ be a reduced representation of f, then $\mu_{f,a}|U = \mu^0_{<\mathfrak{v},\alpha>}$.

2. <u>Definition</u>: The inner product $< , >: V \times V^* \longrightarrow \mathbb{C}$ extends to a bundle map

$$< , >: ((M\times V) \otimes L_f) \oplus (M \times V^*) \longrightarrow L_f \tag{2.48}$$

Thus a section $\langle F,\tilde{\alpha}\rangle \in \Gamma(M,L_f)$ is defined, which is not the zero section. Then $\mu_{f,a} = \mu_{\langle F,\tilde{\alpha}\rangle}$.

3. <u>Definition</u>. Here $\alpha \in V^*$, defines a section $\varepsilon\circ\tilde{\alpha} \in \Gamma(\mathbb{P}(V),H)$ with $\mu_{\varepsilon\circ\tilde{\alpha}} = \mu_{\ddot{E}[a]}$. Then $\mu_{f,a} = f^*(\mu_{\varepsilon\circ\tilde{\alpha}}) = f^*(\mu_{\ddot{E}[a]})$ is the pull back divisor.

4. <u>Definition</u>. The section $\varepsilon\circ\tilde{\alpha}$ defines a pull back section α_f of $f^*(H)$ over $M-I_f$. Then $\mu_{f,a} = \mu_{\alpha_f}$ on $M-I_f$ continues to the divisor $\mu_{f,a}$ on M.

The map f is said to be a <u>linearly by non-degenerate</u> if f is free for all $a \in \mathbb{P}(V^*)$.

Let (M,τ) be a parabolic manifold of dimension m. Let $f: M \longrightarrow \mathbb{P}(V)$ be a meromorphic map. The <u>spherical image</u>

$$(2.49) \qquad A_f(t) = t^{2-2m}\int_{M[t]} f^*(\Omega)\wedge v^{m-1} = A_f(t,H,\ell) \ge 0$$

increases with $t > 0$. The limits

$$(2.50) \qquad 0 \le A_f(0) = \lim_{0<t\to 0} A_f(t) < \infty \qquad 0 \le A_f(\infty) = \lim_{t\to\infty} A_f(t) \le \infty$$

exist. If $t > 0$, then

$$(2.51) \qquad A_f(t) = \int_{M[t]_*} f^*(\Omega)\wedge\omega^{m-1} + A_f(0).$$

The map f is constant, if and only if $A_f(\infty) = 0$. If $A_f(\infty) = \infty$, the map f is said to have <u>transcendental growth</u>. If $(M,\tau) = (\mathbb{C}^m,\tau_o)$ then $A_f(0)$ is a non-negative integer, which vanishes if and only if f is holomorphic at $0 \in \mathbb{C}^m$. Also $A_f(\infty) < \infty$ if and only if f is rational.

Returning to the general case, the <u>characteristic</u> T_f of f is defined by

$$(2.52) \qquad T_f(r,s) = \int_s^r A_f(t)\,\frac{dt}{t} = T_f(r,s;H,\ell) \ge 0$$

for $0 < s < r \in \mathbb{R}$. Here $T_f(r,s)/\log r \longrightarrow A_f(\infty)$ for $r \longrightarrow \infty$. In particular $T_f(r,s) \longrightarrow \infty$ if and only if f is not constant.

Take $a \in \mathbb{P}(V)$. Assume that f is free for a. The <u>counting function</u> and the <u>valence function</u> of the divisor $\mu_{f,a}$ are denoted by $n_{f,a}$ and $N_{f,a}$ respectively. For $r \in \mathbb{C}_\tau$, the <u>compensation function</u> is defined by

$$(2.53) \qquad m_{f,a}(r) = \int_{M[t]} \log \frac{1}{\Box f;a\Box}\, \sigma \geq 0.$$

For $0 < s < r \in \mathbb{C}_\tau$ with $s \in \mathbb{C}_\tau$, (2.32) implies the <u>First Main Theorem</u>

$$(2.54) \qquad T_f(r,s) = N_{f,a}(r,s) + m_{f,a}(r) - m_{f,a}(s).$$

Since T_f and $N_{f,a}$ are continuous, the function $m_{f,a}$ extends to the full positive axis $\mathbb{R}^+$ as a continuous function such that (2.54) holds for all $0 < s < r$. The <u>defect</u>

$$(2.55) \quad 0 \leq \delta_f(a) = \liminf_{r\to\infty} \frac{m_{f,a}(r)}{T_f(r,s)} = 1 - \limsup_{r\to\infty} \frac{N_{f,a}(r,s)}{T_f(r,s)} \leq 1$$

is defined, if f is not constant, and measures the failure of f(M) to intersect $\ddot{E}[a]$. The topic of defects is important, but first another subject shall be considered.

<u>THEOREM 2.3</u>. Let (M,τ) be a parabolic manifold of dimension m. Let V be a hermitian vector space of dimension $n+1 > 1$. Let $f: M \longrightarrow \mathbb{P}(V)$ be a meromorphic map. Let ℓ be the hermitian metric on V and the induced hermitian metric along the fibers of MxV. Let $\varkappa$ be a hermitian metric along the fibers of the hyperplane section bundle L_f of f. Let $\gamma = \ell\otimes\varkappa$ be the induced hermitian metric along the holomorphic vector bundle $W = (M\times V)\otimes L_f$. Let $0 \neq G \in \Gamma(M,W)$ be a representation section of f. Take $0 < s < r \in \mathbb{C}_\tau$ with $s \in \mathbb{C}_\tau$. Then we have

$$(2.56)\qquad T_f(r,s) + N_{\mu_G}(r,s) = T(r,s,L_f,\varkappa) + \int_{M\langle r\rangle} \log\|G\|_\gamma \sigma + \int_{M\langle s\rangle} \log\|G\|_\gamma \sigma .$$

<u>Proof</u>. The Green Residue Theorem (2.27) implies

$$(2.57)\qquad \int_s^r t^{2-2m} \int_{M[t]} dd^c \log\|G\|_\gamma^2\, \upsilon^m \frac{dt}{t} + N_{\mu_G}(r,s) = \int_{M[r]} \log\|G\|_\gamma \sigma - \int_{M\langle s\rangle} \log\|G\|_\gamma \sigma .$$

Take any $x \in M$. Then there is an open neighborhood U of x and there are representations $\mathfrak{v}: U \longrightarrow V$ and $\mathfrak{g}: U \longrightarrow V$ such that $\mathfrak{v}$ is reduced and such that $G|U = \tilde{\mathfrak{g}}\otimes\mathfrak{v}^\Delta$ where $\tilde{\mathfrak{g}}(z) = (z,\mathfrak{g}(z))$ for all $z \in U$. We have

$$\|G\|_\gamma \mid U = \|\mathfrak{g}\|\ \|\mathfrak{v}^\Delta\|_\varkappa$$

$$\begin{aligned} dd^c \log\|G\|_\gamma^2 | U &= dd^c \log\|\mathfrak{g}\|^2 + dd^c \log\|\mathfrak{v}^\Delta\|_\varkappa^2 \\ &= \mathfrak{g}^* \mathbb{P}^*(\Omega) - c(L_f,\varkappa)|U \\ &= (\mathbb{P}\circ\mathfrak{g})^*(\Omega) - c(L_f,\varkappa)|U \\ &= f^*(\Omega)|U - c(L_f,\varkappa)|U. \end{aligned}$$

Therefore

$$(2.58)\qquad \int_s^r t^{2-2m} \int_{M[t]} dd^c \log\|G\|_\gamma^2\, \upsilon^m \frac{dt}{t} = T_f(r,s) - T(r,s,L_f,\varkappa).$$

Now (2.58) and (2.57) implies (2.56); q.e.d.

If F is the reduced representation section of f, then $\mu_F = 0$ and we obtain

$$(2.59)\qquad T_f(r,s) = T(r,s,L_f,\varkappa) + \int_{M\langle r\rangle} \log\|F\|_\gamma \sigma - \int_{M\langle s\rangle} \log\|F\|_\gamma \sigma .$$

If there is a global reduced representation $\mathfrak{v}: M \longrightarrow V$, then there is a global holomorphic frame $\mathfrak{v}^\Delta$ of L_f. Thus L_f is trivial. We can take $\varkappa$ such that $\|\mathfrak{v}^\Delta\|_\varkappa \equiv 1$. Hence $c(L_f,\varkappa) \equiv 0$. Also

$$\|F\|_{\gamma} = \|\mathfrak{v}\| \, \|\mathfrak{v}^{\Delta}\|_{\varkappa} = \|\mathfrak{v}\|.$$

Therefore we obtain

$$T_f(r,s) = \int_{M\langle r\rangle} \log\|\mathfrak{v}\|\sigma - \int_{M\langle s\rangle} \log\|\mathfrak{v}\|\sigma.$$

If in addition there is a global representation $\mathfrak{g}: M \longrightarrow V$, then a representation section $G = \tilde{\mathfrak{g}}\otimes\mathfrak{v}^{\Delta}$ is defined with $\|G\|_{\gamma} = \|\mathfrak{g}\|$ and $\mu_G = \mu_{\mathfrak{g}}$. Therefore we have

$$(2.61) \qquad T_f(r,s) + N_{\mu_{\mathfrak{g}}}(r,s) = \int_{M\langle r\rangle} \log \|\mathfrak{g}\|\sigma - \int_{M\langle s\rangle} \log \|\mathfrak{g}\|\sigma.$$

Of course (2.60) and (2.61) are known a long time (Stoll [20] Satz 9.1) and Theorem 2.3 is the natural extension to the case where M is not a Cousin II domain.

Now, we will outline the Second Main Theorem and the Defect Relation. Let B be a holomorphic form of degree m-1 on the parabolic manifold (M,τ) of dimension m. Then τ is said to <u>majorize</u> B if and only if for every $r > 0$ there is constant $c > 0$ such that

$$(2.61) \qquad m\, i_{m-1}\, B\wedge\bar{B} \le c\, \upsilon^{m-1} \quad \text{on} \quad M[r].$$

The minimum of all those constants $c \ge 1$ is denoted by $Y_0(r)$. Then $Y(r) = \lim_{r<t\to r} Y_0(t) \ge 1$ defines an increasing, semi-continuous function Y called the <u>majorant</u> of B for τ. Let $U \ne \phi$ be an open, connected subset of M. An injective, holomorphic map

$$(2.62) \qquad \mathfrak{z} = (z_1,\dots,z_m): U \longrightarrow \mathbb{C}^m$$

is called a chart and $z_1,\dots, z_m$ are called <u>local</u> <u>coordinates</u> on U (at x if $x \in U$). Define the associated frame of $K(M)$ by

$$(2.63) \qquad \zeta = dz_1\wedge\dots\wedge dz_m.$$

Let $\mathfrak{v}: U \longrightarrow V$ be a holomorphic vector function. Then there exists one and only one holomorphic vector function $\mathfrak{v}': U \longrightarrow V$ called the <u>B-derivative</u> such that

$$\mathfrak{v}'\zeta = d\mathfrak{v}\wedge B. \tag{2.64}$$

The operation can be iterated: $\mathfrak{v}^{(p)} = (\mathfrak{v}^{(p-1)})'$ with $\mathfrak{v}^{(0)} = \mathfrak{v}$. For $0 \le p \in \mathbb{Z}$ define

$$\mathfrak{v}_{:p} = \mathfrak{v}\wedge\mathfrak{v}'\wedge\ldots\wedge\mathfrak{v}^{(p)}: U \longrightarrow \bigwedge_{p+1} V. \tag{2.65}$$

Let $f: M \longrightarrow \mathbb{P}(V)$ be a meromorphic map. Let $\mathfrak{Z} = \{U_\lambda, \mathfrak{v}_\lambda, \mathfrak{z}_\lambda\}_{\lambda\in\Lambda}$ be the family of all triplets $(U_\lambda, \mathfrak{v}_\lambda, \mathfrak{z}_\lambda)$ where $U_\lambda \neq \emptyset$ is an open, connected subset of M, where $\mathfrak{v}_\lambda: U_\lambda \longrightarrow V$ is a reduced representation of f and where $\mathfrak{z}_\lambda: U \longrightarrow \mathbb{C}^m$ is a chart of M with associated K(M)-frame ζ_λ. For $\lambda = (\lambda_0, \ldots, \lambda_p) \in \Lambda^{p+1}$ define

$$U_\lambda = U_{\lambda_0\ldots\lambda_p} = U_{\lambda_0}\cap\ldots\cap U_{\lambda_p} \tag{2.66}$$

$$\Lambda[p] = \{\lambda \in \Lambda^{p+1} \mid U_\lambda \neq \emptyset\}. \tag{2.67}$$

If $(\lambda,\mu) \in \Lambda[1]$, there are holomorphic functions $v_{\lambda\mu}: U_{\lambda\mu} \longrightarrow \mathbb{C}_*$ and $\Delta_{\lambda\mu}: U_{\lambda\mu} \longrightarrow \mathbb{C}_*$ such that

$$\mathfrak{v}_\lambda = v_{\lambda\mu}\mathfrak{v}_\mu, \quad \mathfrak{v}^\Delta_\lambda = v_{\mu\lambda}\,\mathfrak{v}^\Delta_\mu, \quad \zeta_\lambda = \Delta_{\lambda\mu}\,\zeta_\mu \tag{2.68}$$

$$\mathfrak{v}_{\lambda:p} = v^{p+1}_{\lambda\mu}\,\Delta^{(1/2)p(p+1)}_{\mu\lambda}\,\mathfrak{v}_{\mu:p} \tag{2.69}$$

$$v_{\lambda\lambda} = \Delta_{\lambda\lambda} = 1 \quad , \quad v_{\lambda\mu}\,v_{\mu\lambda} = \Delta_{\lambda\mu}\,\Delta_{\mu\lambda} = 1 \tag{2.70}$$

$$v_{\lambda\mu}\,v_{\mu\rho}\,v_{\rho\lambda} = 1 = \Delta_{\lambda\mu}\,\Delta_{\mu\rho}\,\Delta_{\rho\lambda} \text{ on } U_{\lambda\mu\rho} \quad \text{if } (\lambda,\mu,\rho) \in \Lambda[2]. \tag{2.71}$$

Hence there exists one and only one holomorphic section F_p of the bundle

$$(M \times \bigwedge_{p+1} V) \otimes L^p_f \otimes K(M)^{(1/2)p(p+1)} \tag{2.72}$$

such that

$$F_p|U_\lambda = \mathfrak{v}_{\lambda:p} \otimes (\mathfrak{v}^\Delta_\lambda)^{p+1} \otimes \zeta_\lambda^{(1/2)p(p+1)} \tag{2.73}$$

for all $\lambda \in \Lambda$. Here $F_0 = F \not\equiv 0$, but $F_{p+1} \equiv 0$ if $F_p \equiv 0$ and $F_{n+1} \equiv 0$. Hence a so called <u>generality</u> <u>index</u> $\ell_f \in \mathbb{Z}[0,n]$ exists such that $F_p \not\equiv 0$ if $0 \le p \le \ell_f$ but $F_p \equiv 0$ if $p > \ell_f$. If $\lambda \in \Lambda$, then $\mathfrak{v}_{\lambda:p} \not\equiv 0$ if $0 \le p \le \ell_f$ and $\mathfrak{v}_{\lambda:p} \equiv 0$ if $p > \ell_f$. Here F_p is called the pth <u>representation</u> <u>section</u> of f for B. If $0 \le p \le \ell_f$, one and only one meromorphic map $f_p: M \longrightarrow \mathbb{P}(\bigwedge_{p+1} V)$ called the pth <u>associated</u> <u>map</u> exists such that $\mathfrak{v}_{\lambda:p}$ is a representation of f_p not necessarily reduced. Also $F_p \not\equiv F_{f_p}$ in general. The divisor $d_p = \mu_{F_p} \ge 0$ is called the pth <u>Wronski-divisor</u> where $d_p|U_\lambda = \mu_{\mathfrak{v}_{\lambda:p}}$ for $\lambda \in \Lambda$. Put $d_p = 0$ if $p < 0$ or $p > \ell_f$ and observe that $d_0 = 0$. Then the pth <u>stationary-divisor</u> $\iota_p = d_{p-1} - 2d_p + d_{p+1}$ is non-negative. Actually f_p maps into an algebraic subvariety. Define the pth-<u>Grassmann</u> <u>cone</u> of V by

$$\tilde{G}_p(V) = \{\mathfrak{x}_0 \wedge \cdots \wedge \mathfrak{x}_p \mid \mathfrak{x}_j \in V\}. \tag{2.74}$$

Then $\mathbb{P}(\tilde{G}_p(V)) = G_p(V)$ is a connected, smooth, compact submanifold of $\mathbb{P}(\bigwedge_{p+1} V)$ with $\dim G_p(V) = (n-p)(p+1)$. Here $G_p(V)$ is called the pth <u>Grassmann manifold</u> and $f_p: M \longrightarrow G_p(V)$ maps into it. Let $\Omega_{(p)}$ be the Fubini-Study form on $\mathbb{P}(\bigwedge_{p+1} V)$. Then

$$\mathbb{H}_p = m\, i_{m-1}\, f_p^*(\Omega_{(p)}) \wedge B \wedge \bar{B} \ge 0 \qquad \text{on } M - I_{f_p}. \tag{2.75}$$

Define $h_p: M \longrightarrow \mathbb{R}_+$ by $\mathbb{H}_p = h_p^2\, \upsilon^m$ on $M^+ - I_f$ and $h_p = 0$ on $M - M^+$.

For each $r \in \mathfrak{C}^0_\tau$ the integral

$$S_p(r) = \int_{M\langle r\rangle} (\log h_p)\, \sigma \tag{2.76}$$

exists. If $0 < s < r \in \mathfrak{C}^0_\tau$ with $s \in \mathfrak{C}^0_\tau$ and if $0 \le p \le \ell_f$, the <u>PLÜCKER</u> <u>DIFFERENCE</u> <u>FORMULA</u> holds.

$$N_{\ell_p}(r,s) + T_{f_{p+1}}(r,s) - 2T_{f_p}(r,s) + T_{f_{p-1}}(r,s)$$

(2.77)

$$= S_p(r) - S_p(s) + \mathrm{Ric}_\tau(r,s)$$

Take $0 \le s \in \mathbb{R}$. Let g and h be real valued functions on $\mathbb{R}(s,+\infty)$. Write $g \overset{\cdot}{\le} h$ if there is a set E of finite measure in $\mathbb{R}(s,+\infty)$ such that $g(x) \le h(x)$ for all $x \in \mathbb{R}(s,+\infty)-E$.

Take $\emptyset \ne A \subseteq \mathbb{P}(V)$. The intersection of all projective plans containing A is a projective plan called span(A). Here A is said to be linearly independent if and only if $1+\dim \mathrm{span}(A) = \#A$. A subset $G \ne \emptyset$ of $\mathbb{P}(V)$ is said to be in general position if and only if every subset $\emptyset \ne A \subseteq G$ with $\#A \le n+1$ is linearly independent.

Let (M,τ) be a parabolic manifold of dimension m. Let B be a holomorphic form of degree m-1 on M. Assume that τ majorizes B with majorant Y. Let V be a hermitian vector space of dimension $n+1 > 1$. Let $f: M \longrightarrow \mathbb{P}(V)$ be a meromorphic map with generality index $\ell_f = n$ for B. Let G be a finite subset of $\mathbb{P}(V^*)$ in general position. Take $0 < \varepsilon \in \mathbb{R}$ and $0 < s \in \mathbb{R}$. Then we have the Second Main Theorem

$$N_{d_n}(r,s) + \sum_{a\in G} m_{f,a}(r)$$

(2.78)

$$\overset{\cdot}{\le} (n+1)\, T_f(r,s) + \frac{n(n+1)}{2}\mathrm{Ric}_\tau(r,s) + \varepsilon \log r$$

$$+ n(n+1)\varsigma(1+\varepsilon)^2 (\log T_f(r,s) + \log Y(r) + \log^+ \mathrm{Ric}_\tau(r,s)).$$

Here $\ell_f = n$ implies that f is linearly non-degenerate. Let us consider the special case where there is a proper, surjective, holomorphic map

(2.79)

$$\pi = (\pi_1,\cdots,\pi_m): M \longrightarrow \mathbb{C}^m$$

such that $\tau = \|\pi\|^2 = |\pi_1|^2+\cdots+|\pi_m|^2$. Assume that f is linearly non-degenerate, then there exists a holomorphic form B of degree m-1 on M

such that $\ell_f = n$ and such that τ majorizes B with majorant Y where $Y(r) \le 1 + r^{2n-2}$ for $r > 0$. The branching divisor β of π is the divisor of the form $d\pi_1 \wedge \cdots \wedge d\pi_m$ on M. Then

$$\text{(2.80)} \qquad \mathrm{Ric}_\tau(r,s) = N_\beta(r,s) \ge 0 \qquad \text{if} \quad 0 < s < r.$$

Returning to the general situation of the Second Main Theorem define the <u>defect</u> $\delta_f(a)$ by (2.55) and the

<u>Ricci</u> <u>defect</u>: $$R_f = \limsup_{r \to \infty} \frac{\mathrm{Ric}_\tau(r,s)}{T_f(r,s)}$$

<u>Majorization</u> <u>defect</u>: $$Y_f = \limsup_{r \to \infty} \frac{\log Y(r)}{T_f(r,s)} \ge 0.$$

<u>Ramification</u> <u>defect</u>: $$\Theta_f = \liminf_{r \to \infty} \frac{N_{d_n}(r,s)}{T_f(r,s)} \ge 0.$$

If $R_f < \infty$ and $Y_f < \infty$, then $\Theta_f < \infty$ and (2.75) implies the <u>DEFECT</u> <u>RELATION</u>

$$\text{(2.81)} \qquad \Theta_f + \sum_{a \in G} \delta_f(a) \le n+1 + (1/2)\, n(n+1)\, R_f + n(n+1) s Y_f.$$

In the special case of the covering map $\pi: M \longrightarrow \mathbb{C}^m$ we can take B such that $Y_f = 0$ provided f has transcendent growth and is linearly non- degenerate. If also f separates the fibers of π, then $R_f \le 2(q-1)$ where q is the generic sheet number of π as we will see later. If f has transcendental growth and if the (m-1)-dimensional component of $\pi(\mathrm{supp}\,\beta)$ is affine algebraic in $\mathbb{C}^m$ then $R_f = 0$. In this case we obtain

$$\text{(2.82)} \qquad \sum_{a \in G} \delta_f(a) \le n+1.$$

If $M = \mathbb{C}^m$ and $\tau = \tau_0$ is defined by $\tau_0(z_1, \cdots, z_m) = |z_1|^2 + \cdots + |z_m|^2$, then (2.82) holds under the assumption that $f: \mathbb{C}^m \longrightarrow \mathbb{P}(V)$ is linearly non-degenerate and G is in general position. Observe that the assumption and the statement (2.82) do not involve B.

This ends the outline of Nevanlinna theory on parabolic manifolds. More information is contained for instance in Stoll [20], [22], [23], [24], [25], [26], [27], [28], [29], and [30] and in Shabat [18], in Weyl and Weyl [36] and in Wu [37].

3. THE SYMMETRIC TENSOR PRODUCT.

Multilinear algebra permits us to formulate the theory of algebroid reduction of Nevanlinna theory in a simple, coordinate free fashion. The required facts are assembled in this section. Most proofs are left to the reader who may consult Greub [6].

Let V be a complex vector space of dimension n+1. Put $V_* = V-\{0\}$. Take $p \in \mathbb{N}$. Let $\mathfrak{S}_p$ be the group of bijective maps $\pi: \mathbb{N}[1,p] \to \mathbb{N}[1,p]$. Let $\otimes_p V$ be the p-fold tensor product of V. For each $\pi \in \mathfrak{S}_p$ a linear isomorphism $\hat{\pi}: \otimes_p V \longrightarrow \otimes_p V$ is uniquely defined by

$$\hat{\pi}(\mathfrak{z}_1 \otimes \cdots \otimes \mathfrak{z}_p) = \mathfrak{z}_{\psi(1)} \otimes \cdots \otimes \mathfrak{z}_{\psi(p)} \tag{3.1}$$

where $\psi = \pi^{-1}$ and $\mathfrak{z}_j \in V$ for $j=1,\cdots,p$. Then $\hat{\psi} = \hat{\pi}^{-1}$ and $\hat{id} = id$. If $\pi \in \mathfrak{S}_p$ and $\varphi \in \mathfrak{S}_p$, then $(\pi \circ \varphi)\hat{} = \hat{\pi} \circ \hat{\varphi}$. A vector $\mathfrak{z} \in \otimes_p V$ is said to be symmetric if and only if $\hat{\pi}(\mathfrak{z}) = \mathfrak{z}$ for all $\pi \in \mathfrak{S}_p$. The set of all symmetric $\mathfrak{z} \in \otimes_p V$ is a linear subspace $\amalg_p V$ of $\otimes_p V$ with

$$\dim \amalg_p V = \binom{n+p}{p} \tag{3.2}$$

Here $\amalg_p V$ is called the p-fold symmetric tensor product of V.

This linear map

$$\sigma = \frac{1}{p!} \sum_{\pi \in \mathfrak{S}_p} \hat{\pi} \otimes_p V \longrightarrow \otimes_p V. \tag{3.3}$$

is called the symmetrizer of $\otimes_p V$, where $\operatorname{Im}\sigma = \amalg_p V$ and $\sigma \circ \sigma = \sigma$. Hence σ is a projection and

$$(3.4) \qquad \otimes_p V = (\operatorname{Ker} \sigma) \oplus ⨿_p V.$$

Observe $\otimes_1 V = V = ⨿_1 V$ and define $\otimes_\circ V = \mathbb{C} = ⨿_\circ V$ where σ is the identity on both of these. If $p \in \mathbb{Z}_+$ and $q \in \mathbb{Z}_+$ and $\mathfrak{x} \in ⨿_p V$ and $\mathfrak{y} \in ⨿_q V$ a product

$$(3.5) \qquad \mathfrak{x} \cdot \mathfrak{y} = \sigma(\mathfrak{x} \otimes \mathfrak{y}) = \sigma(\mathfrak{y} \otimes \mathfrak{x}) = \mathfrak{y} \cdot \mathfrak{x} \in ⨿_{p+q} V$$

is defined. The direct sums

$$(3.6) \qquad \otimes V = \bigoplus_{p \in \mathbb{Z}_+} \otimes_p V \qquad ⨿ V = \bigoplus_{p \in \mathbb{Z}_+} ⨿_p V$$

became graded algebras, where σ extends to an algebra homomorphism

$$(3.7) \qquad \sigma \colon \otimes V \longrightarrow ⨿ V.$$

The algebra $⨿V$ is associative, commutative and without zero divisors (see below or Greub [6]) and $1 \in \mathbb{C}$ is the unit element.

If $\{\mathfrak{x}_\lambda\}_{\lambda \in \Lambda}$ is a finite family of vectors in $\otimes V$, take any enumeration $\Lambda = \{\lambda_1, \cdots, \lambda_q\}$ with $\lambda_j \neq \lambda_k$ if $j \neq k$. Then

$$(3.8) \qquad \mathfrak{y} = ⨿_{\lambda \in \Lambda} \mathfrak{x}_\lambda = \mathfrak{x}_{\lambda_1} \cdots \mathfrak{x}_{\lambda_q} \in ⨿ V$$

does not depend on the enumeration and is not zero if $\mathfrak{x}_\lambda \neq 0$ for all $\lambda \in \Lambda$. If $\mathfrak{x}_\lambda \in ⨿_{p_\lambda} V$ then $\mathfrak{y} \in ⨿_p V$ with $p = \sum_{\lambda \in \Lambda} p_\lambda$. If $\mathfrak{x}_\lambda = \mathfrak{x}$ for all $\lambda \in \Lambda$ write $\mathfrak{y} = \mathfrak{x}^q$, which is called the <u>symmetric</u> q^{th} <u>power</u> of $\mathfrak{x}$. If $\{\mathfrak{x}_\lambda\}_{\lambda \in \Lambda}$ is a finite family of vectors in V with $\#\Lambda = p$, then

$$(3.9) \qquad \mathfrak{y} = ⨿_{\lambda \in \Lambda} \mathfrak{x}_\lambda = \mathfrak{x}_{\lambda_1} \cdots \mathfrak{x}_{\lambda_p} \in ⨿_p V$$

is said to be <u>decomposable</u>.

Take $p \in \mathbb{N}$. Take $\mathfrak{a} = (\mathfrak{a}_0, \cdots, \mathfrak{a}_{n+1}) \in V^{n+1}$. For $j = (j_0, \cdots, j_n) \in \mathbb{Z}_+^{n+1}$ define

$$(3.10) \qquad j! = (j_0!) \cdots (j_n!) \qquad \mathfrak{a}^j = \mathfrak{a}_0^{j_0} \cdot \mathfrak{a}_1^{j_1} \cdots \mathfrak{a}_n^{j_n}$$

$$(3.11)\qquad S(p,n) = \{j = (j_0,\cdots,j_n) \in \mathbb{Z}_+^{n+1} \mid j_0+\cdots+j_n = p\}.$$

For each $j \in S(p,n)$ define $c_j > 0$ by $c_j^2 = (p!)/(j!)$. If $\mathfrak{a}$ is a base of V, then $\{c_j\mathfrak{a}^j\}_{j\in S(p,n)}$ is a base of $\amalg_p V$ which shows (3.2).

Let V^* be the dual vector space of V. The <u>inner product</u> of $\mathfrak{v} \in V$ and $\mathfrak{w} \in V^*$ is denoted by $\langle\mathfrak{v},\mathfrak{w}\rangle$, where $\langle\mathfrak{v},\mathfrak{w}\rangle = \langle\mathfrak{w},\mathfrak{v}\rangle$ signifies $V^{**} = V$. A bilinear map

$$(3.12)\qquad \langle\ ,\ \rangle:\ \amalg_p V \times \amalg_p V^* \longrightarrow \mathbb{C}$$

is uniquely defined by $\mathfrak{v}_j \in V$ and $\mathfrak{w}_j \in V^*$ for $j = 1,\cdots p$, and

$$\langle\mathfrak{v}_1\cdots\mathfrak{v}_p,\ \mathfrak{w}_1\cdots\mathfrak{w}_p\rangle$$

$$(3.13)\qquad = (1/p!)\sum_{\pi\in\mathfrak{S}_p}\langle\mathfrak{v}_1,\mathfrak{w}_{\pi(1)}\rangle\cdots\langle\mathfrak{v}_p,\mathfrak{w}_{\pi(p)}\rangle$$

$$= (1/p!)\sum_{\pi\in\mathfrak{S}_p}\langle\mathfrak{v}_{\pi(1)},\mathfrak{w}_1\rangle\cdots\langle\mathfrak{v}_{\pi(p)},\mathfrak{w}_p\rangle.$$

Let $\mathfrak{a} = (\mathfrak{a}_0,\cdots,\mathfrak{a}_n)$ be a base of V. Let $\mathfrak{b} = (\mathfrak{b}_0,\cdots,\mathfrak{b}_n)$ be the dual base of V^*. Take $j \in S(p,n)$ and $k \in S(p,n)$. Then

$$(3.14)\qquad \langle c_j\mathfrak{a}^j, c_k\mathfrak{b}^k\rangle = \begin{Bmatrix} 1 & \text{if } j = k \\ 0 & \text{if } j \neq k \end{Bmatrix}$$

Hence we can identify $\amalg_p V^* = (\amalg_p V)^*$ such that (3.12) is the inner product. If $\mathfrak{v}_j$ and $\mathfrak{w}_j$ are taken as above and if $\mathfrak{v} \in V$ and $\mathfrak{w} \in V^*$, then

$$(3.15)\qquad \langle\mathfrak{v}^p\mathfrak{w}_1\cdots\mathfrak{w}_p\rangle = \langle\mathfrak{v},\mathfrak{w}_1\rangle\cdots\langle\mathfrak{v},\mathfrak{w}_p\rangle$$

$$(3.16)\qquad \langle\mathfrak{v}_1,\cdots\mathfrak{v}_p,\mathfrak{w}^p\rangle = \langle\mathfrak{v}_1,\mathfrak{w}\rangle\cdots\langle\mathfrak{v}_p,\mathfrak{w}\rangle$$

$$\langle \mathfrak{v}^p, \mathfrak{w}^p \rangle = \langle \mathfrak{v}, \mathfrak{w} \rangle^p \tag{3.17}$$

If $\mathfrak{a} \in \amalg_p V$ and $\mathfrak{b} \in \amalg_q V$ and $\mathfrak{x} \in V^*$, then

$$\langle \mathfrak{a} \cdot \mathfrak{b}, \mathfrak{x}^{p+q} \rangle = \langle \mathfrak{a}, \mathfrak{x}^p \rangle \langle \mathfrak{b}, \mathfrak{x}^q \rangle \tag{3.18}$$

Here (3.16) implies (3.18) if $\mathfrak{a}$ and $\mathfrak{b}$ are decomposable. Because the decomposable vectors in $\amalg_p V$ (respectively $\amalg_q V$) generate the vector space $\amalg_p V$ (respectively $\amalg_q V$) (3.18) follows in general.

<u>Lemma 3.1</u>. Take $\mathfrak{a} \in \amalg_p V$ such that $\langle \mathfrak{a}, \mathfrak{x}^p \rangle = 0$ for all $\mathfrak{x} \in V^*$. Then $\mathfrak{a} = 0$.

<u>Proof</u>. Let $\mathfrak{e} = (\mathfrak{e}_0, \ldots, \mathfrak{e}_n)$ be a base of V. Let $\mathfrak{c} = (\mathfrak{c}_0, \ldots, \mathfrak{c}_n)$ be the dual base. Then $\mathfrak{x} = \sum_{j=0}^{n} x_j \mathfrak{c}_j$ with $x_j = \langle \mathfrak{e}_j, \mathfrak{x} \rangle$. Also $\{c_j \mathfrak{e}^j\}_{j \in S(p,n)}$ is a base $\amalg_p V$. Then $a_j = \langle \mathfrak{a}, c_j \mathfrak{c}^j \rangle$ for $j \in S(p,n)$ and

$$0 = \langle \mathfrak{a}, \mathfrak{x}^p \rangle = \sum_{j \in S(p,n)} a_j \, c_j \, \langle \mathfrak{e}^j, \mathfrak{x}^p \rangle = \sum_{j \in S(p,n)} a_j \, c_j \, x_0^{j_0} \cdots x_n^{j_n}$$

Hence $a_j = 0$ for all $j \in S(p,n)$ and $\mathfrak{a} = 0$; q.e.d.

Of course, if $\mathfrak{a} \in \amalg_p V^*$ and $\langle \mathfrak{x}^p, \mathfrak{a} \rangle = 0$ for all $\mathfrak{x} \in V$, then $\mathfrak{a} = 0$. This yields the following identification: A holomorphic function $\alpha: V \longrightarrow \mathbb{C}$ is said to be a homogeneous polynomial of degree p on V if and only if $\alpha(\lambda \mathfrak{x}) = \lambda^p \alpha(\mathfrak{x})$ for $\lambda \in \mathbb{C}$ and $\mathfrak{x} \in V$. The vector space $V_{[p]}$ of all homogeneous polynomials of degree p on V has dimension $\binom{n+p}{p}$. Hence a linear isomorphism

$$\iota_0 : \amalg_p V^* \longrightarrow V_{[p]} \tag{3.19}$$

is defined by $\iota_0(\mathfrak{a})(\mathfrak{x}) = \langle \mathfrak{x}^p, \mathfrak{a} \rangle$ for all $\mathfrak{x} \in V$ and $\mathfrak{a} \in \amalg_p V^*$.

There is another identification: Let $\mathcal{O}(-1)$ be the tautological bundle over $\mathbb{P}(V)$ defined by (2.42). Then we define a holomorphic

frame atlas $\{\xi_{\mathfrak{a}}\}_{\mathfrak{a}\in V_*^*}$. For $\mathfrak{a} \in V_*^*$ define $V_{\mathfrak{a}} = \{\mathfrak{z} \in V \mid \langle \mathfrak{z}, \mathfrak{a}\rangle \neq 0\}$. Then $\mathbb{P}(V_{\mathfrak{a}})$ is open and dense in $\mathbb{P}(V)$. Define $\xi_{\mathfrak{a}} : \mathbb{P}(V_{\mathfrak{a}}) \longrightarrow \mathcal{O}(-1)$ by

$$\xi_{\mathfrak{a}}(x) = \frac{\mathfrak{z}}{\langle \mathfrak{z}, \mathfrak{a}\rangle} \qquad \text{for } x = \mathbb{P}(\mathfrak{z}) \in \mathbb{P}(V_{\mathfrak{a}}). \tag{3.20}$$

If $\mathfrak{a} \in V_*^*$ and $\mathfrak{b} \in V_*^*$. Then $g_{\mathfrak{a}\mathfrak{b}} : \mathbb{P}(V_{\mathfrak{a}}) \cap \mathbb{P}(V_{\mathfrak{b}}) \longrightarrow \mathbb{C}$ is a holomorphic function without zeroes defined by

$$g_{\mathfrak{a}\mathfrak{b}}(x) = \frac{\langle \mathfrak{z}, \mathfrak{a}\rangle}{\langle \mathfrak{z}, \mathfrak{b}\rangle} \qquad \text{for } x = \mathbb{P}(\mathfrak{z}) \in \mathbb{P}(V_{\mathfrak{a}}) \cap \mathbb{P}(V_{\mathfrak{b}}). \tag{3.21}$$

Then

$$\xi_{\mathfrak{a}}\, g_{\mathfrak{a}\mathfrak{b}} = \xi_{\mathfrak{b}} \quad \text{on} \quad \mathbb{P}(V_{\mathfrak{a}}) \cap \mathbb{P}(V_{\mathfrak{b}}). \tag{3.22}$$

Thus $\{\xi_{\mathfrak{a}}\}_{\mathfrak{a}\in V_*^*}$ is a holomorphic frame atlas of $\mathcal{O}(-1)$ and $\{g_{\mathfrak{a}\mathfrak{b}}\}_{(\mathfrak{a},\mathfrak{b})\in (V_*^*)^2}$ is the basic cocycle. Let $H = \mathcal{O}(-1)^*$ be the hyperplane section bundle and let $\xi_{\mathfrak{a}}^*$ be the dual frame. Then

$$\xi_{\mathfrak{a}}^* = g_{\mathfrak{a}\mathfrak{b}}\, \xi_{\mathfrak{b}}^* \qquad \text{on} \quad \mathbb{P}(V_{\mathfrak{a}}) \cap \mathbb{P}(V_{\mathfrak{b}}). \tag{3.23}$$

Take $p \in \mathbb{N}$. Then $\left\{(\xi_{\mathfrak{a}}^*)^p\right\}_{\mathfrak{a}\in V_*^*}$ is a holomorphic frame atlas of H^p with

$$(\xi_{\mathfrak{a}}^*)^p = g_{\mathfrak{a}\mathfrak{b}}^p\, (\xi_{\mathfrak{b}}^*)^p \qquad \text{on} \quad \mathbb{P}(V_{\mathfrak{a}}) \cap \mathbb{P}(V_{\mathfrak{b}}) \tag{3.24}$$

For $p \in \mathbb{N}$ define $V_{(p)} = \Gamma(\mathbb{P}(V), H^p)$ as the vector space of global holomorphic sections of H^p over $\mathbb{P}(V)$. A linear isomorphism $\iota : V_{[p]} \longrightarrow V_{(p)}$ is defined as follows: Take $\alpha \in V_{[p]}$. Take $x \in \mathbb{P}(V)$. Take $\mathfrak{z} \in V_*$ with $\mathbb{P}(\mathfrak{z}) = x$. Then $x \in \mathbb{P}(V_{\mathfrak{a}})$ for some $\mathfrak{a} \in V_*^*$. Define

$$\iota(\alpha)(x) = \frac{\alpha(\mathfrak{z})}{\langle \mathfrak{z}, \mathfrak{a}\rangle^p} \left[\xi_{\mathfrak{a}}^*(x)\right]^p \in H_x^p \tag{3.25}$$

If $x \in \mathbb{P}(V_{\mathfrak{a}}) \cap \mathbb{P}(V_{\mathfrak{b}})$ with $\mathfrak{a} \in V_*^*$ and $\mathfrak{b} \in V_*^*$, then

$$\frac{\alpha(\mathfrak{x})}{\langle\mathfrak{x},\mathfrak{a}\rangle^p}\left[\xi_{\mathfrak{a}}^*(x)\right]^p = \frac{\alpha(\mathfrak{x})}{\langle\mathfrak{x},\mathfrak{a}\rangle^p}\, g_{\mathfrak{a}\mathfrak{b}}(x)^p \left[\xi_{\mathfrak{b}}^*(x)\right]^p$$

$$= \frac{\alpha(\mathfrak{x})}{\langle\mathfrak{x},\mathfrak{a}\rangle^p}\,\frac{\langle\mathfrak{x},\mathfrak{a}\rangle^p}{\langle\mathfrak{x},\mathfrak{b}\rangle^p}\left[\xi_{\mathfrak{b}}^*(x)\right]^p$$

$$= \frac{\alpha(\mathfrak{x})}{\langle\mathfrak{x},\mathfrak{b}\rangle^p}\left[\xi_{\mathfrak{b}}^*(x)\right]^p$$

Hence $\iota(\alpha)(x)$ is well-defined and $\iota(\alpha) \in \Gamma(\mathbb{P}(V),H^p)$. If $\alpha \neq 0$, then $\mathbb{P}(\alpha^{-1}(0)) = Z(\iota(\alpha))$ and $\mathbb{P}^*(\mu_{\iota(\alpha)}) = \mu_\alpha^0$. The divisor $\mu_{\iota(\alpha)}$ is a hypersurface of degree p with support $Z(\iota(\alpha))$. Identify

$$V_{(p)} = V_{[p]} = \amalg_p V^* = (\amalg_p V)^*. \tag{3.26}$$

Now we return to (3.18) which permits us to show that the algebra $\amalg V$ has no zero divisors.

<u>THEOREM 3.2</u> Take $\mathfrak{a} \in \amalg V$ and $\mathfrak{b} \in \amalg V$ with $\mathfrak{a}\mathfrak{b} = 0$. Then either $\mathfrak{a} = 0$ or $\mathfrak{b} = 0$.

<u>Proof</u>. First assume that $\mathfrak{a} \in \amalg_p V$ and $\mathfrak{b} \in \amalg_q V$. Homogeneous polynomials $\alpha: V^* \longrightarrow \mathbb{C}$ and $\beta: V^* \longrightarrow \mathbb{C}$ are defined by $\alpha(\mathfrak{x}) = \langle\mathfrak{a},\mathfrak{x}^p\rangle$ and $\beta(\mathfrak{x}) = \langle\mathfrak{b},\mathfrak{x}^q\rangle$ for all $\mathfrak{x} \in V^*$. By (3.18).

$$(\alpha\cdot\beta)(\mathfrak{x}) = \alpha(\mathfrak{x})\,\beta(\mathfrak{x}) = \langle\mathfrak{a},\mathfrak{x}^p\rangle\langle\mathfrak{b},\mathfrak{x}^q\rangle = \langle\mathfrak{a}\cdot\mathfrak{b},\,\mathfrak{x}^{p+q}\rangle = 0.$$

Hence $\alpha\cdot\beta = 0$. Since α and β are holomorphic functions on V, we have $\alpha = 0$ or $\beta = 0$. Lemma 3.1 implies $\mathfrak{a} = 0$ or $\mathfrak{b} = 0$, which proves the theorem in the special case.

Now consider the general case with $\mathfrak{a} \neq 0$ and $\mathfrak{b} \neq 0$ but $\mathfrak{a}\cdot\mathfrak{b} = 0$. Then $\mathfrak{a} = \sum_{\lambda=0}^{\infty} \mathfrak{a}_\lambda$ and $\mathfrak{b} = \sum_{\lambda=0}^{\infty} \mathfrak{b}_\lambda$ where all but finitely many $\mathfrak{a}_\lambda$ and $\mathfrak{b}_\lambda$ are zero with $\mathfrak{a}_\lambda \in \amalg_\lambda V$ and $\mathfrak{b}_\lambda \in \amalg_\lambda V$ for all $\lambda \in \mathbb{Z}_+$. Since $\mathfrak{a} \neq 0$ and $\mathfrak{b} \neq 0$, smallest numbers p and q exist such that $\mathfrak{a}_p \neq 0$ and $\mathfrak{b}_q \neq 0$.

We have

$$0 = \mathfrak{a}\cdot\mathfrak{b} = \mathfrak{a}_p\,\mathfrak{b}_q + \sum_{m=p+q+1}^{\infty} \mathfrak{c}_m \qquad \text{with } \mathfrak{c}_m \in \amalg_m V$$

Because the sum $\amalg V = \bigoplus_{m\in\mathbb{Z}_+} \amalg_m V$ is directed we have $\mathfrak{a}_p\cdot\mathfrak{b}_q = 0$ and $\mathfrak{c}_m = 0$. By the first part of the proof $\mathfrak{a}_p = 0$ or $\mathfrak{b}_q = 0$ which is a contradiction. Hence $\mathfrak{a} = 0$ or $\mathfrak{b} = 0$; q.e.d.

<u>LEMMA 3.3</u> Take $\mathfrak{x} \in V$ and $\mathfrak{y} \in V$ and $p \in \mathbb{N}$. Then $\mathfrak{x}^p = \mathfrak{y}^p$ if and only if there is $\zeta \in \mathbb{C}$ with $\zeta^p = 1$ and $\mathfrak{y} = \zeta\mathfrak{x}$.

<u>Proof</u>. a) Assume that $\zeta \in \mathbb{C}$ with $\zeta^p = 1$ exists such that $\mathfrak{y} = \zeta\mathfrak{x}$. Then $\mathfrak{y}^p = \zeta\mathfrak{x}^p = \mathfrak{x}^p$.

b) Assume that $\mathfrak{x}^p = \mathfrak{y}^p$. If $\mathfrak{x}^p = \mathfrak{y}^p = 0$, then $\mathfrak{x} = \mathfrak{y} = 0$. Pick $\zeta = 1$. Assume that $\mathfrak{x} \neq 0 \neq \mathfrak{y}$. Take a base $\mathfrak{w}_0, \mathfrak{w}_1, \cdots, \mathfrak{w}_n$ of V^* with $\langle\mathfrak{x},\mathfrak{w}_0\rangle = 1$ and $\langle\mathfrak{x},\mathfrak{w}_j\rangle = 0$ for $1 \le j \le n$. Define $\zeta = \langle\mathfrak{y},\mathfrak{w}_0\rangle \in \mathbb{C}$. Then

$$\zeta^p = \langle\mathfrak{y},\mathfrak{w}_0\rangle^p = \langle\mathfrak{y}^p,\mathfrak{w}_0^p\rangle = \langle\mathfrak{x}^p,\mathfrak{w}_0^p\rangle = \langle\mathfrak{x},\mathfrak{w}_0\rangle^p = 1$$

Take any $j \in \mathbb{N}[1,n]$. Then

$$\langle\mathfrak{y},\mathfrak{w}_j\rangle^p = \langle\mathfrak{y}^p,\mathfrak{w}_j^p\rangle = \langle\mathfrak{x}^p,\mathfrak{w}_j^p\rangle = \langle\mathfrak{x},\mathfrak{w}_j\rangle^p = 0$$

Hence $\langle\mathfrak{y},\mathfrak{w}_j\rangle = 0$. Therefore $\langle\mathfrak{y}-\zeta\mathfrak{x},\mathfrak{w}_j\rangle = 0$ for all $j = 0,1,\cdots,n$, which implies $\mathfrak{y} = \zeta\mathfrak{x}$: q.e.d.

Let a and b integers with $0 \le b \le a \le n$. For $\mathfrak{v} \in \bigwedge_{a+1} V$ and $\mathfrak{w} \in \bigwedge_{b+1} V^*$ the <u>interior</u> <u>product</u> $\mathfrak{v} \llcorner \mathfrak{w} \in \bigwedge_{a-b} V$ is defined by

$$(3.27) \qquad \langle\mathfrak{v} \llcorner \mathfrak{w},\mathfrak{x}\rangle = \langle\mathfrak{v},\mathfrak{w}\wedge\mathfrak{x}\rangle \qquad \text{for all } \mathfrak{x} \in \textstyle\bigwedge_{a-b} V^*.$$

The interior product extends to a bilinear map

$$(3.28) \qquad \llcorner: \amalg_p(\textstyle\bigwedge_{a+1} V) \otimes \amalg_p(\bigwedge_{p+1} V^*) \longrightarrow \amalg_p(\bigwedge_{a-b} V^*)$$

such that

$$\mathfrak{v}_1 \cdots \mathfrak{v}_p \llcorner \mathfrak{w}_1 \cdots \mathfrak{w}_p$$

$$= (1/p!) \sum_{\pi \in \mathfrak{S}_p} (\mathfrak{v}_1 \llcorner \mathfrak{w}_{\pi(1)}) \cdots (\mathfrak{v}_p \llcorner \mathfrak{w}_{\pi(p)}) \tag{3.29}$$

$$= (1/p!) \sum_{\pi \in \mathfrak{S}_p} (\mathfrak{v}_{\pi(1)} \llcorner \mathfrak{w}_1) \cdots (\mathfrak{v}_{\pi(p)} \llcorner \mathfrak{w}_p)$$

where $\mathfrak{v}_j \in \bigwedge_{a+1} V$ and $\mathfrak{w}_j \in \bigwedge_{b+1} V^*$ for $j = 1, \cdots, p$. If in addition $\mathfrak{v} \in \bigwedge_{a+1} V$ and $\mathfrak{w} \in \bigwedge_{b+1} V^*$, then

$$\mathfrak{v}^p \llcorner \mathfrak{w}_1 \cdots \mathfrak{w}_p = (\mathfrak{v} \llcorner \mathfrak{w}_1) \cdots (\mathfrak{v} \llcorner \mathfrak{w}_p) \tag{3.30}$$

$$\mathfrak{v}_1 \cdots \mathfrak{v}_p \llcorner \mathfrak{w}^p = (\mathfrak{v}_1 \llcorner \mathfrak{w}) \cdots (\mathfrak{v}_p \llcorner \mathfrak{w}) \tag{3.31}$$

$$\mathfrak{v}^p \llcorner \mathfrak{w}^p = (\mathfrak{v} \llcorner \mathfrak{w})^p \tag{3.32}$$

If $a = b$, then $\mathfrak{v} \llcorner \mathfrak{w} = \langle \mathfrak{v}, \mathfrak{w} \rangle$.

Let ℓ be a hermitian metric on V. Write $(\mathfrak{v}|\mathfrak{w}) = \ell(\mathfrak{v},\mathfrak{w})$ and define $\|\mathfrak{v}\| \geq 0$ by $\|\mathfrak{v}\|^2 = (\mathfrak{v}|\mathfrak{v})$. A hermitian metric ℓ_p on $\sqcup_p V$ is induced by

$$(\mathfrak{x}_1 \cdots \mathfrak{x}_p | \mathfrak{v}_1 \cdots \mathfrak{v}_p)$$

$$= (1/p!) \sum_{\pi \in \mathfrak{S}_p} (\mathfrak{x}_1 | \mathfrak{v}_{\pi(1)}) \cdots (\mathfrak{x}_p | \mathfrak{v}_{\pi(p)}) \tag{3.33}$$

$$= (1/p!) \sum_{\pi \in \mathfrak{S}_p} (\mathfrak{x}_{\pi(1)} | \mathfrak{v}_1) \cdots (\mathfrak{x}_{\pi(p)} | \mathfrak{v}_p)$$

where $\mathfrak{x}_j \in V$ and $\mathfrak{v}_j \in V$ for $j = 1, \cdots, p$. If, in addition, $\mathfrak{x} \in V$ and $\mathfrak{v} \in V$, then

$$(\mathfrak{x}^p | \mathfrak{v}_1 \cdots \mathfrak{v}_p) = (\mathfrak{x}|\mathfrak{v}_1) \cdots (\mathfrak{x}|\mathfrak{v}_p) \tag{3.34}$$

$$(\mathfrak{x}_1 \cdots \mathfrak{x}_p | \mathfrak{v}^p) = (\mathfrak{x}_1|\mathfrak{v}) \cdots (\mathfrak{x}_p|\mathfrak{v}) \tag{3.35}$$

$$(\mathfrak{x}^p | \mathfrak{v}^p) = (\mathfrak{x}|\mathfrak{v})^p \tag{3.36}$$

$$\|\mathfrak{x}^p\| = \|\mathfrak{x}\|^p \tag{3.37}$$

$$\|\mathfrak{x}_1 \cdots \mathfrak{x}_p\| \le \|\mathfrak{x}_1\| \cdots \|\mathfrak{x}_p\|. \tag{3.38}$$

If $\mathfrak{a} = (\mathfrak{a}_0, \cdots, \mathfrak{a}_n)$ is an orthonormal base of V, then $\left\{c_j \mathfrak{a}^j\right\}_{j \in S(p,n)}$ is an orthonormal base of $\amalg_p V$. The hermitian metric ℓ on V induces a hermitian metric ℓ^* on V^* such that the dual bases of orthonormal bases are orthonormal. Then $(\ell_p)^* = (\ell^*)_p$ on $(\amalg_p V)^* = \amalg_p V^*$.

A <u>hermitian</u> <u>vector</u> <u>space</u> V is a complex vector space with a hermitian metric ℓ affixed to V. Thus V^*, $\amalg_p V$, $\amalg_p V^*$, $\Lambda_p V$ and $\Lambda_p V^* = (\Lambda_p V)^*$ are naturally induced hermitian vector spaces again and the induced hermitian metrics are often denoted by ℓ again.

Now we turn our attention to the projective spaces. Take $v_j \in \mathbb{P}(\amalg_{q_j} V)$ for $j = 1,..,p$ where $q_j \in \mathbb{N}$. Define $q = q_1 + \cdots + q_p$. Then $v_j = \mathbb{P}(\mathfrak{v}_j)$ with $0 \ne \mathfrak{v}_j \in \amalg_{q_j} V$. By Theorem 3.2 $0 \ne \mathfrak{v}_1 \cdots \mathfrak{v}_p \in \amalg_q V$. Hence

$$v_1 \cdots v_p = \mathbb{P}(\mathfrak{v}_1 \cdots \mathfrak{v}_p) \in \mathbb{P}(\amalg_q V) \tag{3.39}$$

is well defined. The product is associative and commutative. If $v_1 = \cdots = v_p = v \in \mathbb{P}(\amalg_q V)$ we obtain $v^p \in \mathbb{P}(\amalg_{pq} V)$. A smooth, injective, holomorphic map $\varphi\colon \mathbb{P}(V) \longrightarrow \mathbb{P}(\amalg_p V)$, called the <u>Veronese map</u> is defined by $\varphi(v) = v^p$ for all $v \in \mathbb{P}(V)$. It embedds $\mathbb{P}(V)$ into $\mathbb{P}(\amalg_p V)$ and lifts to $\hat{\varphi}\colon V \longrightarrow \amalg_p V$ by $\hat{\varphi}(\mathfrak{v}) = \mathfrak{v}^p$ for all $\mathfrak{v} \in V$. Then $\mathbb{P} \circ \hat{\varphi} = \varphi \circ \mathbb{P}$ on V_*. Here $\hat{\varphi}$ is holomorphic but not injective (Lemma 3.3).

The Veronese map has another explanation. Some preparations are needed. Let L be a holomorphic line bundle over a complex manifold N. Let W be a finite dimensional linear subspace of the vector space $\Gamma(N,L)$ of all global holomorphic sections of L over N. Then W <u>spans</u> L if for each $x \in N$, there exists $\alpha \in W$ with $\alpha(x) \ne 0_x$. Assume that W

spans L. For each $x \in N$, the linear subspace $W_x = \{\alpha \in W | \alpha(x) = 0_x\}$ has codimension 1 in W. Hence one and only one $\gamma(x) \in \mathbb{P}(W^*)$ exists such that $E[\gamma(x)] = W_x$. The map $\gamma: N \longrightarrow \mathbb{P}(W^*)$ is called the <u>dual classification map</u>.

We will show that γ is holomorphic. Let $U \neq \phi$ be an open subset of N such that there is a holomorphic frame $\beta \in \Gamma(U,L)$ that is $Z(\beta) = \phi$. Then there is a holomorphic function $g: U \times W \longrightarrow \mathbb{C}$ such that

$$\alpha(x) = g(x,\alpha)\, \beta(x) \qquad \text{for all } (x,\alpha) \in U \times W \tag{3.40}$$

If $x \in U$ is fixed, $g(x) = g(x,\square): W \longrightarrow \mathbb{C}$ is linear. Hence $g(x) \in W^*$. The map $g: U \longrightarrow W^*$ is holomorphic with

$$g(x,\alpha) = \langle \alpha, g(x) \rangle \qquad \text{for all } (x,\alpha) \in U \times W \tag{3.41}$$

Since W spans L, $g(x) \neq 0$ with $\gamma(x) = \mathbb{P}(g(x))$ for all $x \in U$. Therefore g is a reduced representation of γ on U. The map γ is holomorphic. If γ embeds, L is called <u>very ample</u>.

Let $H = H(W^*)$ be the hyperplane section bundle on $\mathbb{P}(W^*)$. The evaluation map $\varepsilon: \mathbb{P}(W^*) \times W \longrightarrow H$ is defined as follows. Take $x \in \mathbb{P}(W^*)$ and $a \in W = W^{**}$. Hence $a: W^* \longrightarrow \mathbb{C}$ is linear and E(x) is a one-dimensional linear subspace of W^*. Thus $\varepsilon(x,a) = a | E(x) \in E(x)^* = H_x$. The bundle map ε is surjective. For fixed $a \in W$, a holomorphic section $\tilde{\varepsilon}(a) = \varepsilon(\square, a)$ of H over $\mathbb{P}(W^*)$ is defined. Here $\tilde{\varepsilon}: W \longrightarrow \Gamma(\mathbb{P}(W^*),H)$ is a linear isomorphism by which $W = \Gamma(\mathbb{P}(W^*),H)$ are identified. Then $\tilde{\varepsilon}$ become the identity and ε becomes the evaluation map. The dual classification map γ pulls back H to

$$\gamma^*(H) = \{(x,y) \in N \times H | y \in H_{\gamma(x)}\} \tag{3.42}$$

Let $\pi: \gamma^*(H) \longrightarrow N$ and $\hat{\gamma}: \gamma^*(H) \longrightarrow H$ be the projections. Then $\pi: \gamma^*(H) \longrightarrow N$ is a holomorphic line bundle over N. A holomorphic isomorphism $\rho: \gamma^*(H) \longrightarrow L$ defined by

$$(3.43)\qquad \rho(x,\varepsilon(\gamma(x),\mathfrak{a})) = \mathfrak{a}(x) \in L_x$$

for all $\mathfrak{a} \in W$ and $x \in N$. Identify $\gamma^*(H) = L$ such that ρ becomes the identity.

Return to the complex vector space V given before. Let $H = H(V)$ be the hyperplane section bundle on $\mathbb{P}(V)$. Then $\amalg_p V^* = V_{(p)} = \Gamma(\mathbb{P}(V), H^p)$ are identified and we have an evaluation map

$$(3.44)\qquad \eta: \mathbb{P}(V) \times \amalg_p V^* \longrightarrow H^p.$$

Take $x \in \mathbb{P}(V)$. Then $\alpha \in V_{[p]} = V_{(p)}$ exists such that $\alpha|E(x) \neq 0$. Thus $x \not\in Z(\alpha)$. Therefore $W = \amalg_p V^*$ spans H^p. The dual classification $\gamma: \mathbb{P}(V) \longrightarrow \mathbb{P}(\amalg_p V)$ is defined. Let $\varphi: \mathbb{P}(V) \longrightarrow \mathbb{P}(\amalg_p V)$ be the pth Veronese map. Take any $x = \mathbb{P}(\mathfrak{x}) \in \mathbb{P}(V)$. We have

$$\begin{aligned} E[\varphi(x)] &= E[x^p] \\ &= \{\mathfrak{y} \in \amalg_p V^* \mid \langle \mathfrak{x}^p, \mathfrak{y}\rangle = 0\} \\ &= \{\alpha \in V_{[p]} \mid \alpha(\mathfrak{x}) = 0\} \\ &= \{\alpha \in V_{[p]} \mid \alpha(x) = 0_x\} = E[\gamma(x)]. \end{aligned}$$

Hence $\gamma(x) = \varphi(x)$ for all $x \in \mathbb{P}(V)$. Then the Veronese map φ is the dual classification map of $\Gamma(\mathbb{P}(V), H^p)$. Since φ imbedds, H^p is very ample. Take any $y \in \mathbb{P}(\amalg_p V^*)$. Then $\ddot{E}[y]$ is a hyperplane in $\mathbb{P}(\amalg_p V)$ and $\varphi^{-1}(\ddot{E}[y])$ is define in $\mathbb{P}(V)$. Also $y = \mathbb{P}(\mathfrak{y})$ with $0 \neq \mathfrak{y} \in \amalg_p V^* = V_{[p]} = V_{(p)}$ is a holomorphic section of H^p over $\mathbb{P}(V)$ whose zero set $Z(\mathfrak{y})$ does not depend on the choice of $\mathfrak{y}$ and is denoted by $Z(y)$. Considering $\mathfrak{y} \in V_{[p]}$ as a homogeneous polynomial of degree p on V we have $Z(y) = \mathbb{P}(\mathfrak{y}^{-1}(0))$. We have

$$\begin{aligned} \varphi^{-1}(\ddot{E}[y]) &= \{x \in \mathbb{P}(V) \mid x^p \in \ddot{E}[y]\} \\ &= \mathbb{P}(\{\mathfrak{x} \in V_* \mid \langle \mathfrak{x}^p, \mathfrak{y}\rangle = 0\}) \\ &= \mathbb{P}(\{\mathfrak{x} \in V_* \mid \mathfrak{y}(\mathfrak{x}) = 0\}) \qquad (3.45) \\ &= \mathbb{P}(\mathfrak{y}^{-1}(0)) = Z(y) \end{aligned}$$

Thus we have embedded $\mathbb{P}(V)$ into $\mathbb{P}(\amalg_p V)$, but there is $\mathbb{P}(V) \subseteq \mathbb{P}(V)^{(p)} \subseteq \mathbb{P}(\amalg_p V)$.

A holomorphic map $\rho\colon \mathbb{P}(V)^p \longrightarrow \mathbb{P}(\amalg_p V)$ is defined by

$$\rho(x_1,\cdots,x_p) = x_1\cdots x_p. \tag{3.46}$$

The image set $\mathbb{P}(V)^{(p)} = \rho(\mathbb{P}(V)^p)$ is called the pth symmetric cartesian product of $\mathbb{P}(V)$. It is an irreducible, analytic subset of $\mathbb{P}(\amalg_p V)$ by Remmert's proper mapping theorem.

LEMMA 3.4 Take $x = (x_1,\cdots,x_p)$ and $y = (y_1,\cdots,y_p)$ in $\mathbb{P}(V)^p$ with $\rho(x) = \rho(y)$. Then there is a permutation $\pi\colon \mathbb{N}[1,p] \longrightarrow \mathbb{N}[1,p]$ such that $y_j = x_{\pi(j)}$ for $j = 1,\cdots,p$.

Proof. Take $\mathfrak{x}_j \in V_*$ and $\mathfrak{y}_j \in V_*$ such that $\mathbb{P}(\mathfrak{x}_j) = x_j$ and $\mathbb{P}(\mathfrak{y}_j) = y_j$ for $j = 1,\cdots,p$. Then

$$\mathbb{P}(\mathfrak{x}_1\cdots\mathfrak{x}_p) = x_1\cdots x_p = \rho(x) = \rho(y) = y_1\cdots y_p = \mathbb{P}(\mathfrak{y}_1\cdots\mathfrak{y}_p).$$

A number $\lambda \in \mathbb{C}_*$ exists such that $\mathfrak{x}_1\cdots\mathfrak{x}_p = \lambda(\mathfrak{y}_1\cdots\mathfrak{y}_p)$. For all $\mathfrak{w} \in V^*$ we have

$$\prod_{j=1}^{p} \langle\mathfrak{x}_j,\mathfrak{w}\rangle = \langle\mathfrak{x}_1\cdots\mathfrak{x}_p,\mathfrak{w}^p\rangle = \langle\lambda\mathfrak{y}_1\cdots\mathfrak{y}_p,\mathfrak{w}^p\rangle = \lambda \prod_{j=1}^{p} \langle\mathfrak{y}_j,\mathfrak{w}\rangle.$$

Hence regarding the factorization of homogeneous polynomials, there are complex numbers $\lambda_j \in \mathbb{C}_*$ and a permutation $\pi \in \mathfrak{S}_p$ such that $\langle\lambda_j\mathfrak{y}_j,\mathfrak{w}\rangle = \langle\mathfrak{x}_{\pi(j)},\mathfrak{w}\rangle$ for all $\mathfrak{w} \in V^*$. Hence $\lambda_j\mathfrak{y}_j = \mathfrak{x}_{\pi(j)}$ and $y_j = x_{\pi(j)}$; q.e.d.

Therefore $\rho\colon \mathbb{P}(V)^p \longrightarrow \mathbb{P}(V)^{(p)}$ is an open, light holomorphic map of generic sheet number $p!$ and $\dim \mathbb{P}(V)^{(p)} = n.p$. The symmetric group $\mathfrak{S}_p$ operates on $\mathbb{P}(V)^p$ by permuting the factors. Then $\mathbb{P}(V)^p/\mathfrak{S}_p$ is an irreducible, normal complex space with residual map ρ_o and $\rho\colon \mathbb{P}(V)^p \longrightarrow \mathbb{P}(V)^{(p)}$ factors into a bijective, holomorphic map

$\rho_1 : \mathbb{P}(V)^p/\mathfrak{S}_p \longrightarrow \mathbb{P}(V)^{(p)}$ such that $\rho = \rho_1 \circ \rho_0$. In fact ρ_1 is biholomorphic (Barlet [3] §2 Lemma 1). Hence $\mathbb{P}(V)^{(p)} = \mathbb{P}(V)^p/\mathfrak{S}_p$ can be identified such that ρ_1 becomes the identity and such that $\rho = \rho_0$. Trivially $x \in \mathbb{P}(V)^{(p)}$ and $y \in \mathbb{P}(V)^{(q)}$ implies $x \cdot y \in \mathbb{P}(V)^{(p+q)}$.

Let ℓ be a hermitian metric on V. Write $\ell(\mathfrak{x}, \mathfrak{y}) = (\mathfrak{x} | \mathfrak{y})$ as before. For $x \in \mathbb{P}(V)$ and $a \in \mathbb{P}(V^*)$ define $\square x, a \square$ by (2.46). Then $\square x, a \square = 0$ if and only if $x \in \ddot{E}[a]$. Take $p \in \mathbb{N}$ and $a \in \mathbb{P}(V_{[p]})$. Then $a = \mathbb{P}(\alpha)$ with $0 \neq \alpha \in V_{[p]} = V_{(p)} = \amalg_p V^*$. Let $\varphi : \mathbb{P}(V) \longrightarrow \mathbb{P}(\amalg_p V)$ be the pth Veronese map. The definition extends to

$$0 \leq \square x, a \square = \frac{|\alpha(\mathfrak{x})|}{\|\alpha\| \, \|\mathfrak{x}\|^p} = \frac{|\langle \mathfrak{x}^p, \alpha \rangle|}{\|\mathfrak{x}\|^p \, \|\alpha\|} = \square \varphi(x), a \square \leq 1. \tag{3.47}$$

The evaluation map (3.44) defines a quotient hermitian metric $\varkappa_p$ along the fibers of H^p. For $x \in \mathbb{P}(V)$ and $a = \mathbb{P}(\alpha) \in \mathbb{P}(V_{(p)})$ put $\alpha_x = \alpha | E(x) \in H^p_x$. Then

$$\square x, a \square = \|\alpha_x\|_{\varkappa_p} / \|\alpha\|. \tag{3.48}$$

Let Ω_p be the Fubini Study form on $\mathbb{P}(\amalg_p V)$ defined by ℓ_p. Define $\tau_p : \amalg_p V \longrightarrow \mathbb{R}_+$ by $\tau_p(\mathfrak{v}) = \|\mathfrak{v}\|^2$ for $\mathfrak{v} \in \amalg_p V$. Then

$$dd^c \log \tau_p = \mathbb{P}^*(\Omega_p). \tag{3.49}$$

Here $\tau_p \circ \hat{\varphi} = \tau_1^p$ with $\hat{\varphi}(\mathfrak{x}) = \mathfrak{x}^p$. Then we obtain

$$\mathbb{P}^*(\varphi^*(\Omega_p)) = \hat{\varphi}^*(\mathbb{P}^*(\Omega_p) = \hat{\varphi}^*(dd^c \log \tau_p) = dd^c \log \tau_p \circ \hat{\varphi}$$

$$= p \, dd^c \log \tau_1 = p \, \mathbb{P}^*(\Omega_1) = \mathbb{P}^*(p\Omega_1)$$

Because $\mathbb{P}^*$ is injective, we obtain

$$\varphi^*(\Omega_p) = p\Omega_1 \tag{3.50}$$

where $\varphi : \mathbb{P}(V) \longrightarrow \mathbb{P}(\amalg_p V)$ is the Veronese map $\varphi(x) = x^p$. We have

$$\int_{\mathbb{P}(V)} \Omega_1^n = 1. \tag{3.51}$$

Let $\Omega_1^* = \Omega^*$ be the Fubini Study form on $\mathbb{P}(V^*)$. Take any $x \in \mathbb{P}(V)$. By Weyl [36] page 130, (2.12) we have

$$(3.52) \qquad \int_{a \in \mathbb{P}(V^*)} \log \frac{1}{\Box x, a\Box} (\Omega^*)^n = (1/2) \sum_{\lambda=1}^{n} (1/\lambda).$$

For $x_1, \cdots, x_p$ in $\mathbb{P}(V)$ define the distance $\Box x_1 \bullet \cdots \bullet x_p \Box$ this way: Take $\mathfrak{x}_j \in V_*$ with $\mathbb{P}(\mathfrak{x}_j) = x_j$ for $j = 1, \cdots, p$. Put

$$(3.53) \qquad 0 < \Box x_1 \bullet \cdots \bullet x_p \Box = \frac{\|\mathfrak{x}_1 \cdots \mathfrak{x}_p\|}{\|\mathfrak{x}_1\| \cdots \|\mathfrak{x}_p\|} \le 1.$$

(see 3.38). By continuity

$$0 < c(n,p) = \mathrm{Min}\, \{\Box x_1 \bullet \cdots \bullet x_p \Box \mid x_j \in \mathbb{P}(V) \text{ for } j = 1, \cdots, p\}$$

Trivially $c(n,1) = 1$.

<u>LEMMA 3.5</u> $c(n,2) = 1/\sqrt{2}$ if $n \ge 2$.

<u>PROOF</u>. Take $x \in \mathbb{P}(V)$ and $y \in \mathbb{P}(V)$. Take $\mathfrak{x} \in V$ and $\mathfrak{y} \in V$ with $\|\mathfrak{x}\| = 1 = \|\mathfrak{y}\|$ and $x = \mathbb{P}(\mathfrak{x})$ and $y = \mathbb{P}(\mathfrak{y})$. Then (3.33) implies

$$2\Box x \bullet y\Box^2 = 2\|\mathfrak{x} \cdot \mathfrak{y}\|^2 = 2(\mathfrak{x} \cdot \mathfrak{y} | \mathfrak{x} \cdot \mathfrak{y}).$$

$$= (\mathfrak{x}|\mathfrak{x})\ (\mathfrak{y}|\mathfrak{y}) + (\mathfrak{x}|\mathfrak{y})\ (\mathfrak{y}|\mathfrak{x}) = 1 + |(\mathfrak{x}|\mathfrak{y})|^2 \ge 1.$$

Hence $\Box x \bullet y\Box \ge 1/\sqrt{2}$ where equality occurs if and only if $(\mathfrak{x}|\mathfrak{y}) = 0$. Therefore $c(n,2) = 1/\sqrt{2}$.

<u>LEMMA 3.6</u>. If $a \in \mathbb{P}(V^*)$ and $x_j \in \mathbb{P}(V)$ for $j = 1, \cdots, p$, then

$$(3.55) \qquad \Box x_1 \bullet \cdots \bullet x_p \Box\ \Box x_1 \cdots x_p, a^p \Box = \prod_{j=1}^{p} \Box x_j, a\Box$$

$$(3.56) \qquad \prod_{j=1}^{p} \Box x_j, a\Box \le \Box x_1 \cdots x_p \Box.$$

<u>Proof</u> Take $\alpha \in V_*^*$ and $\mathfrak{x}_j \in V_*$ with $\|\alpha\| = 1 = \|\mathfrak{x}_j\|$ and with $a = \mathbb{P}(\alpha)$ and $x_j = \mathbb{P}(\mathfrak{x}_j)$ for $j = 1, \cdots, p$. Then $\|\alpha^p\| = 1$ and

$$\square x_1 \bullet \cdots \bullet x_p \square \; \square x_1 \cdots x_p, a^p \square$$

$$= \|\mathfrak{z}_1 \cdots \mathfrak{z}_p\| \frac{|\langle \mathfrak{z}_1 \cdots \mathfrak{z}_p, \alpha^p \rangle|}{\|\mathfrak{z}_1 \cdots \mathfrak{z}_p\|}$$

$$= |\langle \mathfrak{z}_1 \cdots \mathfrak{z}_p, \alpha^p \rangle| = \prod_{j=1}^{p} |\langle \mathfrak{z}_j, \alpha_j \rangle| = \prod_{j=1}^{p} \square x_j, a \square .$$

Since $0 \le \square x_1 \cdots x_p, a^p \square \le 1$, the estimate (3.56) follows; q.e.d.

Now the constant $c(n,p)$ can be estimated below:

<u>LEMMA 3.7</u> $$\log \frac{1}{c(n,p)} \le (p/2) \sum_{\nu=1}^{n} \frac{1}{\nu}.$$

<u>Proof.</u> Take $x_j \in \mathbb{P}(V)$ for $j = 1,\cdots,p$ such that $c(n,p) = \square x_1 \bullet \cdots \bullet x_p \square$. Now (3.52) and (3.56) imply.

$$\log \frac{1}{c(n,p)} \le \sum_{j=1}^{p} \int_{a \in \mathbb{P}(V^*)} \log \frac{1}{\square x_j, a \square} \left(\Omega^*\right)^n = \frac{p}{2} \sum_{\nu=1}^{n} \frac{1}{\nu} \qquad \text{q.e.d.}$$

The degree of the projective algebraic variety $\mathbb{P}(V)^{(p)}$ in $\mathbb{P}(\sqcup_p V)$ will be computed.

<u>Theorem 3.8</u>. Let V be a complex vector space of dimension $n+1 > 1$. Take $p \in \mathbb{N}$. The algebraic variety $\mathbb{P}(V)^{(p)}$ in $\mathbb{P}(\sqcup_p V)$ has degree

$$\deg \mathbb{P}(V)^{(p)} = \frac{(np)!}{p!(n!)^p}. \tag{3.57}$$

<u>Proof</u>. Take a hermitian metric on V. Define $\tau\colon V \longrightarrow \mathbb{R}_+$ by $\tau(\mathfrak{z}) = \|\mathfrak{z}\|^2$ for all $\mathfrak{z} \in V$ and $\tau_p\colon \sqcup_p V \longrightarrow \mathbb{R}_+$ by $\tau_p(\mathfrak{z}) = \|\mathfrak{z}\|^2$ for all $\mathfrak{z} \in \sqcup_p V$. Let $\pi_j\colon \mathbb{P}(V)^p \longrightarrow \mathbb{P}(V)$ and $\tilde{\pi}_j\colon V^p \longrightarrow V$ be the projections defined by $\pi_j(x_1,\cdots,x_p) = x_j$ and $\tilde{\pi}_j(\mathfrak{z}_1,\cdots,\mathfrak{z}_p) = \mathfrak{z}_j$ for $j = 1,\cdots,p$. Define $\tilde{\tau} = \tau \circ \pi_j$. Define $\mathbb{P}\colon V_*^p \longrightarrow \mathbb{P}(V)^p$ by $\mathbb{P}(\mathfrak{z}_1,\cdots,\mathfrak{z}_p) = (\mathbb{P}(\mathfrak{z}_1),\cdots,\mathbb{P}(\mathfrak{z}_p))$. Then $\pi_j \circ \mathbb{P} = \mathbb{P} \circ \tilde{\pi}_j$. Define functions $u\colon \mathbb{P}(V)^p \longrightarrow \mathbb{R}^+$

and $v\colon \mathbb{P}(V)^{(p)} \longrightarrow \mathbb{R}_+$ by

$$u(x_1,\cdots,x_p) = \square x_1 \bullet \cdots \bullet x_p \square^2, \quad v = \log \frac{1}{u}.$$

Define ρ by (3.46) and $\tilde{\rho}\colon V^p \longrightarrow \amalg_p V$ by $\tilde{\rho}(\mathfrak{x}_1,\cdots,\mathfrak{x}_p) = \mathfrak{x}_1 \cdots \mathfrak{x}_p$. Then $\mathbb{P}\circ\tilde{\rho} = \rho\circ\mathbb{P}$. Take $\mathfrak{x}_j \in V_*$ for $j = 1,\cdots,p$. Then

$$u(\mathbb{P}(\mathfrak{x}_1,\cdots,\mathfrak{x}_p)) \prod_{j=1}^{p} \tau(\tilde{\pi}_j(\mathfrak{x}_1,\cdots,\mathfrak{x}_p))$$

$$= u(\mathbb{P}(\mathfrak{x}_1),\cdots,\mathbb{P}(\mathfrak{x}_p)) \prod_{j=1}^{p} \tau(\mathfrak{x}_j)$$

$$= \frac{\|\mathfrak{x}_1\cdots\mathfrak{x}_p\|^2}{\|\mathfrak{x}_1\|^2\cdots\|\mathfrak{x}_p\|^2} \prod_{j=1}^{p} \|\mathfrak{x}_j\|^2 = \|\mathfrak{x}_1\cdots\mathfrak{x}_p\|^2 = \tau_p(\hat{\rho}(\mathfrak{x}_1,\cdots,\mathfrak{x}_p))$$

Therefore

$$(u\circ\mathbb{P})(\tau\circ\tilde{\pi}_p)\cdots(\tau\circ\tilde{\pi}_p) = \tau_p\circ\hat{\rho}.$$

We obtain

$$\mathbb{P}^*\rho^*(\Omega_p) = \hat{\rho}^*\mathbb{P}(\Omega_p) = \hat{\rho}^*(dd^c\log \tau_p) = dd^c\log \tau_p\circ\hat{\rho}$$

$$= dd^c\log u\circ\mathbb{P} + \sum_{j=1}^{p} dd^c\log \tau\circ\tilde{\pi}_j$$

$$= \mathbb{P}^*(dd^c\log u) + \sum_{j=1}^{p} \tilde{\pi}_j^*(dd^c\log \tau)$$

$$= \mathbb{P}^*(dd^c\log u) + \sum_{j=1}^{p} \tilde{\pi}_j^*\mathbb{P}^*(\Omega)$$

$$= \mathbb{P}^*(dd^c\log u + \sum_{j=1}^{p} \pi_j^*(\Omega))$$

Because $\mathbb{P}^*$ is injective, we have

$$\rho^*(\Omega_p) = dd^c\log u + \sum_{j=1}^{p} \pi_j^*(\Omega)$$

Because $\partial\Omega = 0$ and $\bar\partial\Omega = 0$, there is a form χ of class C^∞ and bidegree $(np - 1, np - 1)$ such that

$$\rho^*(\Omega_p^{np}) = dd^c\chi + \Big[\sum_{j=1}^{p} \pi_j^*(\Omega)\Big]^{np}$$

Here $\Omega^q = 0$ if $q > n$. Hence

$$\Big[\sum_{j=1}^{p} \pi_j^*(\Omega)\Big]^{np} = \sum_{j_1+\cdots+j_p = np} \frac{(np)!}{j_1!\cdots j_p!}\, \pi_1^*(\Omega^{j_1})\wedge\cdots\wedge\pi_p^*(\Omega^{j_p})$$

$$= \frac{(np)!}{n!\cdots n!}\, \pi_1^*(\Omega^n)\wedge\cdots\wedge\pi_n^*(\Omega^n).$$

The degree of $\mathbb{P}(V)^{(p)}$ is given by

$$d = \deg \mathbb{P}(V)^{(p)} = \int_{\mathbb{P}(V)^{(p)}} (\Omega_p)^{np}$$

Because p has sheet number p!, we obtain

$$d = \frac{1}{p!} \int_{\mathbb{P}(V)^p} \rho^*(\Omega_p^{np}) = \frac{1}{p!} \int_{\mathbb{P}(V)^p} \Big[\sum_{j=1}^{p} \pi_j^*(\Omega)\Big]^{np}$$

$$= \frac{(np)!}{p!(n!)^p} \int_{\mathbb{P}(V)^p} \pi_1^*(\Omega^n)\wedge\cdots\wedge\pi_n^*(\Omega^n)$$

$$= \frac{(np)!}{p!(n!)^p} \Big[\int_{\mathbb{P}(V)} \Omega^n\Big]^p = \frac{(np)!}{p!(n!)^p} \qquad \text{q.e.d.}$$

4. ALGEBROID REDUCTION

a) The extension of meromorphic maps. Some preparations are needed. This is one of them.

PROPOSITION 4.1. Let V be a complex vector space of dimension n+1. Let M be a connected, complex manifold of dimension m. Let S be an analytic subset of M with $\dim S \le m-2$. Then any meromorphic map $f: M-S \longrightarrow \mathbb{P}(V)$ extends to a meromorphic map $f: M \longrightarrow \mathbb{P}(V)$.

<u>Proof</u>. Take $p \in S$. We have to find a representation of f on a neighborhood of p. Take $a \in \mathbb{P}(V^*)$ such that $f(M-(I_f \cup S)) \not\subseteq \ddot{E}[a]$. The divisor $\mu_{f,a}$ is defined on $M-S$ and extends to a non-negative divisor $\mu_{f,a}$ on M. An open, connected neighborhood U of p and a holomorphic function $g \not\equiv 0$ on U exist such that $\mu_{f,a}|U = \mu_g^0$. A family $\{\mathfrak{v}_\lambda\}_{\lambda\in\Lambda}$ of reduced representations $\mathfrak{v}_\lambda: U_\lambda \longrightarrow V$ with U_λ open and connected in $U-S$ exist such that $\bigcup_{\lambda\in\Lambda} U_\lambda = U-S$. For $(\lambda,\mu) \in \Lambda[1]$, a holomorphic function $v_{\lambda\mu}: U_\lambda \cap U_\mu \longrightarrow \mathbb{C}_*$ exists such that $\mathfrak{v}_\lambda = v_{\lambda\mu}\mathfrak{v}_\mu$ on $U_\lambda \cap U_\mu$. Take $\alpha \in V_*^*$ with $\mathbb{P}(\alpha) = a$. Then $\langle \mathfrak{v}_\lambda, \alpha\rangle \not\equiv 0$ and $\mu_g^0 = \mu_{f,a}|U_\lambda = \mu^0_{\langle \mathfrak{v}_\lambda,\alpha\rangle}$. A holomorphic function $h_\lambda: U_\lambda \longrightarrow \mathbb{C}_*$ exists such that $g|U_\lambda = h_\lambda\langle \mathfrak{v}_\lambda,\alpha\rangle$. Take $(\lambda,\mu) \in \Lambda[1]$. On $U_\lambda \cap U_\mu$ we have

$$h_\lambda \langle \mathfrak{v}_\lambda,\alpha\rangle = g = h_\mu\langle \mathfrak{v}_\mu,\alpha\rangle = h_\mu\, v_{\mu\lambda}\langle \mathfrak{v}_\lambda,\alpha\rangle.$$

Therefore $h_\lambda = h_\mu v_{\mu\lambda}$ and $h_\lambda\mathfrak{v}_\lambda = h_\mu v_{\mu\lambda}\mathfrak{v}_\lambda = h_\mu\mathfrak{v}_\mu$ on $U_\lambda \cap U_\mu$. One and only one holomorphic map $\mathfrak{w}: U-S \longrightarrow V$ exists such that $\mathfrak{w}|U_\lambda = h_\lambda\mathfrak{v}_\lambda$ for all $\lambda \in \Lambda$. Because $\dim S \le m-2$, the holomorphic vector function $\mathfrak{w}: U-S \longrightarrow V$ extends to a holomorphic vector function $\mathfrak{w}: U \longrightarrow V$. Take any $x \in U-(S \cup I_f)$. Then $x \in U_\lambda$ for some $\lambda \in \Lambda$ and $\mathfrak{v}_\lambda(x) \ne 0$. Thus $\mathfrak{w}(x) = h_\lambda(x)\mathfrak{v}_\lambda(x) \ne 0$. Hence $\mathbb{P}(\mathfrak{w}(x)) = \mathbb{P}(\mathfrak{v}(x)) = f(x)$. Thus $\mathfrak{w}$ is a representation of f on U. The map f is meromorphic on M; q.e.d.

b) <u>The branching divisor</u>. Let M and N be connected, complex manifolds of dimension m. Let $\mathfrak{T}(M)$ be the holomorphic tangent bundle of M and let $K(M) = \bigwedge_m \mathfrak{T}^*(M)$ be the canonical bundle of M. Similar $\mathfrak{T}(N)$ and $K(N)$ for N. Let $\pi: M \longrightarrow N$ be a holomorphic map. For $x \in M$, the differential $d\pi(x): \mathfrak{T}_x(M) \longrightarrow \mathfrak{T}_{\pi(x)}(N)$ is a linear map. The set B of all $x \in M$ such that $d\pi(x)$ is not an isomorphism is an analytic subset of M called the <u>branching set of π</u>. Here $B \ne M$ if and only rank $\pi = m$. If so, one and only one divisor $\beta \ge 0$ on M, called <u>branching divisor</u> of π, exists with this property: Let $U \ne \emptyset$ be an open,

connected subset of M and let $W \neq \phi$ be an open connected subset of N with $\pi(U) \subseteq W$. Let ζ be a holomorphic form of bidegree $(m,0)$ and without zeroes on W. Thus ζ is a holomorphic frame of $K(N)$ over W. Then $\pi^*(\zeta)|U \neq 0$ and $\pi^*(\zeta)$ is a holomorphic section of $K(M)$ over U. Then $\beta|U = \mu_{\pi^*(\zeta)}$. Obviously $\operatorname{supp} \beta = B$.

Now we make a fundamental assumption maintained for the remainder of the paper:

<u>FUNDAMENTAL ASSUMPTION</u>: M and N are connected complex manifolds of dimension m and $\pi: M \longrightarrow N$ is a proper, surjective holomorphic map. Here B is the branching set and β is the branching divisor. If $A \subseteq M$, put $A' = \pi(A)$ and $\hat{A} = \pi^{-1}(A')$. If $C \subseteq N$, define $\tilde{C} = \pi^{-1}(C)$. Then $\pi: M-\hat{B} \longrightarrow N-B'$ is a proper, surjective, locally biholomorphic map, hence a covering space of finite sheet member q. Here q is called the <u>sheet number</u> of π.

By a theorem of Remmert [15] (see also Andreotti-Stoll [2]) the <u>exceptional set</u>

$$E = \{x \in M | \dim_x \pi^{-1}(\pi(x)) > 0\} \tag{4.1}$$

is analytic with $E \subseteq B$ and $\dim E' \leq m-2$. Here $E' \subseteq B'$ and $\hat{E} \subseteq \hat{B}$. Also $\dim B' \leq m-1$ and $\dim_x B' = m-1$ if $x \in B' - E'$. If $B \neq \phi$, then B has pure dimension m-1.

<u>LEMMA 4.2</u>. $\#\tilde{z} \leq q$ for all $z \in N-E'$.

<u>Proof</u>. Take $a \in N-E'$. Then $\tilde{a} = \{b_1, \cdots, b_s\}$ with $b_j \neq b_k$ for $j \neq k$. Hence $s = \#\tilde{a}$. For each $j \in \mathbb{N}[1,s]$ there is an open, connected neighborhood Y_j of b_j in $M-\hat{E}$ such that $Y_j \cap Y_k = \phi$ if $j \neq k$. An open, connected neighborhood U of a in N-E' exists such that $\tilde{U} \subseteq Y_1 \cup \cdots \cup Y_s$, because π is proper. Then $U_j = Y_j \cap \tilde{U}$ is an open neighborhood of b_j in $M-\hat{E}$ such that $\tilde{U} = U_1 \cup \cdots \cup U_s$. The restriction $\pi: U_j \longrightarrow U$ is proper,

open and light, hence surjective. Take $z \in U-B'$. The $x_j \in U_j \cap \tilde{z}$ exists and $x_j \neq x_k$ with $j \neq k$. Hence $s = \#\{x_j | j \in \mathbb{N}[1,s]\} \subseteq \#\tilde{z} = q$;

q.e.d.

Take $x \in M-E$. The <u>mapping degree</u> $\nu_\pi(x)$ of π at x is defined as follows: Let U be an open, relative compact neighborhood of x in $M-E$ such that $\bar{U} \cap \pi^{-1}(\pi(x)) = \{x\}$, then

$$1 \leq \nu_\pi(x) = \limsup_{z \to x} \#(\bar{U} \cap \pi^{-1}(\pi(z))) < \infty \tag{4.2}$$

where the definition does not depend on U (Stoll [21]). Here $x \in M-B$ if and only if $\nu_\pi(x) = 1$.

<u>PROPOSITION 4.3</u>. Take $a \in N-E'$. Then

$$\sum_{x \in \tilde{a}} \nu_\pi(x) = q. \tag{4.3}$$

<u>Proof</u>. $\tilde{a} = \{b_1, \cdots, b_s\}$ where $b_j \neq b_k$ if $j \neq k$. For each $j \in \mathbb{N}[1,s]$ there is an open, connected, relative compact neighborhood Y_j of b_j such that $\bar{Y}_j \cap \bar{Y}_k = \phi$ if $j \neq k$ and such that $\tilde{a} \cap \bar{Y}_j = \{\bar{b}_j\}$. Define $q_j = \nu_\pi(b_j)$. An open neighborhood V_j of b_j in $\bar{Y}_j - E$ exists such that $\#(\bar{Y}_j \cap \pi^{-1}(\pi(z)) \leq q_j$ for all $z \in V_j$. An open, connected neighborhood U of a in $N-E'$ exists such that $\tilde{U} \subset V_1 \cup \cdots \cup V_s$. Define $U_j = \tilde{U} \cap V_j$. Then $\tilde{U} = U_1 \cup \cdots \cup U_s$ is a disjoint union. The restriction $\pi\colon U_j \longrightarrow U$ is proper, light and open, hence surjective. Let $\mathring{U}_j$ be a connectivity component of U_j. Then $\mathring{U}_j$ is open and closed in U_j. Hence $\pi\colon \mathring{U}_j \longrightarrow U$ is proper, light and open, hence surjective. Thus $\phi \neq \tilde{a} \cap \mathring{U}_j \subseteq \tilde{a} \cap \bar{Y}_j = \{b_j\}$. Therefore $b_j \in \mathring{U}_j$. Thus $\mathring{U}_j = U_j$ and U_j is connected. Also $\pi\colon U_j - \hat{B} \longrightarrow U-B'$ is a proper, locally biholomorphic map. Hence a covering space of finite sheet number p_j. Take any $z \in U-B'$. Then $\#\tilde{z} \cap U_j = \tilde{z} \cap (U_j - \hat{B}) = p_j$ and

$$q = \#\tilde{z} = \sum_{j=1}^{s} \#\tilde{z} \cap U_j = \sum_{j=1}^{s} p_j.$$

By Lemma 4.2 $\# \tilde{z} \cap U_j \le p_j$ for all $z \in U$. A sequence $\{z_\lambda\}_{\lambda \in \mathbb{N}}$ with $z_\lambda \in U_j$ and $z_\lambda \longrightarrow b_j$ for $\lambda \longrightarrow \infty$ exists such that $\#\pi^{-1}(\pi(z_\lambda)) \cap U_j \longrightarrow q_j$ for $\lambda \longrightarrow \infty$. Here $\pi(z_\lambda) \in U$ and $\#\pi^{-1}(\pi(z_\lambda)) \cap U_j \le p_j$. Hence $q_j \le p_j$ for $j = 1,\cdots,s$. A sequence $\{y_\lambda\}_{\lambda \in \mathbb{N}}$ with $y_\lambda \in U_j - \hat{B}$ and $y_\lambda \longrightarrow b_j$ for $\lambda \longrightarrow \infty$ exists. Then $\pi(y_\lambda) \in U-B'$ and $\#\pi^{-1}(\pi(y_\lambda)) \cap U_j = p_j$ for all λ. By (4.2) we have $q_j \ge p_j$. Hence $q_j = p_j$ for $j = 1,\cdots,s$. We obtain

$$\sum_{z \in \tilde{a}} \nu_\pi(x) = \sum_{j=1}^{s} \nu_\pi(b_j) = \sum_{j=1}^{s} q_j = \sum_{j=1}^{s} p_j = q \qquad \text{q.e.d.}$$

c) <u>The algebroid reduction of a vector function</u>. Let V be a hermitian vector space of dimension n + 1. Let $U \neq \phi$ be an open subset of N. Let $\mathfrak{v}: \tilde{U} \longrightarrow V$ be a continuous vector function. For $z \in U-E'$ <u>the algebroid reduction</u> $\mathfrak{y}: U-E' \longrightarrow \amalg_q V$ is defined by

$$\mathfrak{y}(z) = \amalg_{x \in \tilde{z}} \mathfrak{v}(x)^{\nu_\pi(x)} . \tag{4.4}$$

<u>LEMMA 4.4</u>. $\mathfrak{x}_1,\cdots,\mathfrak{x}_q$ and $\mathfrak{y}_1,\cdots,\mathfrak{y}_q$ be vectors in V. Let $\| \ \|$ be a norm on V. Take $0 \le \eta \in \mathbb{R}$ and $0 \le T \in \mathbb{R}$. Suppose that $\|\mathfrak{x}_j - \mathfrak{y}_j\| \le \eta$, $\|\mathfrak{x}_j\| \le T$, and $\|\mathfrak{y}_j\| \le T$ for all $j = 1,\cdots,q$. Then

$$\| \amalg_{j=1}^{q} \mathfrak{x}_j - \amalg_{j=1}^{q} \mathfrak{y}_j \| \le q\eta T^{q-1} \tag{4.5}$$

The proof is left to the reader. Compare Stoll [21] Lemma 3.3.

<u>PROPOSITION 4.5</u>. The algebroid reduction $\mathfrak{y}: U-E' \longrightarrow \amalg_q V$ defined by (4.4) is continuous. If $\mathfrak{v}: \tilde{U} \longrightarrow V$ is holomorphic, then $\mathfrak{y}: U-E' \longrightarrow \amalg_q V$ is holomorphic and extends to a holomorphic vector function $\mathfrak{y}: U \longrightarrow \amalg_q V$.

<u>Proof</u>. Take $a \in U-E'$. Then $\tilde{a} = \{b_1,\cdots,b_s\}$ with $b_j \neq b_k$ if $j \neq k$. An open neighborhood Y of a exists such that $\bar{Y}$ is compact and

contained in $U-E'$. Then $\pi^{-1}(\bar{Y})$ is compact in $\tilde{U}-\hat{E}$. A constant $L > 1$ exists such that $\|\mathfrak{v}\| \le L$ on $\pi^{-1}(\bar{Y})$. Take any $\varepsilon > 0$. Then there is an open neighborhood Y_j of b_j with $\bar{Y}_j \subseteq \pi^{-1}(\bar{Y})$ such that

$$(4.6) \qquad \|\mathfrak{v}(x) - \mathfrak{v}(b_j)\| < \frac{\varepsilon}{q\cdot s} L^{-q(s+1)}$$

for all $x \in Y_j$. As in the proof of Lemma 4.4 we construct open, connected neighborhoods W of a and W_j of b_j such that $W_j \subseteq Y_j$ and $W \subseteq Y_j$ and $\tilde{W} \subseteq Y$ and $\tilde{W} = W_1 \cup \cdots \cup W_s$ with $W_j \cap W_k = \phi$ for $j \neq k$. Moreover $\pi\colon W_j \longrightarrow W$ is proper, light, surjective with sheet number $q_j = \nu_\pi(b_j)$. If $z \in W$ and $\tilde{z}_j = \tilde{z} \cap W_j$, Proposition 4.3 implies

$$(4.7) \qquad q_j = \sum_{x\in\tilde{z}_j} \nu_\pi(x) \quad , \quad \sum_{j=1}^{s} q_j = q$$

Therefore Lemma 4.4 implies

$$(4.8) \qquad \| \coprod_{z\in\tilde{z}} \mathfrak{v}(x)^{\nu_\pi(x)} - \mathfrak{v}(b_j)^{q_j}\| \le q_j \frac{\varepsilon}{q\cdot s} L^{-q(s+1)} L^{q_j-1} \le \frac{\varepsilon}{s} L^{-qs}.$$

Also we have

$$(4.9) \qquad \| \coprod_{x\in\tilde{z}} \mathfrak{v}(x)^{\nu_\pi(x)} \| \le L^{q_j} \le L^{q}.$$

Again Lemma 4.4 implies

$$\|\mathfrak{y}(z) - \mathfrak{y}(a)\| = \| \coprod_{x\in\tilde{z}} \mathfrak{v}(x)^{\nu_\pi(x)} - \coprod_{x\in\tilde{z}} \mathfrak{v}(x)^{\nu_\pi(x)} \|$$

$$= \| \coprod_{j=1}^{s} \left[\coprod_{x\in\tilde{z}} \mathfrak{v}(x)^{\nu_\pi(x)} \right] - \coprod_{j=1}^{s} \mathfrak{v}(b_j)^{q_j} \|$$

$$\le s\cdot\frac{\varepsilon}{s} L^{-qs}\cdot (L^q)^{s-1} < \varepsilon$$

Therefore $\mathfrak{y}$ is continuous on $U-E'$.

Assume that $\mathfrak{v}$ is holomorphic on $\tilde{U}$. Take any point $a \in U-B'$. An open, connected neighborhood W of a in $U-B'$ exists such that $\tilde{W} = W_1 \cup \cdots \cup W_q$ is an disjoint union of open, connected set W_j such

that $\pi_j = \pi: W_j \longrightarrow W$ is biholomorphic for $j = 1,\cdots,q$. Then $\rho_j = \pi_j^{-1}: W \longrightarrow W_j$ is biholomorphic. Hence $\mathfrak{v}\circ\rho_j: W \longrightarrow V$ is holomorphic with

$$\mathfrak{y}|W = (\mathfrak{v}\circ\rho_1)\cdots(\mathfrak{v}\circ\rho_q). \tag{4.10}$$

Hence $\mathfrak{y}$ is holomorphic on W. Therefore $\mathfrak{y}: U-E' \longrightarrow \amalg_q V$ is holomorphic. Because $\dim E' \le m-2$, the vector function $\mathfrak{y}$ uniquely extends to a holomorphic vector function $\mathfrak{y}: U \longrightarrow \amalg_q V$, q.e.d. (compare Stoll [21] Proposition 3.6).

Take $\mathfrak{w} \in V$. The algebroid reduction $\mathfrak{p}$ of $\mathfrak{w}-\mathfrak{v}: \tilde{U} \longrightarrow V$ is given by

$$\mathfrak{p}(z,\mathfrak{w}) = \amalg_{x\in\tilde{z}} (\mathfrak{w}-\mathfrak{v}(x))^{\nu_\pi(x)} = \sum_{j=0}^{q-1} \mathfrak{p}_{q-j}(z)\mathfrak{w}^j + \mathfrak{w}^q, \tag{4.11}$$

for all $z \in U-E'$ and $\mathfrak{w} \in V$. For fixed $z \in U-E'$, $\mathfrak{p}$ is a polynomial in the coordinates of $\mathfrak{w}$. Hence $\mathfrak{p}(z,\square): V \longrightarrow \amalg_q V$ is holomorphic.

<u>LEMMA 4.6</u>. Take $s \in \mathbb{Z}[0,q-1]$. If $\mathfrak{v}: \tilde{U} \longrightarrow V$ is continuous, $\mathfrak{p}_{q-s}: U-E' \longrightarrow \amalg_{q-s} V$ is continuous. If $\mathfrak{v}: \tilde{U} \longrightarrow V$ is holomorphic, $\mathfrak{p}_{q-s}$ extends uniquely to a holomorphic map $\mathfrak{p}_{q-s}: U \longrightarrow \amalg_{q-s} V$.

<u>Proof</u>. If $s = 0$, then $\mathfrak{p}_q = (-1)^q\mathfrak{y}$ and the statement follows from Proposition 4.5. Assume that $1 \le s \le q-1$. Replacing $\mathfrak{w}$ by $t\mathfrak{w}$ with $t \in \mathbb{C}$ and differentiating for t implies that $\mathfrak{p}_{q-s}\cdot\mathfrak{w}^s: U-E' \longrightarrow V$ is continuous respectively holomorphic for each fixed $\mathfrak{w} \in V$. Abbreviate $q-s = p$. Let $(\mathfrak{a}_0,\cdots,\mathfrak{a}_n) = \mathfrak{a}$ be a base of V and let $\mathfrak{b} = (\mathfrak{b}_0,\ldots,\mathfrak{b}_n)$ be the dual base. We use the notation of (3.10) and (3.11). Then $\{c_j(p)\mathfrak{a}^j\}_{j\in S(p,n)}$ is a base of $\amalg_p V$ and $\{c_j(p)\mathfrak{b}^j\}_{j\in S(p,n)}$ is the dual base of $\amalg_p V^*$. Define $P_j = \langle \mathfrak{p}_{q-s}, c_j(p)\mathfrak{b}^j\rangle: U-E' \longrightarrow \mathbb{C}$ for all $j\in S(p.n)$. Thus

$$\mathfrak{p}_p = \sum_{j\in S(p,n)} P_j\, c_j(p)\mathfrak{a}^j = \sum_{j\in S(p,n)} P_j\, c_j(p)\mathfrak{a}_0^{j_0}\cdots\mathfrak{a}_n^{j_n}. \tag{4.12}$$

Take $\mathfrak{w} = \mathfrak{a}_n^s$ Then

$$\mathfrak{p}_p \, \mathfrak{a}_n^s = \sum_{j \in S(p,n)} P_j \, c_j(p) \mathfrak{a}_0^{j_0} \cdots \mathfrak{a}_{n-1}^{j_{n-1}} \mathfrak{a}_n^{j_n+s}.$$

Take any $j \in S(p,n)$. Define $k = (j_0, \cdots, j_{n-1}, j_n+s) \in S(q,n)$. Then

$$P_j = \frac{c_k(q)}{c_j(p)} \langle \mathfrak{p}_p \, \mathfrak{a}_n^s, c_k(q) \mathfrak{b}^k \rangle. \tag{4.13}$$

Hence P_j is continuous respectively holomorphic on U-E'. By (4.12) the same holds for $\mathfrak{p}_p$. If $\mathfrak{v}$ is holomorphic then P_j and $\mathfrak{p}_p$ extend trivially to U as holomorphic functions, q.e.d.

The coefficients $\mathfrak{p}_j$ are called the <u>elementary symmetric functions</u> of $\mathfrak{v}$. In particular

$$\mathfrak{p}_1(z) = - \sum_{x \in \tilde{z}} \nu_\pi(x) \, \mathfrak{v}(x) \in V \tag{4.14}$$

is continuous on U-E' if $\mathfrak{v}$ is continuous and holomorphic on U if $\mathfrak{v}$ is holomorphic on $\tilde{U}$.

Thus $\mathfrak{p}$ seems to be the direct analogon to (1.1). Clearly $\mathfrak{p}(\pi(x), \mathfrak{v}(x)) = 0$ for all $x \in \tilde{U}-\tilde{E}$. However, if $\mathfrak{v}$ is a representation of a meromorphic map, the construction ought to be invariant under multiplication of $\mathfrak{v}$ with a holomorphic function. This is not the case. Hence we introduce another construction. Assume that $\mathfrak{v}$ is holomorphic.

Take $p \in \mathbb{N}[1,n+1]$. A holomorphic vector function

$$\psi_p : U \times \bigwedge_p V^* \longrightarrow \amalg_q \bigwedge_{p-1} V^* \tag{4.15}$$

is defined for $z \in U$ and $\mathfrak{w} \in \bigwedge_p V^*$, by

$$\psi_p(z, \mathfrak{w}) = \mathfrak{w}^q \llcorner \mathfrak{y}(z). \tag{4.16}$$

If $z \in$ U-E' and $\mathfrak{w} \in \bigwedge_p V^*$, we have

$$(4.17)\qquad \psi_p(z,\mathfrak{w}) = \amalg_{x\in\tilde{z}}(\mathfrak{w} \llcorner \mathfrak{v}(x))^{\nu_\pi(x)}$$

$$(4.18)\qquad \psi_p(z,\mathfrak{w}) = \amalg_{x\in\tilde{z}} \langle\mathfrak{v}(x),\mathfrak{w}\rangle^{\nu_\pi(x)} \in \mathbb{C}.$$

Here ψ_p is called the pth <u>interior reduction polynomial</u> of $\mathfrak{v}$ for π.

Take $0 \neq e \in \bigwedge_{n+1} V^*$. The <u>dual</u> of $\mathfrak{v}$ is the holomorphic vector function

$$(4.19)\qquad \mathfrak{v}^* = e \llcorner \mathfrak{v} : \tilde{U} \longrightarrow \bigwedge_n V^*$$

<u>PROPOSITION 4.7</u>. $\psi_n(\pi,\mathfrak{v}^*) = 0$ on $\tilde{U}$.

<u>Proof</u>. Take $y \in \tilde{U}-\hat{E}$. Then $z = \pi(y) \in U-E'$ and $y \in \tilde{z}$. Hence

$$\psi_n(\pi(y),\mathfrak{v}^*(y)) = \amalg_{x\in\tilde{z}} (\mathfrak{v}^*(y) \llcorner \mathfrak{v}(x))^{\nu_\pi(x)}$$

$$= \amalg_{x\in\tilde{z}} \Big[(e \llcorner \mathfrak{v}(y)) \llcorner \mathfrak{v}(x)\Big]^{\nu_\pi(x)}$$

$$= \amalg_{x\in\tilde{z}} \Big[(e \llcorner \mathfrak{v}(y)) \wedge \mathfrak{v}(x)\Big]^{\nu_\pi(x)} = 0 \qquad \text{q.e.d.}$$

If $n = 1$, let c_0, c_1 be a base of V. Let e_0, e_1 be the dual base. Put $e = e_0 \wedge e_1$. Then

$$(4.20)\qquad \mathfrak{v} = v_0 c_0 + v_1 c_1 \quad , \quad \mathfrak{w} = w_0 e_0 + w_1 e_1$$

$$(4.21)\qquad \mathfrak{v}^* = -v_1 e_0 + v_0 e_1 \ , \ \langle\mathfrak{v},\mathfrak{w}\rangle = v_0 w_0 + v_1 w_1$$

$$(4.22)\qquad \psi_1(z,\mathfrak{w}) = \Pi_{z\in\tilde{z}}(v_0(x)w_0 + v_1(x)w_1)^{\nu_\pi(x)} = \sum_{j=0}^{q} g_j(z) w_0^{q-j} w_1^j$$

If we regard v_0, v_1 as homogeneous coordinates and put $f = -v_0/v_1$ and $w = w_1/w_0$ and $h_j = g_j/g_q$, we obtain

$$(4.23)\qquad \ddot{\psi}_1(z,w) = \frac{\psi_1(z,\mathfrak{w})}{w_0^q\, g_q(z)} = \Pi_{x\in\tilde{z}} (w-f(x))^{\nu_\pi(x)} = \sum_{j=0}^{q} h_j(z) w^j$$

which corresponds to (1.1).

d) <u>The reduction of a meromorphic map</u>. Let V be a hermitian vector space of dimension $n+1 > 1$. Let $f: M \longrightarrow \mathbb{P}(V)$ be a meromorphic map with indeterminacy I_f. Then $I'_f = \pi(I_f)$ is analytic in N with $\dim I'_f \le m-2$ since $\dim I_f \le m-2$.

THEOREM 4.8. A meromorphic map $g: N \longrightarrow \mathbb{P}(\amalg_q V)$ with $I_g \subseteq I'_f \cup E'$ is defined by

$$(4.24) \qquad g(z) = \amalg_{x \in \tilde{z}} (f(x))^{\nu_\pi(x)} \qquad \text{for all } z \in N-(I'_f \cup E').$$

If $b \in N-E'$, an open, connected neighborhood U of b exists such that there is a reduced representation $\mathfrak{v}: \tilde{U} \longrightarrow V$. If $U \neq \emptyset$ is an open, connected subset of $N-E'$, and if $\mathfrak{v}: \tilde{U} \longrightarrow V$ is a representation of f on $\tilde{U}$, then the algebroid reduction $\mathfrak{y}: U \longrightarrow \amalg_q V$ of $\mathfrak{v}$ for π is a representation of g. If $\mathfrak{v}$ is reduced, $\mathfrak{y}$ is reduced.

Proof. Take $b \in N-E'$. For each $c \in \tilde{b} = \pi^{-1}(b)$ there is an open neighborhood $W_0(c)$ of c in $M-\hat{E}$ such that $W_0(c) \cap W_0(d) = 0$ if $c \in \tilde{b}$ and $d \in \tilde{b}$ with $c \neq d$ and such that there is a reduced representation $\mathfrak{v}_c: W_0(c) \longrightarrow V$ of f. Then $W_0 = \bigcup_{c \in \tilde{b}} W_0(c)$ is an open neighborhood of $\tilde{b}$. Since π is proper, an open connected neighborhood U of b exists such that $\tilde{U} \subseteq W_0$. Then $U(c) = \tilde{U} \cap W_0(c)$ is an open neighborhood of $c \in \tilde{b}$ and $\tilde{U} = \bigcup_{c \in \tilde{b}} U(c)$ is a disjoint union. A reduced representation $\mathfrak{v}: \tilde{U} \longrightarrow V$ is defined by $\mathfrak{v}|U(c) = \mathfrak{v}_c$ for all $c \in \tilde{b}$.

Now let $U \neq \emptyset$ be any open connected subset of $N-E'$ with a representation $\mathfrak{v}: \tilde{U} \longrightarrow V$ of f. Let $\mathfrak{y}: U \longrightarrow \amalg_q V$ be the algebroid reduction of $\mathfrak{v}$ for π. Then $\mathfrak{y}$ is holomorphic and $\mathfrak{y}(z) \neq 0$ if $z \in U-\pi(\mathfrak{v}^{-1}(0))$. If so, then

$$\mathbb{P}(\mathfrak{y}(z)) = \mathbb{P}\left(\amalg_{x \in \tilde{z}} \mathfrak{v}(x)^{\nu_\pi(x)}\right) = \amalg_{x \in \tilde{z}} \mathbb{P}(\mathfrak{v}(x))^{\nu_\pi(x)} = \amalg_{x \in \tilde{z}} f(x)^{\nu_\pi(x)} = g(z).$$

Thus $\mathfrak{y}$ is a representation of g. If $\mathfrak{v}$ is reduced, then $\mathfrak{v}^{-1}(0) = I_f \cap \tilde{U}$ and $\mathfrak{y}(z) \neq 0$ for all $z \in U - I'_f$ where $\dim I'_f \leq m-2$. Hence $\mathfrak{y}$ is a reduced representation of g. We have shown, that g is holomorphic on $N-(E' \cup I'_g)$ and meromorphic on N with $I_g \subseteq E' \cup I'_f$ (Proposition 4.1); q.e.d.

From the definition we see that g actually maps into the algebraic subvariety $\mathbb{P}(V)^{(q)}$ of $\mathbb{P}(\amalg_q V)$, which corresponds to the fact that the qth associated map $f_q: M \longrightarrow \mathbb{P}(\bigwedge_{q+1} V)$ actually maps into the Grassmann manifold $G_q(V)$.

The pth interior reduction polynomial can be globalized to a section in a twisted holomorphic vector bundle and if $p = 2$ into a meromorphic map, but this line of investigation shall not be further pursued here.

If $h: N \longrightarrow \mathbb{P}(V)$ is a meromorphic map, then h lifts to a meromorphic map $f = h \circ \pi: M \longrightarrow \mathbb{P}(V)$ whose algebroid reduction is $g = h^q: N \longrightarrow \mathbb{P}(\amalg_q V)$.

There is also the question: Given any meromorphic map $g: N \longrightarrow \mathbb{P}(V)^{(q)}$ does there exist a connected complex manifold M of dimension m, a proper, surjective, holomorphic map $\pi: M \longrightarrow N$ of sheet number q and a meromorphic map $f: M \longrightarrow \mathbb{P}(V)$ such that g is the algebroid reduction of f for π. In order not to interrupt the line of investigation here we will study this lifting problem later in section 7.

e) THE REDUCTION OF THE FUBINI-STUDY FORM. As before let $f: M \longrightarrow \mathbb{P}(V)$ be a meromorphic map where V is a hermitian vector space of dimension n+1. Take $z \in N-(E' \cup I'_f)$. Enumerate

$$\tilde{z} = \{x_1, \cdots, x_q\} \tag{4.25}$$

such that

(4.26) $$\nu_\pi(x) = \#\{j \in \mathbb{N}[1,q] \mid x_j = x\}$$

for all $x \in \tilde{z}$. Such an enumeration shall be called a <u>multiplicity enumeration</u> of $\tilde{z}$. Then

(4.27) $$u_f(z) = \Box f(x_1)\bullet\cdots\bullet f(x_q)\Box$$

does not depend on the choice of the multiplicity enumeration of $\tilde{z}$. A function $u_f\colon N-(E'\cup I'_q) \longrightarrow \mathbb{R}^+$ is defined. (3.54) implies

(4.28) $$0 < c(n,q) \le u_f \le 1.$$

Recall that $\pi\colon M-\hat{B} \longrightarrow N-B'$ is a covering space of sheet number q. Take $a \in N-B'$. Then $\tilde{a} = \{b_1,\cdots,b_q\}$ with $b_j \ne b_k$ if $j \ne k$. Then there exist open connected neighborhoods U of a and U_j of b_j and reduced representations $\mathfrak{v}_j\colon U_j \longrightarrow V$ such that

(4.29) $$\tilde{U} = \pi^{-1}(U) = U_1 \cup\cdots\cup U_q \subset M-\hat{B}$$

(4.30) $$U_j \cap U_k = \emptyset \text{ if } 1 \le j < k \le q.$$

(4.31) $\pi_j := \pi|U_j\colon U_j \longrightarrow U$ is biholomorphic for $j = 1,\cdots,q$.

A reduced representation $\mathfrak{v}\colon \tilde{U} \longrightarrow V$ is defined by $\mathfrak{v}|U_j = \mathfrak{v}_j$ for $j = 1,\cdots,q$. Then $\rho_j = \pi_j^{-1}\colon U \longrightarrow U_j$ is biholomorphic for $j = 1,\cdots,q$. We obtain

(4.32) $$\mathfrak{y} = (\mathfrak{v}\circ\rho_1)\cdots(\mathfrak{v}\circ\rho_q)\colon U \longrightarrow V$$

(4.33) $$u_f = \Box(f\circ\rho_1\bullet\cdots\bullet f\circ\rho_q)\Box \qquad \text{on } U-I'_f$$

(4.34) $$u_f = \frac{\|(\mathfrak{v}\circ\rho_1)\cdots(\mathfrak{v}\circ\rho_q)\|}{\|(\mathfrak{v}\circ\rho_1)\|\cdots\|\mathfrak{v}\circ\rho_q\|} = \frac{\|\mathfrak{y}\|}{\prod_{j=1}^{q}\|\mathfrak{v}\circ\rho_j\|}$$

on $U-I'_f$. Thus u_f is of class C^∞ on $N-(B'\cup I'_f)$.

Let φ be a differential form on $\dot{M}$. The direct image $\pi_*(\varphi)$ is a form of the same class and degree on N-B′ such that

$$(4.35)\qquad \pi_*(\varphi)|U = \sum_{j=1}^{q} \rho_j^*(\varphi)$$

for any possible choice of U, U_j as in (4.29)-(4.31). If $\varphi = h$ is of degree 0, that is if h is a function, $\pi_*(h)$ extends to N-E′ by

$$(4.36)\qquad \pi_*(h)(z) = \sum_{x\in\tilde{z}} \nu_\pi(x)h(x) \qquad \text{for } z \in N-E'$$

which is the first elementary symmetric function of h. By Lemma 4.6 and (4.14), $\pi_*(h)$ is continuous on N-E′ if h is continuous on $M-\hat{E}$.

<u>THEOREM 4.9.</u> The function u_f is positive and of class C^∞ an $N-(B'\cup I_f')$. Let Ω be the Fubini-Study form on $\mathbb{P}(V)$. Let Ω_q be the Fubini-Study form on $\mathbb{P}(\amalg_q V)$. Let g be the algebraid reduction of f. Then

$$(4.37)\qquad g^*(\Omega_q) - \pi_* f^*(\Omega) = 2\, dd^c \log u_f \quad \text{on } N-(B'\cup I_f')$$

<u>Proof</u>. Take $a \in N-(B'\cup I_f')$. Use the construction of (4.29)-(4.31) with $U \subseteq N-(B'\cup I_f')$. On U we have

$$2\, dd^c \log u_f = dd^c \log \|\mathfrak{v}\|^2 - \sum_{j=1}^{q} \rho_j^*(dd^c \log \|\mathfrak{v}_j\|^2)$$

$$= g^*(\Omega_q) - \sum_{j=1}^{q} \rho_j^* f^*(\Omega) = g^*(\Omega_q) - \pi_* f^*(\Omega) \qquad \text{q.e.d.}$$

<u>THEOREM 4.10</u>. f is free for $a \in \mathbb{P}(V^*)$ if and only if the algebroid reduction of f is free for $a^q \in \mathbb{P}(\amalg_q V)$. If so, then

$$(4.38)\qquad \log \square g, a^q \square = \pi_*(\log \square f, a \square) - \log u_f$$

on $N-(E'\cup I_f \cup f^{-1}(\ddot{E}[a]'))$. Here $(f^{-1}(\ddot{E}[a]))' = g^{-1}(\ddot{E}[a^q])$.

<u>Proof</u>. Take $\alpha \in V_*^*$ with $\mathbb{P}(\alpha) = a$. Then $\mathbb{P}(\alpha^q) = a^q$. Take $z \in N-(E' \cup I_q')$. Take a multiplicity enumeration $\tilde{z} = \{x_1, \cdots, x_q\}$. An open connected neighborhood U of z in $N-(E' \cup I_f')$ and a reduced representation $\mathfrak{v}: \tilde{U} \longrightarrow V_*$ of f exist. Let $\mathfrak{y}: U \longrightarrow \amalg_q V$ be the algebroid reduction of $\mathfrak{v}$ for π. Then

$$u_f(z) \,\square g(z), a^q \square$$

$$= \square f(x_1) \bullet \cdots \bullet f(x_q) \square \; \frac{|\langle \mathfrak{y}(z), \alpha^q \rangle|}{\|\mathfrak{y}(z)\| \; \|\alpha^q\|}$$

$$= \frac{\|\mathfrak{v}(x_1) \cdots \mathfrak{v}(x_q)\|}{\|\mathfrak{v}(x_1)\| \cdots \|\mathfrak{v}(x_q)\|} \; \frac{|\langle \mathfrak{v}(x_1) \cdots \mathfrak{v}(x_q), \alpha^q \rangle|}{\|\mathfrak{v}(x_1) \cdots \mathfrak{v}(x_q)\| \; \|\alpha^q\|}$$

$$= \frac{|\langle \mathfrak{v}(x_1), \alpha \rangle \cdots \langle \mathfrak{v}(x_q), \alpha \rangle|}{\|\mathfrak{v}(x_1)\| \cdots \|\mathfrak{v}(x_q)\| \; \|\alpha\|^q}$$

$$= \square f(x_1); a \square \cdots \square f(x_q); a \square$$

Here f is free for a iff $\square f, a \square \not\equiv 0$ iff $\square g, a^q \square \not\equiv 0$ iff g is free for a^q. Take $z \in N-E' \cup I_f'$, then $z \in g^{-1}(\ddot{E}[a^q])$ iff $\square g(z), a^q \square = 0$ that is iff $x_j \in \tilde{z}$ exists such that $\square f(x_j), a \square = 0$ that is iff $z \in (f^{-1}(\ddot{E}[a]))'$. Hence $(f^{-1}(\ddot{E}[a]))' = g^{-1}(\ddot{E}[a^q])$ by taking the closure. If $z \in N-(E' \cup I_g' \cup g^{-1}(\ddot{E}[a^q])$, we obtain

$$\log \square g(z), a^q \square + \log u_f(z) = \sum_{j=1}^{q} \log \square f(x_j), a \square$$

$$= \pi_* \,(\log \square f, a \square)(z) \qquad \text{q.e.d.}$$

<u>COROLLARY 4.11.</u> u_f is continuous on $N-E' \cup I'$.

<u>Proof</u>. Take $z_0 \in N-(E' \cup I_f')$. Then $\tilde{z}_0 \cap I_f = \phi$. Then $a \in \mathbb{P}(V^*)$ exists such that $f(\tilde{z}_0) \cap \ddot{E}[a] = \phi$ since $\#f(\tilde{z}_0) \leq q$. Therefore an open neighborhood U of z_0 in $N-(E' \cup I_f')$ exists such that $\square f(x), a \square > 0$ for all $x \in \tilde{U}$ and $\square g(z), a^q \square > 0$ for all $z \in U$. Hence $\log \square g, a^q \square$ and $\pi_*(\log \square f, a \square)$ are continuous on U. Thus u_f is continuous on U, q.e.d.

If $h: N \longrightarrow \mathbb{P}(V)$ is a meromorphic map, then $f = h \circ \pi: M \longrightarrow \mathbb{P}(V)$ is a meromorphic map with $u_f \equiv 1$.

f) <u>THE ALGEBROID REDUCTION OF A DIVISOR</u> Let $\nu: M \longrightarrow \mathbb{Z}$ be a divisor. A <u>direct image divisor</u> $\pi_*(\nu): N \longrightarrow \mathbb{Z}$ will be defined.

The collection $\mathfrak{C} = (U, \tilde{U}, g, h, G, H)$ is said to be a <u>construction</u> for ν (at a if $a \in U$) if the following properties are satisfied:

1) $U \neq \phi$ is an open, connected subset of $N-E'$ and $\tilde{U} = \pi^{-1}(U)$.

2) g and h are holomorphic functions on $\tilde{U}$ and $g^{-1}(o)$ and $h^{-1}(o)$ do not contain any non-empty open subset of $\tilde{U}$.

3) $$\nu | \tilde{U} = \mu_g^0 - \mu_h^0.$$

4) $$\dim g^{-1}(0) \cap h^{-1}(0) \leq m-2.$$

5) G is the algebroid reduction of g for π and H is the reduction of h for π.

By Proposition 4.5 (with $V = \mathbb{C}$), the functions G and H are holomorphic. If $z \in U$, then

$$G(z) = \prod_{x \in \tilde{z}} g(x)^{\nu_\pi(x)} \qquad H(z) = \prod_{x \in \tilde{z}} h(x)^{\nu_\pi(x)} \tag{4.39}$$

$$G^{-1}(0) = \pi(g^{-1}(0)) \qquad H^{-1}(0) = \pi(h^{-1}(0)) \tag{4.40}$$

<u>LEMMA 4.12</u>. Let $\nu: M \longrightarrow \mathbb{Z}$ be a divisor. Take $a \in N-E'$. Then there is a construction for ν at a.

<u>Proof</u>. $\tilde{a} = \{b_1, \cdots, b_s\}$ with $b_j \neq b_k$ for $j \neq k$. For each $j \in \mathbb{N}[1,s]$ there exists an open, connected neighborhood Y_j of b_j in $M-\hat{E}$ with $Y_j \cap Y_k = \phi$ if $j \neq k$ and holomorphic functions $g_j \neq 0$ and $h_j \neq 0$ on Y_j such that

$$\dim g_j^{-1}(0) \cap h_j^{-1}(0) \leq m-2 \quad , \quad \nu | Y_j = \mu_{g_j}^0 - \mu_{h_j}^0 .$$

An open, connected neighborhood U of a in N-E′ exists such that $\tilde{U} \subseteq \bigcup_{j=1}^{s} Y_j$. Then $U_j = Y_j \cap \tilde{U}$ is an open neighborhood of b_j and $\tilde{U} = U_1 \cup \cdots \cup U_s$ is a disjoint union. Define $g: \tilde{U} \longrightarrow \mathbb{C}$ and $h: \tilde{U} \longrightarrow \mathbb{C}$ by $g|U_j = g_j|U_j$ and $h|U_j = h_j|U_j$. Then g and h are holomorphic. Let G and H be their algebroid reductions for π. Then $\mathfrak{C} = (U,\tilde{U},g,h,G,H)$ is a construction for ν at a; q.e.d.

LEMMA 4.13. Let $\nu: M \longrightarrow \mathbb{Z}$ be a divisor. Let $\mathfrak{C}_\lambda = (U_\lambda,\tilde{U}_\lambda,g_\lambda,h_\lambda,G_\lambda,H_\lambda)$ be constructions for ν for $\lambda \in \Lambda$. Then for each $(\varkappa,\lambda) \in \Lambda^2$ with $U_\varkappa \cap U_\lambda \neq \phi$ there are unique holomorphic functions $k_{\varkappa\lambda}: \tilde{U}_\varkappa \cap \tilde{U}_\lambda \longrightarrow \mathbb{C}_*$ and $\ell_{\varkappa\lambda}: \tilde{U}_\varkappa \cap \tilde{U}_\lambda \longrightarrow \mathbb{C}_*$ without zeroes such that $g_\varkappa = k_{\varkappa\lambda} g_\lambda$ and $h_\varkappa = \ell_{\varkappa\lambda} g_\lambda$ on $\tilde{U}_\varkappa \cap \tilde{U}_\lambda$. Let $K_{\varkappa\lambda}$ be the algebroid reduction of $k_{\varkappa\lambda}$ and let $L_{\varkappa\lambda}$ be the algebroid reduction of $\ell_{\varkappa\lambda}$. Then $K_{\varkappa\lambda}$ and $L_{\varkappa\lambda}$ are holomorphic and without zeroes on $U_\varkappa \cap U_\lambda$ and

(4.41) $$G_\varkappa = K_{\varkappa\lambda} G_\lambda \quad , \quad H_\varkappa = L_{\varkappa\lambda} H_\lambda \qquad \text{on } U_\varkappa \cap U_\lambda .$$

Moreover, if $(\varkappa,\lambda,\mu) \in \Lambda^3$ with $U_\varkappa \cap U_\lambda \cap U_\mu \neq \phi$ then

(4.42) $$k_{\varkappa\lambda} k_{\lambda\mu} k_{\mu\varkappa} = 1 = \ell_{\varkappa\lambda} \ell_{\lambda\mu} \ell_{\mu\varkappa} \qquad \text{on } \tilde{U}_\varkappa \cap \tilde{U}_\lambda \cap \tilde{U}_\mu$$

(4.43) $$K_{\varkappa\lambda} K_{\lambda\mu} K_{\mu\varkappa} = 1 = L_{\varkappa\lambda} L_{\lambda\mu} L_{\mu\varkappa} \qquad \text{on } U_\varkappa \cap U_\lambda \cap U_\mu .$$

PROOF. Take $(\varkappa,\lambda) \in \Lambda^2$ with $U_\varkappa \cap U_\lambda \neq \phi$. Then $\mu_{g_\varkappa} - \mu_{h_\varkappa} = \mu_{g_\lambda} - \mu_{h_\lambda}$ on $\tilde{U}_\varkappa \cap \tilde{U}_\lambda$. Define $\rho = \mu_{g_\varkappa} - \mu_{g_\lambda} = \mu_{h_\varkappa} - \mu_{h_\lambda}$ on $\tilde{U}_\varkappa \cap \tilde{U}_\lambda$. Assume that ρ is not the null divisor. Then supp $\rho \neq \phi$. Let x_0 be a simple point of supp ρ. One and only one branch R of supp ρ exists with $x_0 \in R$. By perhaps switching $\varkappa$ and λ we can assume that $\rho(x_0) > 0$. An open neighborhood W of x_0 exists such that $\rho(x) > 0$ for all $x \in R \cap W$. Hence $\mu_{g_\varkappa}(x) > \mu_{g_\lambda}(x) \geq 0$ and $\mu_{h_\varkappa}(x) > \mu_{h_\lambda}(x) \geq 0$ for all $x \in R \cap W$. Hence $R \cap W \subseteq g_\varkappa^{-1}(0) \cap h_\varkappa^{-1}(0)$ with dim $R \cap W = m-1$. This contradicts property 4) of the construc-

tion. Hence $\rho \equiv o$. Thus holomorphic functions $k_{\varkappa\lambda}$ and $\ell_{\varkappa\lambda}$ without zeroes exist uniquely on $\tilde{U}_\varkappa \cap \tilde{U}_\lambda$ such that $g_\varkappa = k_{\varkappa\lambda} q_\lambda$ and $h_\varkappa = \ell_{\varkappa\lambda} h_\lambda$ on $\tilde{U}_\varkappa \cap \tilde{U}_\lambda$. Also (4.41) follows easily.

Take $(\varkappa,\lambda,\mu) \in \Lambda^3$ with $U_\varkappa \cap U_\lambda \cap U_\mu \neq \phi$. On $\tilde{U}_\varkappa \cap \tilde{U}_\lambda \cap \tilde{U}_\mu$ we have

$$g_\varkappa = k_{\varkappa\lambda}\, g_\lambda = k_{\varkappa\lambda}\, k_{\lambda\mu}\, g_\mu = k_{\varkappa\lambda}\, k_{\lambda\mu}\, k_{\mu\varkappa}\, g_\varkappa.$$

Hence $k_{\varkappa\lambda}\, k_{\lambda\mu}\, k_{\mu\varkappa} = 1$. Similar $\ell_{\varkappa\lambda}\, \ell_{\lambda\mu}\, \ell_{\mu\varkappa} = 1$ and (4.43) follows; q.e.d.

In Lemma 4.13, we can assume that $N\text{-}E' = \bigcup_{\lambda\in\Lambda} U_\lambda$. Thus (4.41) implies

$$(4.44) \quad \mu^0_{G_\varkappa} | U_\varkappa \cap U_\lambda = \mu^0_{G_\lambda} | U_\varkappa \cap U_\lambda \qquad \mu^0_{H_\varkappa} | U_\varkappa \cap U_\lambda = \mu^0_{H_\lambda} | U_\varkappa \cap U_\lambda$$

$$(4.45) \qquad (\mu^0_{G_\varkappa} - \mu^0_{H_\varkappa}) | U_\varkappa \cap U_\lambda = (\mu^0_{G_\lambda} - \mu^0_{H_\lambda}) | U_\varkappa \cap U_\lambda$$

$$(4.46) \quad \dim \operatorname{supp} \mu^0_{G_\varkappa} \cap \operatorname{supp} \mu^0_{H_\varkappa} = \dim G_\varkappa^{-1}(0) \cap H_\varkappa^{-1}(0) \le m-2.$$

Hence there are unique divisors $\pi_*(\nu)_1 \ge 0$ and $\pi_*(\nu)_2 \ge 0$ and $\pi_*(\nu)$ on N-E' such that

$$(4.47) \qquad \pi_*(\nu)_1 | U_\varkappa = \mu^0_{G_\varkappa} \qquad \pi_*(\nu)_2 | U_\varkappa = \mu^0_{H_\varkappa}$$

$$(4.48) \qquad \pi_*(\nu) | U_\varkappa = \mu^0_{G_\varkappa} - \mu^0_{H_\varkappa}$$

$$(4.49) \qquad \pi_*(\nu) = \pi_*(\nu)_1 - \pi_*(\nu)_2$$

$$(4.50) \qquad \dim (\operatorname{supp} \pi_*(\nu)_1 \cap \operatorname{supp} \pi_*(\nu)_2) \le m-2$$

Also $\pi_*(\nu)$, $\pi_*(\nu)_1$ and $\pi_*(\nu)_2$ extend uniquely onto W such that $\pi_*(\nu)_1$ and $\pi_*(\nu)_2$ remain non-negative. Also (4.49) and (4.50) hold on N. By (2.15)-(2.17) we conclude that

$$(4.51) \qquad \pi_*(\nu)_1 = \pi_*(\nu)^+ \quad , \quad \pi_*(\nu)_2 = \pi_*(\nu)^-.$$

Also we have $\nu = \nu^+ - \nu^-$ on M with $\nu^+ = \mu^0_{g_\varkappa}$ on $\tilde{U}_\varkappa$ and $\nu^- = \nu^0_{h_\varkappa}$ on $\tilde{U}_\varkappa$ by property 4) of the construction. Thus (4.44), (4.45), (4.46), (4.47), (4.51) imply

$$(4.52) \qquad \pi_*(\nu^+) = \pi_*(\nu)^+ \qquad \pi_*(\nu^-) = \pi_*(\nu)^-.$$

Also $(-\nu)^+ = \nu^-$ and $(-\nu)^- = \nu^+$ implies

$$\pi_*(-\nu) = \pi_*(-\nu)^+ - \pi_*(-\nu)^- = \pi_*((-\nu)^+) - \pi_*((-\nu)^-)$$

$$= \pi_*(\nu^-) - \pi_*(\nu^+) = \pi_*(\nu)^- - \pi_*(\nu)^+ = - \pi_*(\nu)$$

Hence $\quad \pi_*(-\nu) = -\pi_*(\nu)$.

<u>LEMMA 4.14</u>. Let ν be a divisor on M. Let $U \neq \phi$ be an open, connected subset of N-E'. Let g and h be holomorphic functions on $\tilde{U}$ such that $g^{-1}(0)$ and $h^{-1}(0)$ do not contain non-empty open subsets of $\tilde{U}$ and such that $\nu | \tilde{U} = \mu^0_g - \mu^0_h$. Let G and H be the algebroid reduction of g respectively h for π. Then $\mu^0_G - \mu^0_H = \pi_*(\nu) | U$.

<u>Remarks</u>. The Lemma is not a tautology, because condition 4) for a construction is not satisfied.

<u>Proof</u>. Take $a \in U$. Take a construction $\mathfrak{C} = (W, \tilde{W}, g_1, h_1, G_1, H_1)$ at a with $W \subseteq U$. Hence $\tilde{W} \subseteq \tilde{U}$. Define $\rho = \mu^0_g - \mu^0_{g_1} = \mu^0_h - \mu^0_{h_1}$, on $\tilde{W}$. Assume that ρ is not non-positive. Then there exists a simple point x_0 of supp ρ with $\rho(x_0) < 0$. One and only one branch R of supp ρ with $x_0 \in R$ exists. There is an open neighborhood Y of x_0 with $Y \subseteq \tilde{W}$ such that $\rho(x) < 0$ for all $x \in R \cap Y$ and $\rho(x) = 0$ for all $x \in Y-R$. Hence $\mu^0_{g_1}(x) > \mu^0_g(x) \geq 0$ and $\mu^0_{h_1}(x) > \mu^0_h(x) \geq 0$ for all $x \in R \cap Y$. Hence $R \cap Y \subseteq g_1^{-1}(0) \cap h_1^{-1}(0)$ with $\dim R \cap Y = m-1$. This contradicts property 4) of the construction. Hence $\rho \geq 0$. Hence one and only one holomorphic function $p: \tilde{W} \longrightarrow \mathbb{C}$ exists such that $g = pg_1$. Here $\rho = \mu^0_p$. Hence $\mu^0_h = \mu^0_{h_1} + \mu^0_p = \mu^0_{ph_1}$. One and only one holomorphic function

$k\colon \tilde{W} \longrightarrow \mathbb{C}_*$ exists such that $h = kph_1$. Let K and P be the algebroid reductions of k and p respectively. Then K and $P \not\equiv 0$ are holomorphic with $G = PG_1$ and $H = KPH_1$ and K has no zeroes. Therefore

$$\mu_G - \mu_H = \mu_P + \mu_{G_1} - \mu_P - \mu_{H_1} = \mu_{G_1} - \mu_{H_1} = \pi_*(\nu)$$

on W. Thus $\mu_G - \mu_H = \pi_*(\nu)|U$; q.e.d.

<u>Proposition 4.15.</u> Let $\nu_1,\cdots,\nu_p$ be divisors on M. Then

$$\pi_*(\nu_1+\cdots+\nu_p) = \pi_*(\nu_1)+\cdots+\pi_*(\nu_p)$$

<u>Proof</u>. Take $a \in N-E'$. Let $\mathfrak{C}_j = (U_j,\tilde{U}_j,g_j,h_j,G_j,H_j)$ be a construction for ν_j at a. Let U be an open, connected neighborhood of a in $\bigcap_{j=1}^{p} U_j$. Then $\tilde{U} \subseteq \bigcap_{j=1}^{p} \tilde{U}_j$. Then $g = g_1\cdots g_p$ and $h = h_1\cdots h_p$ are holomorphic functions on $\tilde{U}$ and $g^{-1}(0)$ and $h^{-1}(0)$ do not contain interior points. Moreover

$$\mu_g^0 - \mu_h^0 = \sum_{j=1}^{p} \mu_{g_j}^0|\tilde{U} - \sum_{j=1}^{p} \mu_{h_j}^0|\tilde{U} = \sum_{j=1}^{p} (\mu_{g_j}^0-\mu_{h_j}^0)|\tilde{U} = \sum_{j=1}^{p} \nu_j|\tilde{U}$$

Let G and H be the algebroid reductions of g and h respectively. By Lemma 4.14 we have $\pi_*(\nu_1+\cdots+\nu_p)|U = \mu_G^0 - \mu_H^0$. Also $G = \prod_{j=1}^{p} G_j|U$ and $H = \prod_{j=1}^{p} H_j|U$. Hence

$$\pi_*(\nu_1+\cdots+\nu_p)|U = \mu_G^0 - \mu_H^0 = \sum_{j=1}^{p} \mu_{G_j}^0|U - \sum_{j=1}^{p} \mu_{H_j}^0|U$$

$$= \sum_{j=1}^{p} (\mu_{G_j}^0 - \mu_{H_j}^0)|U = \sum_{j=1}^{p} \pi^*(\nu_j)|U \qquad \text{q.e.d.}$$

<u>PROPOSITION 4.16</u>. Let ν_1 and ν_2 be divisors on M with $\nu_1 \le \nu_2$. Then $\pi_*(\nu_1) \le \pi_*(\nu_2)$.

<u>Proof</u>. $\rho = \nu_2 - \nu_1 \ge 0$. Hence $\rho = \rho^+$. Therefore

$$0 \le \pi^*(\rho)^+ = \pi^*(\rho^+) = \pi^*(\rho) = \pi^*(\nu_2) - \pi^*(\nu_1) \qquad \text{q.e.d.}$$

Let $\mathfrak{D}_M$ be the module of divisors on M. We have shown that $\pi_*: \mathfrak{D}_M \longrightarrow \mathfrak{D}_N$ is an increasing homomorphism.

THEOREM 4.17. Let V be a complex vector space of dimension $n+1 > 1$. Let $f: M \longrightarrow \mathbb{P}(V)$ be a meromorphic map. Let $g: N \longrightarrow \mathbb{P}(\amalg_q V)$ be the algebroid reduction of f for π. Take $a \in \mathbb{P}(V^*)$. Define $b = a^q \in \mathbb{P}(\amalg_q V^*)$. Assume that f is free for a. Hence g is free for b. Then

$$\pi_*(\mu_{f,a}) = \mu_{g,b}. \tag{4.53}$$

Proof. Take $z_0 \in N-E'$. An open, connected neighborhood U of z_0 exists in N-E′ such there is a reduced representation $\mathfrak{v}: \tilde{U} \longrightarrow V$ of f. Let $\mathfrak{y}$ be the algebroid reduction of $\mathfrak{v}$. Then $\mathfrak{y}$ is a reduced representation of g on U. If $z \in U$, then

$$\mathfrak{y}(z) = \amalg_{x \in \tilde{z}} \mathfrak{v}(x)^{\nu_\pi(x)}.$$

Take $\alpha \in V_*^*$ with $\mathbb{P}(\alpha) = a$. Then $b = a^q = \mathbb{P}(\alpha^q)$ and

$$\langle \mathfrak{y}(z), \alpha^q \rangle = \prod_{x \in \tilde{z}} \langle \mathfrak{v}(x), \alpha \rangle^{\nu_\pi(x)}.$$

Thus $(U, \tilde{U}, \langle \mathfrak{v}, \alpha \rangle, 1, \langle \mathfrak{y}, \alpha^q \rangle, 1)$ is a construction for $\mu_{f,a}$ at z_0. We have

$$\pi^*(\mu_{f,a})|U = \mu^0_{\langle \mathfrak{y}, \alpha^q \rangle} = \mu_{g,b}|U. \qquad \text{q.e.d.}$$

PROPOSITION 4.18. Let ν be a divisor on M. Take $z \in N-B'$. Then

$$\pi_*(\nu)(z) = \sum_{x \in \tilde{z}} \nu(x). \tag{4.54}$$

Proof. Take $a \in N-B'$. Let $\mathfrak{C} = (U, \tilde{U}, g, h, G, H)$ be a construction of ν for π with $a \in U \subseteq N-B'$ and where U is simply connected. Then $\tilde{U} = U_1 \cup \cdots \cup U_q$ is a disjoint union of open subset U_j such that $\pi_j := \pi: U_j \longrightarrow U$ is biholomorphic. Define $\rho_j = \pi_j^{-1}$. Then

$$G = \prod_{j=1}^{q} g\circ\rho_j \qquad H = \prod_{j=1}^{q} h\circ\rho_j$$

$$\mu_G^0 = \sum_{j=1}^{q} \mu_g^0\circ\rho_j \qquad \mu_H^0 = \sum_{j=1}^{q} \mu_h^0\circ\rho_j .$$

Define $a_j = \rho_j(a)$. Then $a_j \neq a_k$ for $j \neq k$ and $\tilde{a} = \{a_1,\cdots,a_q\}$. We obtain

$$\pi_*(\nu)(a) = \mu_G^0(a) - \mu_H^0(a) = \sum_{j=1}^{q} \mu_g^0(a_j) - \sum_{j=1}^{q} \mu_h^0(a_j) = \sum_{j=1}^{q} \nu(a_j)$$

q.e.d.

If χ is a form of bidegree (m-1,m-1) and continuous on N, if $\tilde{\chi} = \pi^*(\chi)$, if K is compact in N, if S = supp ν does not have a branch contained in $\hat{B}$, then Proposition 4.18 implies

$$\int_{\tilde{K}} \nu\, \tilde{\chi} = \int_{K} \pi_*(\nu)\, \chi \tag{4.55}$$

which is most helpful, except that the condition $\dim (S \cap \hat{B}) \leq m-2$ is too restrictive. Thus we shall prove (4.54) for all $z \in N-P$ where P is analytic with $\dim P \leq m-2$. Then (4.55) will hold for any divisor on M. However this extension of Proposition 4.18 will require considerable effort.

Let X be a complex space. $\Sigma(X)$ is the set of singular points of X and $\Re(X) = X - \Sigma(X)$ is the set of simple (regular) points. Assume that X is pure m-dimensional. Let Y be a complex space and let $\varphi: X \longrightarrow Y$ be a holomorphic map. For the properties of the rank of φ see Andreotti-Stoll [2]. Since X is pure dimensional the <u>rank</u> of φ can be defined by

$$\operatorname{rank}_x \varphi = m - \dim_x \varphi^{-1}(\varphi(x)) \qquad \text{for } x \in X \tag{4.56}$$

$$\operatorname{rank}_X \varphi = \operatorname{Max} \{\operatorname{rank}_x \varphi \mid x \in X\}. \tag{4.57}$$

For each $p \in \mathbb{N}$ the set $R_p(\varphi) = \{x \in X \mid \operatorname{rank}_x \varphi < p\}$ is analytic in X. The map φ is said to be <u>locally injective</u> at $x \in X$ if there is an open

neighborhood U of x such that $\varphi|U$ is injective. The map φ is said to be light if $\dim_x \varphi^{-1}(\varphi(x)) = 0$ for all $x \in X$.

<u>LEMMA 4.19</u>. Let X and Y be complex spaces. Assume that X is pure m-dimensional. Let $\varphi: X \longrightarrow Y$ be a holomorphic map. Then

$$T = \{x \in X|\ \varphi \text{ is not locally injective at } x\}$$

is an analytic subset of X. If φ is light, then $\dim T \leq m-1$. If φ is proper, then $\dim \varphi(T) \leq m-1$.

<u>Proof</u>. The diagonal $\Delta = \{(x,x)|x \in X\}$ is analytic in $X{\times}X$. A biholomorphic map $\lambda: X \longrightarrow \Delta$ is defined by $\lambda(x) = (x,x)$ for all $x \in X$. The set

$$Z = \{(x_1,x_2) \in X{\times}X|\varphi(x_1) = \varphi(x_2)\}$$

is analytic in $X{\times}X$. Define $A = Z-\Delta$. Then $\bar{A}$ is the union of all branches of Z not contained in Δ. Thus $\bar{A}$ is analytic. Hence $S = \Delta \cap \bar{A}$ is analytic in $X{\times}X$ and in Δ.

<u>1. Claim</u>: $\lambda(T) = S$.

<u>Proof of the 1. Claim</u> $\{b_j\}_{j\in\mathbb{N}}$ and $\{c_j\}_{j\in\mathbb{N}}$ converging to a with $b_j \neq c_j$ but $\varphi(b_j) = \varphi(c_j)$ for all $j \in \mathbb{N}$. Hence $(b_j,c_j) \in Z-\Delta$ with $(b_j,c_j) \longrightarrow (a,a)$ for $j \longrightarrow \infty$. Thus $\lambda(a) = (a,a) \in S$.Now take $a \in X$ with $(a,a) \in S$. A sequence $\{(b_j,c_j)\}_{j\in\mathbb{N}}$ with $(b_j,c_j) \in A$ converges to a. Hence $b_j \neq c_j$ and $\varphi(b_j) = \varphi(c_j)$ and $b_j \longrightarrow a$ and $c_j \longrightarrow a$ for $j \longrightarrow \infty$. Hence φ is not locally injective at a and $a \in T$ and $\lambda(a) = (a,a) \in \lambda(T)$. Thus $\lambda(T) = S$. The 1. Claim is proved.

Because $\lambda: X \longrightarrow \Delta$ is biholomorphic, T is analytic in X.

<u>2. Claim</u>. If φ is light, then $\dim T \leq m-1$.

<u>Proof of the 2. Claim</u>. Assume that $\dim T = m$. Then there is a branch R_0 of X with $R_0 \subseteq T$. Then $R = \lambda(R_0) \subseteq S$ and R is a branch of Δ. The intersection B of R with all other branches of Δ is thin

analytic in R. Hence R-B is open in Δ, and open and dense in R. Because φ is light, R is a branch of Z by Lemma 1.28 of Andreotti-Stoll [2]. The intersection C of R with all the other branches of Z is analytic and thin in R. Hence R-C is open in Z and in R and dense in R. Thus $R-(B\cup C)$ is open and dense in R. Take (x_0,x_0) in $R-(B\cup C)$. Then $(x_0,x_0) \in S$ and R-C is an open neighborhood of (x_0,x_0) in the space Z. Hence $(x_1,x_2) \in (R-C)\cap(Z-\Delta) \subseteq \Delta\cap(Z-\Delta) = \phi$ exists, which is impossible. Hence $\dim T \le m-1$. The 2. Claim is proved.

<u>3. Claim</u>. If φ is proper, $\dim \varphi(T) \le m-1$.

<u>Proof of the 3. Claim</u>. By the Remmert proper mapping theorem, $\varphi(T)$ is analytic in Y. Let $\mathfrak{B}$ be the set branches of X. Take any $R \in \mathfrak{B}$. Then $R\cap T$ is analytic in X and $\varphi(R\cap T)$ is analytic in Y. We claim $\dim \varphi(R\cap T) \le m-1$. If $R \not\subseteq T$, then $\dim R\cap T \le m-1$. Hence $\dim \varphi(R\cap T) \le m-1$. Assume $R\subseteq T$. The set

$$P = \{x \in R \mid \operatorname{rank}_x(\varphi|R) < m\}$$

is analytic in R. Assume that $R_1 = R - P \ne \phi$. Define $\mathfrak{B}_1 = \mathfrak{B} - \{R\}$. Then

$$Q = R\cap \bigcup_{L\in\mathfrak{B}_1} L$$

is analytic in X and analytic and thin in R. Here R-Q is open in X and open in R and dense in R. Then $R_2 = R-(P\cup Q)$ is open and dense in R, open in X and R_2 is an irreducible complex space of dimension m. By the definition of P the map $\varphi: R_2 \longrightarrow Y$ is light. The 2. Claim provides us with a thin analytic subset T_2 of R_2 such that $\varphi: R_2-T_2 \longrightarrow Y$ is locally injective at every point of R_2-T_2. Since R_2-T_2 is open in X, we see that $(R_2-T_2)\cap T = \phi$. But $R_2-T_2 \subseteq R_2 \subseteq R \subseteq T$. Hence $R_2-T_2 = \phi$. Thus $R_2 = T_2$, while T_2 is thin analytic in R_2. We have a contradiction. We conclude that $R_1 = \phi$ and $R = P$. Hence $\operatorname{rank}(\varphi|R) \le m-1$ and $\dim\varphi(R\cap T) = \dim\varphi(R) \le m-1$. Because $\mathfrak{B}$ is locally finite, $\mathfrak{B}$ is at most countable and

$$\varphi(T) = \bigcup_{R \in \mathfrak{B}} \varphi(R \cap T)$$

has at most dimension m-1; q.e.d.

It may be helpful to recapitulate the present set up:

(S1) M and N are connected complex manifolds of dimension m.

(S2) $\pi\colon M \longrightarrow N$ is a surjective, proper holomorphic map of sheet number q.

(S3) $E = \{x \in N \mid \mathrm{rank}_x \pi \leq m-1\}$ is the exceptional set, which is analytic in M with $\dim E \leq m-1$. Here $E' = \pi(E)$ is analytic in N with $\dim E' \leq m-2$. Also $\hat{E} = \pi^{-1}(E')$ is analytic with $\dim \hat{E} \leq m-1$ and $E \subseteq \hat{E}$. If $x \in \hat{E}-E$, then $\dim_x \hat{E} \leq m-2$.

(S4) β is the branching divisor of π and $B = \mathrm{supp}\, \beta$ is the branching set of π. Either $B = \phi$ or B is pure (m-1)-dimensional. Also $E \subseteq B$. The set $B' = \pi(B)$ is analytic in N with $\dim B' \leq m-1$ and $E' \subseteq B'$. If $x \in B'-E'$, then $\dim_x B' = m-1$. Also $\hat{B} = \pi^{-1}(B')$ is analytic in M with $\hat{E} \subseteq \hat{B}$ and $\dim \hat{B} \leq m-1$. If $x \in \hat{B}-\hat{E}$, then $\dim_x \hat{B} = m-1$.

(S5) The map $\pi\colon M-E \longrightarrow N$ is light and open. The map $\pi\colon M-\hat{E} \longrightarrow N-E'$ is proper, light, open, surjective. The map $\pi\colon M-\hat{B} \longrightarrow N-B'$ is a q-sheeted covering space and locally biholomorphic.

(S6) $B = \{x \in M \mid \pi$ is not locally biholomorphic at $x\}$

$B = \{x \in M \mid \pi$ is not locally injective at $x\}$.

Now, we introduce additional features.

(S7) By Lemma 4.19

(4.58) $$T = \{x \in B \mid \pi|B \text{ is not locally injective at } x\}$$

is an analytic subset of B and $T' = \pi(T)$ is analytic in N with

$T' \subseteq B'$ and $\dim T' \le m-2$. Also $E \subseteq T$ and $E' \subseteq T'$. If $x \in T-E$, then $\dim_x(T-E) \le m-2$. The set $\hat{T} = \pi^{-1}(T')$ is analytic in M with $\hat{E} \subseteq \hat{T} \subseteq \hat{B}$ and $T \subseteq \hat{T}$. Also $\dim \hat{T} \le m-1$. If $x \in \hat{T}-E$, then $\dim_x \hat{T} \le m-2$.

(S8) Let $\nu : M \longrightarrow \mathbb{Z}$ be a divisor. Assume that ν is not the null divisor. Then $P = \operatorname{supp} \nu$ is a pure $(m-1)$-dimensional analytic subset of M. Also $P' = \pi(P)$ is analytic in N with $\dim P' \le m-1$. If $x \in P' - E'$ then $\dim_x P' = m-1$. Also $\hat{P} = \pi^{-1}(P')$ is analytic in M with $P \subseteq \hat{P}$ and $\dim \hat{P} \le m-1$. If $x \in \hat{P}-\hat{E}$, then $\dim_x \hat{P} = m-1$.

(S9) The analytic set $Q = P \cup B$ is pure $(m-1)$-dimensional in M and $Q' = \pi(Q) = P' \cup B'$ is analytic in N with $\dim Q' \le m-1$. If $x \in Q'-E'$, then $\dim_x Q' = m-1$. The set $\hat{Q} = \pi^{-1}(Q') = \hat{P} \cup \hat{B}$ is analytic in M with $\dim \hat{Q} \le m-1$ and $Q \subseteq \hat{Q}$. If $x \in \hat{Q}-\hat{E}$, then $\dim_x \hat{Q} = m-1$.

(S10) The set

$$S = E' \cup T' \cup \Sigma(Q') \cup \pi(\Sigma(\hat{Q})) \tag{4.59}$$

is analytic with $\dim S \le m-2$. Also $\tilde{S} = \pi^{-1}(S)$ is analytic with $\dim \tilde{S} \le m-1$ and $\dim(\tilde{S}-\hat{E}) \le m-2$. The map $\pi : M-\tilde{S} \longrightarrow N-S$ is proper, light, open, surjective and holomorphic. (In (4.59), E' is mentioned for psychological purposes only since $E' \subseteq T'$).

<u>LEMMA 4.20</u>. Assume (S1)-(S10) holds. Take $a \in N-S$. Let Y be an open neighborhood of a in N-S. Enumerate $\tilde{a} = \pi^{-1}(a) = \{b_1, \cdots, b_s\}$ with $b_j \neq b_k$ for $j \neq k$. Define $q_j = \nu_\pi(b_j) \in \mathbb{N}$. Then there is a disjoint union

$$\mathbb{N}[1,s] = I_1 \cup I_2 \cup I_3 \cup I_4 \tag{4.60}$$

such that

$$b_j \in P \cap B \qquad q_j > 1 \qquad \text{if } j \in I_1 \tag{4.61}$$

(4.62) $b_j \in B-P \qquad q_j > 1 \qquad \text{if} \quad j \in I_2$

(4.63) $b_j \in P-B \qquad q_j = 1 \qquad \text{if} \quad j \in I_3$

(4.64) $b_j \in M-(P \cup B) \qquad q_j = 1 \qquad \text{if} \quad j \in I_4$

There is an open, connected neighborhood U of a in Y and there are open, connected neighborhoods U_j of b_j in $M-\hat{S}$ such that we have these properties:

(A1) $\hat{U} = \pi^{-1}(U) = U_1 \cup \cdots \cup U_s$ is a disjoint union.

(A2) $\pi: U_j \longrightarrow U$ is proper, light, open and surjective for $j = 1,\cdots,s$.

(A3) $B \cap U_j = \phi$ if $j \in I_3 \cup I_4$ and $P \cap U_j = \phi$ if $j = I_2 \cup I_4$.

(A4) Let $D = \{z \in \mathbb{C} \mid |z| < 1\}$ be the unit disc. Then there are biholomorphic maps $\varphi_j: U_j \longrightarrow D^m$ and $\alpha: U \longrightarrow D^m$ and holomorphic maps $\psi_j: D^m \longrightarrow D^m$ such that $\psi_j o \varphi_j = \alpha o \pi | U_j$, such that $\varphi_j(b_j) = 0$ and $\alpha(a) = 0$ where

(4.65) $$\psi_j(x_1,\cdots,x_m) = (x_1,\cdots,x_{m-1},x_m^{q_j})$$

for all $(x_1,\cdots,x_m) \in D^m$ and $j \in \mathbb{N}[1,s]$.

(A5) Abbreviate $W = D^{m-1} \times \{0\}$. Then

$\alpha(U \cap Q') = W = \varphi_j(\hat{Q} \cap U_j) \qquad \text{if } j \in \mathbb{N}[1,s]$

$\varphi_j(B \cap P \cap U_j) = \varphi_j(Q \cap U_j) = W \qquad \text{if } j \in I_1$

$\varphi_j((B-P) \cap U_j) = \varphi_j(Q \cap U_j) = W \qquad \text{if } j \in I_2$

$\varphi_j((P-B) \cap U_j) = \varphi_j(Q \cap U_j) = W \qquad \text{if } j \in I_3$

$\varphi_j(Q \cap U_j) = \phi \qquad \text{if } j \in I_4$

(A6) There are holomorphic functions $g_j \not\equiv 0$ and $h_j \not\equiv 0$ on U_j such that

$$\dim g_j^{-1}(0) \cap h_j^{-1}(0) \le m-2 \quad \nu|U_j = \mu_{g_j}^0 - \mu_{h_j}^0 \quad \text{for } j = 1,\cdots,s.$$

<u>Remark</u>. Holomorphic functions $g: \tilde{U} \longrightarrow \mathbb{C}$ and $h: \tilde{U} \longrightarrow \mathbb{C}$ are defined by $g|U_j = g_j$ and $h|U_j = h_j$. Then $\dim g^{-1}(0) \cap h^{-1}(0) \le m-2$ and $\nu|\tilde{U} = \mu_g^0 - \mu_h^0$. Let G and H be the algebroid reductions of G and H respectively, then $\mathfrak{C} = (U,\tilde{U},g,h,G,H)$ is a construction for ν which we shall call a <u>winding construction of ν for π at a</u>.

<u>Proof</u>. The statement is trivial if $a \in N-Q'$. Thus we assume that $a \in Q'$. Then a is a simple point of Q' since $a \in Q'-S$. Therefore there exists an open, connected neighborhood U^0 of a in $N-S$ such that $U^0 \cap Q'$ is a connected, complex manifold of dimension $m-1$ and such that $U^0 \cap B' = \phi$ if $a \in N-B'$ and such that $U^0 \cap P' = \phi$ if $a \in N-B'$ and $U^0 \cap B' = U^0 \cap P' = U^0 \cap Q'$ if $a \in B' \cap P'$. Also we can take U^0 such that there is open, connected neighborhood W^0 of $0 \in \mathbb{C}^m$ and a biholomorphic map

$$\alpha = (\alpha_1,\cdots,\alpha_m): U^0 \longrightarrow W^0 \qquad \text{with } \alpha(a) = 0$$

$$\alpha_m^{-1}(0) = \{z \in U^0 | \alpha_m(z) = 0\} = U^0 \cap Q'.$$

Define I_1, I_2, I_3 and I_4 by (4.61)-(4.64). Then the disjoint union (4.60) holds. For each $j \in \mathbb{N}[1,s]$ we pick an open, connected, simply connected neighborhood Y_j of b_j in $\tilde{U}^0$ such that we have the following properties:

(B1) $Y_j \cap Y_k = \phi$ if $1 \le j < k \le s$

(B2) $Y_j \cap \hat{Q}$ is a connected, complex manifold of dimension $m-1$ (possible since $b_j \in \hat{Q}-\Sigma(\hat{Q})$) for all $j \in \mathbb{N}[1,s]$.

(B3) If $j \in I_1 \cup I_2$, then $\pi|(Y_j \cap B)$ is injective (possible since $b_j \in B-T$). If $j \in I_3 \cup I_4$, then $\pi|Y_j$ is injective (possible since $b_j \in M-B$).

(B4) If $j \in I_1$, then $Y_j \cap \hat{Q} = Y_j \cap Q = Y_j \cap P = Y_j \cap B$ (possible since $b_j \in P \cap B$ and $b_j \in \hat{Q}-\Sigma(\hat{Q})$). If $j \in I_2$, then $Y_j \cap P = \phi$ and $Y_j \cap \hat{Q} = Y_j \cap B$. (possible since $b_j \in B - P \cup \Sigma(\hat{Q})$). If $j \in I_3$, then $Y_j \cap B = \phi$ and $Y_j \cap \hat{Q} = Y_j \cap B$ (possible since $b_j \in P - B \cup \Sigma(\hat{Q})$. If $j \in I_4$, then $Y_j \cap P = Y_j \cap B = \phi$ (possible since $b_j \in M-(B \cup P)$).

(B5) Holomorphic functions $g_j \not\equiv 0$ and $h_j \not\equiv 0$ exist on Y_j such that $\nu | Y_j = \mu^0_{g_j} - \mu^0_{h_j}$ and $\dim (g_j^{-1}(0) \cap h_j^{-1}(0)) \leq m-2$ for $j = 1,\cdots,s$.

(B6) A holomorphic function Δ_j exists on Y_j such that $\Delta_j^{-1}(0) = Y_j \cap \hat{Q}_j$ and such that $d\Delta_j(x) \neq 0$ for all $x \in Y_j$ for $j = 1,\cdots,s$.

For $n = 1,\cdots,m$ holomorphic functions $\pi_n = \alpha_n \circ \pi$ are defined on $\tilde{U}^0$. Take $x \in Y_j \cap \hat{Q}$. Then $\pi(x) \in U^0 \cap Q'$. Hence $\pi_m(x) = \alpha_m(\pi(x)) = 0$. Take $x \in Y_j$ with $\pi_m(x) = 0$. Then $\alpha_m(\pi(x)) = \pi_m(x) = 0$. Hence $\pi(x) \in U^0 \cap Q'$ and $x \in \hat{Q} \cap Y_j$. Thus $\pi_m^{-1}(0) \cap Y_j = \hat{Q} \cap Y_j$. By (B2) and (B6) there exists a holomorphic function $f_j\colon Y_j \longrightarrow \mathbb{C}_*$ and an integer $p_j \in \mathbb{N}$ such that $\pi_m = f_j \Delta_j^{p_j}$ on Y_j. Since Y_j is simply connected, there is a holomorphic function $\chi_j\colon Y_j \longrightarrow \mathbb{C}_*$ such that $f_j = \chi_j^{p_j}$. Then $\varphi_{m,j} = \chi_j \Delta_j$ is a holomorphic function on Y_j such that

$$\pi_m = (\varphi_{m,j})^{p_j} \quad \text{on } Y_j, \quad (\varphi_{m,j})^{-1}(0) = Y_j \cap \hat{Q}.$$

Take any $x \in Y_j$. If $x \in Y_j \cap \hat{Q}$, then $d\varphi_{m,j}(x) = \chi_j(x) d\Delta_j(x) \neq 0$. If $x \in Y_j - \hat{Q}$, then $x \in M-B$ and $d\pi_1(x) \wedge \cdots \wedge d\pi_m(x) \neq 0$. Hence

$$0 \neq d\pi_m(x) = p_j\, \varphi_{m,j}(x)^{p_j - 1}\, d\varphi_{m,j}(x).$$

Therefore $d\varphi_{m,j}(x) \neq 0$. Thus $d\varphi_{m,j}(x) \neq 0$ for all $x \in Y_j$. Define $\varphi_{n,j} = \pi_n | Y_j$ for $n = 1,\cdots,m-1$. We obtain a holomorphic map

$$\varphi_j = (\varphi_{1j},\cdots,\varphi_{m,j})\colon Y_j \longrightarrow \mathbb{C}^m.$$

<u>1. CLAIM</u>: The map $\varphi_j\colon Y_j \longrightarrow \mathbb{C}^m$ is locally biholomorphic for $j \in \mathbb{N}[1,s]$.

<u>PROOF OF THE 1. CLAIM</u>: Take $x \in Y_j - \hat{Q}$. Then

$$0 \neq d\pi_1(x)\wedge\cdots\wedge d\pi_m(x) = p_j\varphi_{m,j}(x)^{p_j-1}\, d\varphi_{1,j}(x)\wedge\cdots\wedge d\varphi_{m,j}(x).$$

Hence $d\varphi_{1,j}(x)\wedge\cdots\wedge d\varphi_{m,j}(x) \neq 0$. Thus φ_j is locally biholomorphic at x.

Take $x \in Y_j \cap \hat{Q}$. By (B3) and (B4), the map $\pi\colon Y_j \cap \hat{Q} \longrightarrow U^0 \cap Q'$ is injective where $Y_j \cap \hat{Q}$ and $U^0 \cap Q$ are (m-1)-dimensional, complex manifolds. Also $\alpha\colon U^0 \cap Q' \longrightarrow \mathbb{C}^{m-1}x\{0\}$ is injective. Hence the map

$$(\pi_1,\cdots,\pi_{m-1})\colon Y_j \cap \hat{Q} \longrightarrow \mathbb{C}^{m-1}.$$

is injective, where $\varphi_{n,j} = \pi_n | Y_j$ for $j = 1,\cdots,m-1$. Let $\xi_1,\ldots,\xi_m$ be a base of the holomorphic tangent space $\mathfrak{T}_x(M)$ such that $\xi_1,\cdots,\xi_{m-1}$ is a base of $\mathfrak{T}_x(Y_j \cap \hat{Q})$ with

$$d\varphi_{n,j}(x,\xi_k) = d\pi_n(x,\xi_k) = \begin{cases} 1 \text{ if } n = k \\ 0 \text{ if } n \neq k \end{cases}$$

where $n \in \mathbb{N}[1,m-1]$ and $k \in \mathbb{N}[1,m-1]$. Now $\varphi_{m,j}|(Y_j \cap \hat{Q}) = 0$ implies $d\varphi_{m,j}(x,\xi_k) = 0$ for all $k \in \mathbb{N}[1,m-1]$. Because $d\varphi_{m,j}(x) \neq 0$, we obtain $d\varphi_{m.j}(x,\xi_m) \neq 0$. Therefore

$$(d\varphi_{1,j}(x)\wedge\cdots\wedge d\varphi_{m,j}(x))(\xi_1\wedge\cdots\wedge\xi_m) = d\varphi_m(x,\xi_m) \neq 0.$$

Hence φ_j is locally biholomorphic at x. The 1. Claim is proved.

For $r > 0$ define $\mathbb{C}(r) = \{z \in \mathbb{C} \mid |z| < r\}$. Abbreviate $D = \mathbb{C}(1)$. For each $j \in \mathbb{N}[1,s]$ there is an open, connected neighborhood Y_j^* of b_j in Y_j such that $\varphi_j\colon Y_j^* \longrightarrow \mathbb{C}^m$ is injective. Because $\pi\colon M \longrightarrow N$ is proper, there is a number $r > 0$ such that $(\mathbb{C}(r))^m \subset W_0$ and such that $U = \alpha^{-1}((\mathbb{C}(r))^m)$ is an open, connected neighborhood of a with $\tilde{U} \subseteq Y_1^* \cup \cdots \cup Y_s^*$. By multiplying α with $1/r$, we can assume that

$r = 1$. Then $\alpha: U \longrightarrow D^m$ is a biholomorphic map with $\alpha(a) = 0$ and with $\alpha(U \cap Q') = D^{m-1} x \{0\} = W$. Also $U_j = Y_j^* \cap \tilde{U}$ is an open neighborhood of b_j such that $\tilde{U} = U_1 \cup \cdots \cup U_s$ is a disjoint union. Because $\varphi_j: Y_j^* \longrightarrow \mathbb{C}^m$ is holomorphic and injective, $Z_j = \varphi_j(U_j)$ is open in $\mathbb{C}^m$ and $\varphi_j: U_j \longrightarrow Z_j$ is biholomorphic.

Because $\pi: M \longrightarrow N$ is proper and because $\tilde{U} = \pi^{-1}(U)$, the map $\pi: \tilde{U} \longrightarrow U$ is proper. Because $\tilde{U} = U_1 \cup \cdots \cup U_s$ is a disjoint union of open sets, the sets U_j are closed in the topology of $\tilde{U}$. Therefore $\pi: U_j \longrightarrow U$ is proper. The map $\pi: U_j \longrightarrow U$ is proper, light, open and holomorphic, where U_j and U are pure m-dimensional complex manifolds and where U is connected. Hence $\pi(U_j) = U$. Let U_j^* be a connectivity component of U_j. Then U_j^* is open and closed in U_j. Thus $\pi: U_j^* \longrightarrow U$ is light, open, proper and holomorphic. Hence $\pi(U_j^*) = U$. Thus $\emptyset \neq \pi^{-1}(a) \cap U_j^* \subseteq \pi^{-1}(a) \cap U_j = \{b_j\}$. We obtain $b_j \in U_j^*$. Thus there is only one connectivity component of U_j and $U_j = U_j^*$ is connected.

<u>2. CLAIM</u>: Take $j \in \mathbb{N}[1,s]$. Then $Z_j = D^m$. Define $\psi_j = \alpha \circ \pi \circ \varphi_j^{-1}$. For $x = (x_1, \cdots, x_m) \in D^m$ the holomorphic map $\psi_j: D^m \longrightarrow D^m$ is given by

$$\psi_j(x_1, \cdots, x_m) = (x_1, \cdots, x_{m-1}, x_m^{p_j}).$$

<u>PROOF OF THE 2. CLAIM</u>: The map $\psi_j = \alpha \circ \pi \circ \varphi_j^{-1}: Z_j \longrightarrow D^m$ is holomorphic. Take $x = (x_1, \cdots, x_m) \in Z_j$. Then $y = \varphi_j^{-1}(x) \in U_j$ and $\varphi_j(y) = x$, which implies $\pi_n(y) = \varphi_{n,j}(y) = x_n$ for $n = 1, \cdots, m-1$ and

$$\pi_m(y) = \varphi_{m,j}(y)^{p_j} = x_m^{p_j}$$

$$\psi_j(x) = \alpha(\pi(y)) = (\pi_1(y), \cdots, \pi_m(y)) = (x_1, \cdots, x_{m-1}, x_m^{p_j}) \in D^m.$$

Hence $|x_n| < 1$ for all $n \in \mathbb{N}[1,m]$. Therefore $Z_j \subseteq D^m$. Take $v \in D^{m-1}$. Then

$$Z_j(v) = \{x_m \in \mathbb{C} \mid (v, x_m) \in Z_j\}$$

is an open subset of D such that

$$Z_j = \{(v,x_m) \mid x_m \in Z_j(v) \text{ and } v \in D^{m-1}\}.$$

Take any $v \in D^{m-1}$ and $z \in D$. Because $\psi_j = \alpha\circ\pi\circ\varphi_j^{-1}\colon Z_j \longrightarrow D^m$ is surjective, there exists $x = (x_1,\cdots,x_m) \in \mathcal{Z}_j$ with $\psi_j(x) = (v,z)$. Therefore $(x_1,\cdots,x_{m-1}) = v$ and $x_m^{p_j} = z$. Thus $x_m \in Z_j(v)$. Hence $Z_j(v) \neq \phi$. Take $\zeta \in \partial Z_j(v)$. A sequence $\{z_\lambda\}_{\lambda\in\mathbb{N}}$ exists with $z_\lambda \in Z_j(v)$ and $z_\lambda \longrightarrow \zeta$ for $\lambda \longrightarrow \infty$. Because $\pi\colon U_j \longrightarrow U$ is proper, light, surjective also $\psi_j = \alpha\circ\pi\circ\varphi_j^{-1}\colon Z_j \longrightarrow D^m$ is proper. Hence

$$(v,\zeta^{p_j}) = \lim_{\lambda\to\infty} \psi_j(v,z_\lambda) \in \partial D^m.$$

Since $v \in D^{m-1}$ we obtain $\zeta^{p_j} \in \partial D$. Thus $|\zeta| = 1$. Hence $\partial Z_j(v) \subseteq \partial D$. Thus $Z_j(v) \neq \phi$ is open and closed in D. Therefore $Z_j(v) = D$ for all $v \in D^{m-1}$. Consequently $Z_j = D^m$. The 2. Claim is proved.

We have proved (A1)-(A4) except for $p_j = q_j$ and (A6) is a consequence of (B6). Take $y \in U_j$. If $y \in \hat{Q}\cap Y_j$, then $\pi_m(y) = 0$. Hence $\varphi_{m,j}(y)^{p_j} = \pi_m(y) = 0$. Thus $\varphi_{m,j}(y) = 0$ and $\varphi_j(y) \in D^{m-1}x\{0\} = W$. If $\varphi_j(y) \in W$, then $\varphi_{m,j}(y) = 0$. Hence $\pi_m(y) = 0$ and $y \in \hat{Q}\cap Y_j$. Thus $\varphi_j(U_j\cap\hat{Q}) = W = \alpha(U\cap Q')$. Also (B4) implies

$$U_j\cap\hat{Q} = U_j\cap Q = U_j\cap P = U_j\cap B \qquad \text{if } j \in I_1$$

$$U_j\cap\hat{Q} = U_j\cap Q = U_j\cap B\,, \quad U_j\cap P = \phi \qquad \text{if } j \in I_2$$

$$U_j\cap\hat{Q} = U_j\cap Q = U_j\cap B\,, \quad U_j\cap B = \phi \qquad \text{if } j \in I_3$$

$$U_j\cap Q = U_j\cap P = U_j\cap B = \phi \qquad \text{if } j \in I_4$$

which implies (A5). Thus $p_j = q_j$ remains to be proven.

3. CLAIM. For each $j \in \mathbb{N}[1,s]$ and $x \in U_j\cap\hat{Q}$, we have $\nu_\pi(x) = p_j = q_j$.

<u>PROOF OF THE 3. CLAIM</u>: Take any $x \in U_j \cap \hat{Q}$. Then $\varphi_j(x) = (v,0)$ with $v \in D^{m-1}$. Because φ_j and α are biholomorphic with $\alpha \circ \pi | U_j = \psi_j \circ \varphi_j$. We have $\nu_\pi(x) = \nu_{\psi_j}(\varphi_j(x)) = \nu_{\psi_j}(v,0)$. Define $\zeta_j = \exp(2\pi i/p_j)$. Take $z = (z_1,\ldots,z_m) \in D^m$. Then

$$\psi_j^{-1}(\psi_j(z)) = \{(z_1,\cdots,z_{m-1}, \zeta_j^\lambda z_m) | \lambda \in \mathbb{N}[1, p_j]\}$$

$$\#\psi_j^{-1}(\psi_j(z)) = \begin{cases} p_j & \text{if} \quad z \in D^m - W \\ 1 & \text{if} \quad z \in W. \end{cases}$$

Therefore

$$\nu_\pi(x) = \nu_{\psi_j}(v,0) = \limsup_{z \to (v,0)} \#\psi_j^{-1}(\psi_j(z)) = p_j.$$

Since $b_j \in U_j \cap \hat{Q}$ we obtain $p_j = \nu_\pi(b_j) = q_j$; q.e.d.

<u>THEOREM 4.21</u>. Let M and N be connected, complex manifolds of dimension m. Let $\pi: M \longrightarrow N$ be a surjective, proper, holomorphic map with branching divisor β. Let $\nu: M \longrightarrow \mathbb{Z}$ be a divisor. Assume that ν is not the null divisor. Determine S by (4.59). Take $a \in N-S$. Then

$$\pi_*(\nu)(a) = \sum_{x \in \tilde{a}} \nu(x) \tag{4.66}$$

<u>Proof</u>. If $a \in N-P'$, the statement is trivial. Assume that $a \in P'-S$. Take a winding construction $(U,\tilde{U},g,h,G,H)$ of ν at a. Then

$$\nu | \tilde{U} = \mu_g^0 - \mu_h^0 \qquad \pi_*(\nu) | U = \mu_G^0 - \mu_H^0.$$

The function $f = g/h$ is meromorphic on $\tilde{U}$ and holomorphic without zeroes on $\tilde{U}-P$. The function $F = G/H$ is meromorphic on U and holomorphic without zeroes on $U-Q'$. For each $j \in \mathbb{N}[1,s]$ the function $f_j = f \circ \varphi_j^{-1}$ is meromorphic on D^m and holomorphic without zeroes on $D^{m-1} x (D-\{0\})$ then

$$p_j = \nu(b_j) = \mu_g^0(b_j) - \mu_h^0(b_j) = \nu_{f_j,0}(0) - \nu_{f_j,\infty}(0) \in \mathbb{Z}.$$

By Rouché's Theorem $p_j = \nu(x) = \nu_{f_j,0}(x) - \nu_{f_j,\infty}(x)$ for $x \in D^{m-1} x \{0\}$. Therefore there exists a holomorphic function $k_j\colon D^m \longrightarrow \mathbb{C}_*$ without zeroes such that

$$f_j(z_1,\cdots,z_m) = z_m^{p_j}\, k_j(z_1,\cdots,z_m)$$

for all $(z_1,\cdots,z_m) \in D^m$ with $z_m \neq 0$. Define $p = \pi_*(\nu)(a)$. Then $p = \pi_*(\nu)(x)$ for all $x \in U \cap Q'$. The function $F_0 = F \circ \alpha^{-1}$ is meromorphic on D^m and holomorphic without zeroes on $D^{m-1} x (D-\{0\})$. Here $p = \mu_{F_0,0}(z) - \mu_{F_0,\infty}(z)$ for all $z \in D^{m-1} x (D-\{0\})$ A holomorphic function $K\colon D^m \longrightarrow \mathbb{C}_*$ without zeroes exists such that

$$F_0(z_1,\cdots,z_m) = z_m^p\, K(z_1,\cdots,z_m)$$

for all $(z_1,\cdots,z_m) \in D^m$ with $z_m \neq 0$.

Take $r \in \mathbb{R}(0,1)$. Take $0 \in D^{m-1}$. Then $z(r) = \alpha^{-1}(0,r) \in U-Q'$. Define $q_j = \nu_\pi(b_j)$ and

$$\zeta_j = e^{\frac{2\pi i}{q_j}} \qquad x_{j\lambda}(r) = r^{\frac{1}{q_j}} \zeta_j^\lambda \qquad w_{j\lambda}(r) = \varphi_j^{-1}(0, x_{j\lambda}(r))$$

for $j \in \mathbb{N}[1,s]$ and $\lambda \in \mathbb{N}[1,q_j]$. Then $w_{j\lambda}(r) \neq w_{p,\mu}(r)$ if $(j,\lambda) \neq (k,\mu)$. Therefore

$$\# \{w_{j,\lambda}(r) \mid j \in \mathbb{N}[1,s], \lambda \in \mathbb{N}[1,q_j]\} = \sum_{j=1}^{s} q_j = q$$

$$\alpha(\pi(w_{j\lambda}(r)) = \psi_j(\varphi_j(w_{j\lambda}(r))) = \psi_j(0, x_{j\lambda}(r))$$

$$= (0, x_{j\lambda}(r)^{q_j}) = (0,r) = \alpha(z(r)).$$

Hence $\pi(w_{j\lambda}(r)) = z(r)$. Because $q = \#\pi^{-1}(z(r))$ we obtain

$$\pi^{-1}(z(r)) = \{w_{j,\lambda}(r) \mid j \in \mathbb{N}[1,s],\ \lambda \in \mathbb{N}[1,q_j]\}$$

Also we have

$$F(z(r)) = F(\alpha^{-1}(0,r)) = F_0(0,r) = K(0,r)r^P,$$

Here $x_{j\lambda}(r) \longrightarrow 0$ and $k_j(0,x_{j\lambda}(r)) \longrightarrow k_j(0,0) \neq 0$ and $K(0,r) \longrightarrow K(0,0) \neq 0$ for $r \longrightarrow 0$. Hence

$$F(z(r)) = \prod_{j=1}^{s} \prod_{\lambda=1}^{q_j} f(w_{j\lambda}(r))$$

$$= \prod_{j=1}^{s} \prod_{\lambda=1}^{q_j} f(\varphi_j^{-1}(0,x_{j\lambda}(r)))$$

$$= \prod_{j=1}^{s} \prod_{\lambda=1}^{q_j} f_j(0,x_{j\lambda}(r))$$

$$= \prod_{j=1}^{s} \prod_{\lambda=1}^{q_j} x_{j\lambda}(r)^{p_j} k_j(0,x_{j\lambda}(r)))$$

$$= \prod_{j=1}^{s} \prod_{\lambda=1}^{q_j} r^{(p_j/q_j)} \zeta_j^{\lambda p_j} k_j(0,x_{j\lambda}(r))$$

$$= r^{p_1+\cdots+p_s} \prod_{j=1}^{s} \prod_{\lambda=1}^{q_j} \zeta_j^{\lambda p_j} k_j(0,x_{j\lambda}(r))$$

which implies

$$r^{p-p_1-\cdots-p_s} = \frac{1}{|K(0,r)|} \prod_{j=1}^{s} \prod_{\lambda=1}^{q_j} |k_j(0,x_{j\lambda}(r))|$$

$$r^{p-p_1-\cdots-p_s} \longrightarrow \frac{1}{|K(0,0)|} \prod_{j=1}^{s} |k_j(0,0)| \in \mathbb{R}-\{0\} \text{ for } r \longrightarrow 0,$$

which implies $p = p_1+\cdots+p_s$. Therefore we have

$$\pi_*(\nu)(a) = p = p_1+\cdots+p_s = \nu(b_1)+\cdots+\nu(b_s) \qquad \text{q.e.d.}$$

THEOREM 4.22. Let M and N be connected, complex manifolds of dimension m. Let $\pi: M \longrightarrow N$ be a surjective, holomorphic map of sheet

number q. Let $\nu: M \longrightarrow \mathbb{Z}$ be a divisor with support P. Denote $R = \operatorname{supp} \pi_*(\nu)$. Let K be a compact subset of N. Let χ be a continuous form of bidegree (m-1-m-1) on K. Abbreviate $\tilde{K} = \pi^{-1}(K)$ and $\tilde{\chi} = \pi^*(\chi)$. Then

$$\int_{P\cap\tilde{K}} \nu\tilde{\chi} = \int_{R\cap K} \pi_*(\nu)\chi \tag{4.67}$$

<u>PROOF</u>. Assume the notations of (S1)-(S10). Define S by (4.59). Let $\mathfrak{B}_0$ be the set of all branches of P not contained in E. Let $\mathfrak{B}_1$ be the set of all branches of P contained in E. Then $\mathfrak{B} = \mathfrak{B}_0 \cup \mathfrak{B}_1$ is the disjoint union of all branches of P. Then $P_j = \bigcup_{C\in\mathfrak{B}_j} C$ is pure (m-1) dimensional or empty with $P = P_0 \cup P_1$ and with $\dim (P_0 \cap P_1) \le m-2$. Also $P_1 \subseteq E$ and $P_1' \subseteq E'$. Hence $\dim P_1' \le m-2$. Let $\iota : P_1 \longrightarrow M$ and $j: P_1' \longrightarrow N$ be the inclusion maps. Let $\pi_1 = \pi: P_1 \longrightarrow P_1'$ be the restriction. Then $\pi\circ\iota = j\circ\pi_1$. For degree reasons we have $j^*(\chi) = 0$. Therefore

$$\iota^*(\tilde{\chi}) = \iota^*\pi^*(\chi) = (\pi\circ\iota)^*(\chi) = (j\circ\pi_1)^*(\chi) = \pi_1^*(j^*(\chi)) = \pi_1^*(0) = 0.$$

Take $C \in \mathfrak{B}_0$. Then $\dim (C \cap E) \le m-2$. Hence $\dim (C \cap \tilde{S}) \le m-2$. Thus $\dim (P_0 \cap \tilde{S}) \le m-2$. Also $\pi: P_0 - \tilde{S} \longrightarrow P_0' - \tilde{S} = P' - S$ is a locally biholomorphic, proper, surjective holomorphic map between pure (m-1)-dimensional, complex manifolds if $\mathfrak{B}_0 \ne \phi$. Hence Theorem 4.20 implies

$$\int_{P\cap\tilde{K}} \nu\tilde{\chi} = \int_{P_0\cap\tilde{K}-\tilde{S}} \nu\tilde{\chi} = \int_{z\in K\cap(P'-S)} \sum_{x\in\tilde{z}} \nu(x)\chi$$

$$= \int_{P'\cap k} \pi_*(\nu)\chi = \int_{r\cap K} \pi_*(\nu)\chi$$

since $R \subseteq P'$ and $\pi_*(\nu)|(P'-R) = 0$. If $\mathfrak{B}_0 = \phi$, then $R \subseteq P' \subseteq E'$ and $R = \phi$. Hence $\pi_*(\nu) = 0$. Also $\tilde{\chi} = 0$ on $P = P_1$. Hence (4.67) verifies as 0 = 0; q.e.d.

Now, we apply (S1)-(S10) and Lemma 4.20 to the case $\nu = \beta$. Then $B = P = Q$ and $I_2 = I_3 = \phi$.

<u>THEOREM 4.23</u>. Let M and N be connected, complex manifolds of dimension m. Let $\pi: M \longrightarrow N$ be a proper, surjective, holomorphic map of sheet number q. Let E be the exceptional set. Let β be the branching divisor with $B = \operatorname{supp} \beta$. Let $\nu_\pi(x)$ the local mapping degree of π at $x \in M-E$. Then there exists an analytic subset S of N with $\dim S \leq m-2$ such that $E' \subseteq S$ and

$$(4.68) \qquad \beta(x) = \nu_\pi(x)-1 \qquad \text{if } x \in M-\tilde{S}$$

$$(4.69) \qquad \pi_*(\beta)(z) = q-\#\tilde{z} \qquad \text{if } z \in N-S.$$

<u>Proof</u>. Apply (S1)-(S10) with $\nu = \beta$. Determine S by (4.59). If $x \in M-B$, then $\beta(x) = 0 = 1-1 = \nu_\pi(x)-1$. Thus take $x \in B-\tilde{S}$. Define $a = \pi(x)$. Then $x = b_j$ for some $j \in I_1$. We have $\nu_\pi(x) = \nu_\pi(b_j) = q_j$. On U_j we have

$$d\pi_1\wedge\cdots\wedge d\pi_m = q_j(\varphi_{m,j})^{q_j-1}\, d\varphi_{1,j}\wedge\cdots\wedge d\varphi_{m,j}.$$

Here $d\varphi_{m,j}(b_j) \neq 0$ and $d\varphi_{1,j}(b_j)\wedge\cdots\wedge d\varphi_{m,j}(b_j) \neq 0$. Hence

$$\beta(x) = \beta(b_j) = q_j-1 = \nu_\pi(x)-1.$$

Thus (4.68) is proved. Then take $z \in N-S$. This yields

$$\pi_*(\beta)(z) = \sum_{z\in\tilde{z}} \beta(x) = \sum_{x\in\tilde{z}} (\nu_\pi(x)-1) = \sum_{x\in\tilde{z}} \nu_\pi(x) - \#\tilde{z} = q-\#\tilde{z}$$

q.e.d.

Because $\pi_*(\beta)$ is a divisor, $\#\tilde{z}$ is locally constant on $B-S = \Re(B)-S$. Define the proper surjective light map $\pi: \mathbb{C}^2 \longrightarrow \mathbb{C}^2$ by $\pi(z,w) = (z^2,w^2)$. Then $\nu_\pi(z,w)-1 = \beta(z,w) = 1$ if $z \neq 0 = w$ or $z = 0 \neq w$. But $\nu_\pi(0)-1 = 3 \neq 2 = \beta(0)$. Thus $\nu_\pi-1$ is not a divisor.

5. REDUCTION OF NEVANLINNA THEORY

The following general assumptions will be made:

(G1) M and N are connected, complex manifolds of dimension m.

(G2) $\pi: M \longrightarrow N$ is a proper, surjective holomorphic map of sheet number q with branching divisor β. Here $B = \operatorname{supp} \beta$ is the branching set and E the exceptional set.

(G3) τ is a parabolic exhaustion of N and $\tilde{\tau} = \tau \circ \pi$.

(G4) V is a hermitian vector space of dimension $n+1 > 1$.

(G5) $f: M \longrightarrow \mathbb{P}(V)$ is a meromorphic map and $g: N \longrightarrow \mathbb{P}(\amalg_q V)$ is the algebroid reduction of f for π.

Then $\tilde{\tau}$ is a parabolic exhaustion of M with $\mathfrak{C}_{\tilde{\tau}} = \mathfrak{C}_{\tau}$ and

(5.1) $$\tilde{\upsilon} = dd^c\tilde{\tau} = \pi^*(\upsilon) \qquad \tilde{\omega} = dd^c \log \tilde{\tau} = \pi^*(\omega).$$

(5.2) $$\tilde{\sigma} = d^c \log \tilde{\tau} \wedge \tilde{\omega}^{m-1} = \pi^*(\sigma).$$

(5.3) $$M[r] = \pi^{-1}(N[r]) \qquad M(r) = \pi^{-1}(N(r)) \qquad M\langle r\rangle = \pi^{-1}(N\langle r\rangle).$$

(5.4) $$\tilde{\varsigma} = \int_{M\langle r\rangle} \tilde{\sigma} = \int_{M\langle r\rangle} \pi^*(\sigma) = \int_{N\langle r\rangle} q\sigma = q\varsigma$$

THEOREM 5.1. The Riemann-Hurwitz Formula. Take $0 < s < r \in \mathfrak{C}^0_{\tau}$ with $s \in \mathfrak{C}^0_{\tau}$. Then $s \in \mathfrak{C}^0_{\tilde{\tau}}$ and $r \in \mathfrak{C}^0_{\tilde{\tau}}$ and

(5.5) $$\operatorname{Ric}_{\tilde{\tau}}(r,s) = q \operatorname{Ric}_{\tau}(r,s) + N_{\beta}(r,s)$$

Remark: (5.5) is the analogon to the classical Riemann-Hurwitz formula for proper, surjective, holomorphic maps from one compact Riemann surface onto another. If τ is strictly parabolic, that is, if (M,τ) is biholomorphically isometric to $(\mathbb{C}^m, \| \ \|^2)$, then $\operatorname{Ric}_{\tau}(r,s) = 0$ and $\operatorname{Ric}_{\tilde{\tau}}(r,s) = N_{\beta}(r,s)$. This special case was proved earlier Stoll [24] Theorem 15.4.

<u>Proof</u>. Let $\chi > 0$ be a positive form of degree 2m and class C^∞ on M. Then χ defines a hermitian metric along the fibers of K(M) whose norm shall be denoted by $\| \ \|_\chi$. Similar, let $\psi > 0$ be a positive form of degree 2m and class C^∞ on N which defines the norm $\| \ \|_\psi$ along the fibers of K(N) and the dual norm $\| \ \|_\psi^*$ along the dual bundle $K(N)^*$ which pulls back to $\pi^*(K(N)^*)$. These hermitian metrics induce a hermitian metric $\varkappa$ with norm $\| \ \|_\varkappa$ along the fibers of the <u>Jacobian bundle</u> $K(\pi) = K(M) \otimes \pi^*(K(N)^*)$. By Stoll [24] page 110 there exists one and only one <u>Jacobian</u> <u>section</u> $P \in \Gamma(M, K(\pi))$ with this property: Let $U \neq \phi$ be an open, connected subset of N with a holomorphic frame $\zeta \in \Gamma(U, K(N))$ of K(N) over U (ζ is a holomorphic form of bidegree (m,0) without zeroes on U). Let $\zeta^* \in \Gamma(U, K(N)^*)$ be the dual frame. Let ζ_π^* be the pull back <u>section</u> $\zeta_\pi^* \in \Gamma(\tilde{U}, \pi^*(K(N)^*))$. Let $\pi^*(\zeta) \in \Gamma(\tilde{U}, K(M))$ be the pull back <u>form</u> of the form ζ by π. Then

(5.6) $$P|\tilde{U} = \pi^*(\zeta) \otimes \zeta_\pi^*.$$

Recall that $\beta|\tilde{U} = \mu_{\pi^*(\zeta)}$ is the divisor of $\pi^*(\zeta)$ by the definition of the branching divisor. Since ζ_π^* is without zeroes, $\mu_P|\tilde{U} = \mu_{\pi^*(\zeta)}$. Therefore $\beta = \mu_P$ globally. The First Main Theorem (2.31) implies

(5.7) $$T(r,s,K(\pi),\varkappa) + \int_{M\langle r\rangle} \log\|P\|_\varkappa \, \sigma - \int_{M\langle s\rangle} \log\|P\|_\varkappa \, \sigma = N_\beta(r,s)$$

Locally we have

$$\|P\|_\varkappa|\tilde{U} = \|\pi^*(\zeta)\|_\chi \, \|\zeta_\pi^*\|_\psi = \frac{\|\pi^*(\zeta)\|_\chi}{\|\zeta\|_\psi \circ \pi}$$

$$c(K(\pi),\varkappa)|\tilde{U} = - dd^c \log\|P\|_\varkappa^2$$

$$= - dd^c \log\|\pi^*(\zeta)\|_\chi^2 + dd^c \log(\|\zeta\|_\psi^2 \circ \pi)$$

$$= \operatorname{Ric} \chi|\tilde{U} - \pi^*(\operatorname{Ric} \psi)|\tilde{U}$$

Hence

(5.8) $$c(K(\pi),\varkappa) = \operatorname{Ric} \chi - \pi^*(\operatorname{Ric} \psi)$$

which implies

$$(5.9)\qquad T(r,s,K(\pi),\varkappa) = \mathrm{Ric}(r,s,\chi) - \int_s^r \int_{M[t]} \pi^*(\mathrm{Ric}\ \psi)\wedge\tilde{\upsilon}^{m-1}\ \frac{dt}{t^{2m-1}}$$

Here

$$\int_s^r t^{1-2m} \int_{M[t]} \pi^*(\mathrm{Ric}\ \psi)\wedge\tilde{\upsilon}^{m-1} dt$$

$$= \int_s^r t^{1-2m} \int_{M[t]} \pi^*(\mathrm{Ric}\psi\wedge\upsilon^{m-1}) dt$$

$$= q \int_s^r t^{1-2m} \int_{N[t]} \mathrm{Ric}\ \psi\wedge\upsilon^{m-1} dt = q\ \mathrm{Ric}(r,s,\psi).$$

Therefore we obtain

$$(5.10)\qquad T(r,s,K(\pi),\varkappa) = \mathrm{Ric}(r,s,\chi) - q\ \mathrm{Ric}(r,s,\psi).$$

Define $v\colon N \longrightarrow \mathbb{R}_+$ by $\upsilon^m = v\psi$. Then $\tilde{\upsilon}^m = \pi^*(\upsilon^m) = (v\circ\pi)\cdot\pi^*(\psi)$. We have

$$(5.11)\qquad \mathrm{Ric}_\tau(r,s) = \int_{N\langle r\rangle} (\log v)\ \sigma - \int_{N\langle s\rangle} (\log v)\ \sigma + \mathrm{Ric}(r,s,\psi).$$

Again we return to the local situation (5.6):

$$i_m \zeta\wedge\bar{\zeta} = \|\zeta\|_\psi^2\ \psi|U$$

$$i_m \pi^*(\zeta)\wedge\pi^*(\bar{\zeta}) = \|\pi^*(\zeta)\|_\chi^2\ \chi|\tilde{U}$$

$$\tilde{\upsilon}^m = (v\circ\pi)\ (\|\zeta\|_\psi\circ\pi)^{-2}\ i_m\pi^*(\zeta)\wedge\pi^*(\bar{\zeta})$$

$$= (v\circ\pi)\ (\|\zeta\|_\psi\circ\pi)^{-2}\ \|\pi^*(\zeta)\|_\chi^2\cdot\chi \qquad \text{on } \tilde{U}$$

$$= (v\circ\pi)\ \|P\|_\varkappa^2\ \chi.$$

Since $\log \|P\|_\varkappa^2\ \tilde{\sigma}$ and $\log(v\circ\pi)\ \tilde{\sigma}$ are integrable over $M\langle r\rangle$ and $M\langle s\rangle$ with

$$\int_{M\langle r\rangle} \log(v\circ\pi)\ \tilde{\sigma} = q \int_{N\langle r\rangle} (\log v)\ \sigma$$

We obtain

$$\mathrm{Ric}_{\tilde\tau}(r,s) = \int_{M\langle r\rangle} \log(v\circ\pi)\,\|P\|_{\varkappa}^{2}\,\tilde\sigma - \int_{M\langle s\rangle} \log(v\circ\pi)\|P\|_{\varkappa}^{2}\,\tilde\sigma + \mathrm{Ric}(r,s,\chi)$$

$$= \int_{M\langle r\rangle} \log\|P\|_{\varkappa}\,\tilde\sigma - \int_{M\langle s\rangle} \log\|P\|_{\varkappa}\,\tilde\sigma + \mathrm{Ric}(r,s,\chi)$$

$$+ q\int_{N\langle r\rangle} (\log v)\,\sigma - q\int_{N\langle s\rangle} (\log v)\,\sigma$$

$$= N_{\beta}(r,s) - T(r,s,K(\pi),\varkappa) + \mathrm{Ric}(r,s,\chi) + q\,\mathrm{Ric}_{\tau}(r,s)$$

$$- q\,\mathrm{Ric}\,(r,s,\psi)$$

$$= N_{\beta}(r,s) + q\,\mathrm{Ric}_{\tau}(r,s) \qquad \text{q.e.d.}$$

Now Theorem 4.22 implies immediately.

<u>THEOREM 5.2</u>. Assume that (G1)-(G3) hold. Let $\nu: M \longrightarrow \mathbb{Z}$ be a divisor and let $\mu = \pi_{*}(\nu)$ be the direct image divisor. Then $n_{\mu}(t) = n_{\nu}(t)$ for $t > 0$. If $0 < s < r$, then $N_{\mu}(r,s) = N_{\nu}(r,s)$.

<u>THEOREM 5.3</u>. Assume that (G1)-(G5) hold. Take $a \in \mathbb{P}(V^{*})$ and define $b = a^{q} \in \mathbb{P}(\amalg_{q} V^{*})$. Then f is free for a if and only if g is free for b. Assume that f is free for a. Define c(n,q) by (3.54). Define u_f by (4.27). Take $t > 0$ and $0 < s < r$. Then

$$n_{g,b}(t) = n_{f,a}(t) \qquad N_{g,b}(r,s) = N_{f,a}(r,s) \tag{5.12}$$

$$m_{f,a}(r) = m_{g,b}(r) + \int_{N\langle r\rangle} \log\frac{1}{u_f}\,\sigma \tag{5.13}$$

$$0 \le m_{g,b}(r) \le m_{f,a}(r) \le m_{g,b}(r) + \varsigma \log\frac{1}{c(n,q)} \tag{5.14}$$

Remarks: Here (5.13) and (5.14) hold originally for $r \in \mathfrak{E}_\tau = \mathfrak{E}_{\tilde{\tau}}$ only, but the compensation functions extend to continuous functions on $\mathbb{R}_+$. Thus the integral $\displaystyle\int_{N<r>} \log 1/u_f \, \sigma$ extends to a continuous function on $\mathbb{R}_+$ and the "integral" is to be understood in this sense.

Proof. Since $\pi_*(\nu_{f,a}) = \mu_{g,b}$ by Theorem 4.17, Theorem 5.2 implies (5.12). Because $\pi: M-\tilde{B} \longrightarrow N-B'$ is a holomorphic covering space of sheet number q, and because $(B' \cup I'_f) \cap M<r>$ and $(\hat{B} \cup \hat{I}_f) \cap M<r>$ have measure zero on $N<r>$ and $M<r>$ respectively, Theorem 4.10 implies

$$-m_{f,a}(r) = \int_{M<r>} \log \square f, a \square \tilde{\sigma} = \int_{N<r>} \pi_*(\log \square f, a \square)\sigma$$

$$= \int_{N<r>} \log \square g, b \square \sigma + \int_{N<r>} \log u_f \, \sigma$$

$$= - m_{g,b}(r) + \int_{N<r>} \log u_f \, \sigma$$

which implies (5.13). Now (4.28) and (5.13) imply (5.14); q.e.d.

THEOREM 5.4. Assume that (G1)-(G5) hold. Take $0 < s < r$. Then

$$(5.15) \qquad T_f(r,s) = T_g(r,s) + \int_{N<r>} \log \frac{1}{u_f} \, \sigma - \int_{N<s>} \log \frac{1}{u_f} \, \sigma$$

$$(5.16) \qquad |T_f(r,s) - T_g(r,s)| \le s \log \frac{1}{c(n,q)} \le \frac{sq}{2} \sum_{\nu=1}^{n} \frac{1}{\nu}$$

$$(5.17) \quad 2 \int_s^r \int_{N[t]} (dd^c \log \frac{1}{u_f}) \wedge v^{m-1} \, t^{1-2m} dt = \int_{N<r>} \log \frac{1}{u_f} \, \sigma - \int_{N<s>} \log \frac{1}{u_f} \, \sigma.$$

Proof. Take $a \in \mathbb{P}(V^*)$ such that f is free for a. Then g is free for $b = a^q$ and

$$T_f(r,s) = N_{f,a}(r,s) + m_{f,a}(r) - m_{f,a}(s)$$

$$= N_{g,b}(r,s) + m_{g,b}(r) + \int_{N\langle r\rangle} \log \frac{1}{u_f}\,\sigma - m_{g,b}(s) - \int_{N\langle s\rangle} \log \frac{1}{u_f}\,\sigma$$

$$= T_g(r,s) + \int_{N\langle r\rangle} \log \frac{1}{u_f}\,\sigma - \int_{N\langle s\rangle} \log \frac{1}{u_f}\,\sigma$$

which proves (5.15). Also

$$0 \le \int_{N\langle r\rangle} \log \frac{1}{u_f}\,\sigma \le \int_{N\langle r\rangle} \log \frac{1}{c(n,q)}\,\sigma = \varsigma \log \frac{1}{c(n,q)}$$

$$\le \frac{\varsigma q}{2} \sum_{\nu=1}^{n} \frac{1}{\nu}$$

which implies (5.16).

For $t > 0$ we have

$$\int_{M[t]} f^*(\Omega)\wedge\tilde{\upsilon}^{m+1} = \int_{M[t]} f^*(\Omega)\wedge\pi^*(\upsilon^{m+1}) = \int_{N[t]} \pi_*(f^*(\Omega))\wedge\upsilon^{m+1}$$

$$= \int_{N[t]} g^*(\Omega_q)\wedge\upsilon^{m+1} + 2\int_{N[t]} dd^c\log\frac{1}{u_f}\wedge\upsilon^{m-1}$$

which implies

$$T_f(r,s) = T_g(r,s) + 2\int_s^r \int_{N[t]} dd^c\log\frac{1}{u_f}\wedge\upsilon^{m-1}\; t^{1-2m}dt.$$

Comparison with (5.15) yields (5.17); q.e.d.

Observe that u_f is of class C^∞ on $N-(B'\cup I_f)$. Thus if f is holomorphic and $\beta = 0$, then (5.17) follows from Stokes Theorem. Since we only know that u_f is continuous on $B'-(I_f'\cup E')$, such a direct method may be more difficult in the general case.

It is remarkable that the bound in (5.16) depends on ς, n and q only, but not on the maps involved. Here $\tilde{\varsigma} = \varsigma.q$.

If f is not constant, then g is not constant and the defects are defined with

$$\delta_f(a) = \delta_g(a^q) \tag{5.18}$$

which invites a transfer of the defect relation from f to g or from g to f. However, this requires the preservation of linear non-degeneracy of the map and of general position of the hyperplane, but such an invariance does not exist in general. Clearly if $f: M \longrightarrow \mathbb{P}(W)$ where W is a proper, linear subspace of V, then $g: N \longrightarrow \mathbb{P}(\amalg_q W)$ where $\amalg_q W$ is a proper linear subspace of $\amalg_q V$, but the inverse is not true as the following example shows:

<u>Example 1</u>: Let $M = N = \mathbb{C}$ and $\pi: \mathbb{C} \longrightarrow \mathbb{C}$ be given by $\pi(z) = z^2$. Then $q = 2$. Let $h: \mathbb{C} \longrightarrow \mathbb{C}$ be any entire function. Let V be a complex vector space of dimension 3. Hence $n = 2$. Let $\mathfrak{e}_0, \mathfrak{e}_1, \mathfrak{e}_2$ be a base of V. A holomorphic map $\mathfrak{v}: \mathbb{C} \longrightarrow V$ is defined by

$$\mathfrak{v}(z) = \mathfrak{e}_0 + z\mathfrak{e}_1 + h(z^2)\,\mathfrak{e}_2.$$

Then $f = \mathbb{P}\circ\mathfrak{v}: \mathbb{C} \longrightarrow \mathbb{P}(V)$ is a holomorphic map. Let $\mathfrak{y}: \mathbb{C} \longrightarrow \amalg_2 V$ be the algebroid reduction of $\mathfrak{v}$. Then

$$\mathfrak{y}(w) = \mathfrak{v}(\sqrt{w})\,\mathfrak{v}(-\sqrt{w}) = \mathfrak{e}_0^2 - w\mathfrak{e}_1^2 + h(w)^2\mathfrak{e}_2 + 2h(w)\mathfrak{e}_0\mathfrak{e}_2$$

Since $\mathfrak{e}_0^2, \mathfrak{e}_1^2, \mathfrak{e}_2^2, \mathfrak{e}_0\mathfrak{e}_1, \mathfrak{e}_0\mathfrak{e}_2, \mathfrak{e}_1\mathfrak{e}_2$ is a base of $\amalg_2 V$, the algebroid reduction $g = \mathbb{P}\circ\mathfrak{y}$ of f is linearly degenerate. However, if h is not constant, f is linearly non-degenerate.

<u>Example 2</u>. Here $a_j \in \mathbb{P}(V^*)$ for $j = 1,\cdots,p$ will not be in general position but $a_j^q \in \mathbb{P}(\amalg_q V^*)$ will be in general position for $j = 1,\cdots,p$.

Take $q = 2$ and $p = 6$ and $n+1 = \dim V = 3$. Let $\epsilon_1, \epsilon_2, \epsilon_3$ be a base of V^*. Then $\epsilon_1^2, \epsilon_2^2, \epsilon_3^2, \epsilon_1\epsilon_2, \epsilon_1\epsilon_3, \epsilon_2\epsilon_3$ is a base of $\amalg_2 V^*$. Define $a_j = \mathbb{P}(\alpha_j)$ for $j = 1,2,\cdots,6$ by $\alpha_j = \epsilon_j$ for $j = 1,2,3$ and

$$\alpha_4 = \epsilon_1 + \epsilon_3 \qquad \alpha_5 = \epsilon_1 + 2\epsilon_2 \qquad \alpha_3 = \epsilon_1 + 3\epsilon_2 + \epsilon_3$$

Then $\alpha_4 = \alpha_1 + \alpha_3$. Hence $a_1, a_2, \cdots, a_6$ are <u>not in general position</u> in $\mathbb{P}(V^*)$. We have $\alpha_j^2 = e_j^2$ for $j = 1,2,3$ and

$$\alpha_4^2 = e_1^2 + e_3^2 + 2e_1e_3$$

$$\alpha_5^2 = e_1^2 + 4e_2^2 + 4e_1e_2$$

$$\alpha_6^2 = e_1^2 + 9e_2^2 + e_3^2 + 6e_1e_2 + 2e_1e_3 + 6e_2e_3$$

$$\bigwedge_{j=1}^{6} \alpha_j^2 = -48e_1^2 \wedge e_2^2 \wedge e_3^2 \wedge e_1e_2 \wedge e_1e_3 \wedge e_2e_3 \neq 0$$

Hence $a_1, a_2, \cdots, a_6$ are in <u>general position</u> in $\mathbb{P}(\amalg_2 V^*)$.

<u>Example 3</u>. Here $a_j \in \mathbb{P}(V^*)$ are in general position but $a_j^q \in \mathbb{P}(\amalg_q V)$ are not. Take $q = 2$ and $p = 6$ and $n+1 = \dim V^* = 3$. Let e_1, e_2, e_3 be a base of V^*. Then $e_1^2, e_2^2, e_3^2, e_1e_2, e_1e_3, e_2e_3$ is a base of $\amalg_2 V^*$. Define $a_j = \mathbb{P}(\alpha_j)$ for $j = 1, \cdots, 6$ by $\alpha_j = e_j$ for $j = 1,2,3$ and

$$\alpha_4 = e_1 + 2e_2 + 4e_3 \qquad \alpha_5 = e_1 - e_2 + 2e_3 \qquad \alpha_6 = e_1 + e_2 + 6e_3$$

Any 3 vectors from $\alpha_1, \cdots, \alpha_6$ are linearly independent. Hence $a_1, \cdots, a_6$ are in general position. We have $\alpha_j^2 = e_j^2$ for $j = 1,2,3$ and

$$\alpha_4^2 = e_1^2 + 4e_2^2 + 16e_3^2 + 4e_1e_2 + 8e_1e_3 + 16\,e_2e_3$$

$$\alpha_5^2 = e_1^2 + e_2^2 + 4e_3^2 - 2e_1e_2 + 4e_1e_3 - 4e_2e_3$$

$$\alpha_6^2 = e_1^2 + e_2^2 + 36e_3^2 + 2e_1e_2 + 12e_1e_3 + 12e_2e_3$$

Hence

$$\alpha_1^2 + 4\alpha_2^2 - 16\alpha_3^2 - \alpha_4^2 - \alpha_5^2 + \alpha_6^2 = 0.$$

Therefore $a_1^2, a_2^2, \cdots, a_6^2$ are not in general position in $\mathbb{P}(\amalg_2 V^*)$.

6. THE DISCRIMINANT

The concept of the discriminant shall be extended to meromorphic maps $f: M \longrightarrow \mathbb{P}(V)$. A first Main Theorem for discriminants will be proved. The discriminant divisor majorizes the direct image of the branching divisor. As an application we obtain a theorem of Noguchi. We assume that (G1)-(G5) holds and abbreviate

$$W = \amalg_{q(q-1)} V \wedge V. \tag{6.1}$$

a) THE CONSTRUCTION OF THE DISCRIMINANT SECTION. Let $U \neq \emptyset$ be an open, connected subset of N-E′ such that there is a reduced representation $\mathfrak{v}: \tilde{U} \longrightarrow V$ of f. A holomorphic vector function $\mathfrak{d}: U \longrightarrow W$ called the discriminant of $\mathfrak{v}$ is defined as follows: Take any $z \in U$. Enumerate $\tilde{z} = \{x_1, \cdots, x_q\}$ such that

$$\nu_\pi(x) = \#\{j \in \mathbb{N}[1,q] \mid x_j = x\} \qquad \text{for all } x \in \tilde{z}. \tag{6.2}$$

Define

$$P(z) = \{(j,k) \in \mathbb{N}[1,q]^2 \mid j \neq k\} \tag{6.3}$$

$$\mathfrak{d}(z) = \amalg_{(j,k) \in P(z)} \mathfrak{v}(x_j) \wedge \mathfrak{v}(x_k) \tag{6.4}$$

Then $\mathfrak{d}(z)$ does not depend on the choice of the enumeration.

LEMMA 6.1. $\mathfrak{d}: U \longrightarrow W$ is holomorphic and $\mathfrak{d}(z) = 0$ for all $z \in U \cap B'$.

Proof. Take any $a \in U-B'$. Then there is an open, connected neighborhood Y of a in U-B′ such that $\tilde{Y} = Y_1 \cup \cdots \cup Y_q$ is a disjoint union of open, connected subsets Y_j of M such that $\pi_j = \pi: Y_j \longrightarrow Y$ is biholomorphic with $\sigma_j = \pi_j^{-1}$ for all $j \in \mathbb{N}[1,q]$. Therefore

$$\mathfrak{d} | Y = \amalg_{j=1}^{q} \amalg_{j \neq k=1}^{q} (\mathfrak{v} \circ \sigma_j) \wedge (\mathfrak{v} \circ \sigma_k): Y \longrightarrow W \tag{6.5}$$

is holomorphic. Thus $\mathfrak{d}$ is holomorphic on U-B′.

Take any $z \in B' \cap U$. Then $\#\tilde{z} < q$. Enumerate $\tilde{z}$ as in (6.2) and define $P(z)$ by (6.3). Then $(j,k) \in P(z)$ exists with $x_j = x_k$. Hence $\mathfrak{v}(x_j) \wedge \mathfrak{v}(x_k) = 0$. By (6.4) we obtain $\delta(z) = 0$.

By the Riemann extension theorem, it suffices to show that δ is continuous at every point $a \in U \cap B'$. Here $s = \#\tilde{a} < q$ and $\tilde{a} = \{b_1, \cdots, b_q\}$ with $b_j \neq b_k$ if $j \neq k$. Let Y be an open, connected neighborhood of a such that $\bar{Y}$ is compact and contained in U. Then $K = \pi^{-1}(\bar{Y})$ is compact in $\tilde{U}$. A constant $c > 1$ exists such that $\|\mathfrak{v}(x)\| \leq c$ for all $x \in K$. Define $c_0 = c^{q(q-1)-1}$. Take any $\varepsilon > 0$. For each $j \in \mathbb{N}[1,s]$ there is an open neighborhood Z_j of b_j in K such that

$$\|\mathfrak{v}(x) - \mathfrak{v}(b_j)\| < \frac{\sqrt{\varepsilon}}{2c_0} \quad \text{for all } x \in Z_j.$$

If $x \in Z_j$ and $y \in Z_j$, then

$$\|\mathfrak{v}(x) \wedge \mathfrak{v}(y)\| = \|(\mathfrak{v}(x) - \mathfrak{v}(b_j)) \wedge \mathfrak{v}(y) + \mathfrak{v}(b_j) \wedge (\mathfrak{v}(y) - \mathfrak{v}(b_j))\|$$

$$\leq \|\mathfrak{v}(x) - \mathfrak{v}(b_j)\| \, \|\mathfrak{v}(y)\| + \|\mathfrak{v}(b_j)\| \, \|\mathfrak{v}(y) - \mathfrak{v}(b_j)\| < \frac{c\sqrt{\varepsilon}}{c_0}.$$

There is an open neighborhood Z of a in Y such that $\tilde{Z} \subseteq Z_1 \cup \cdots \cup Z_s$. Take any $z \in Z$. If $z \in B'$, then $\delta(z) = 0 = \delta(a)$ and $\|\delta(z) - \delta(a)\| = 0 < \varepsilon$. Assume that $z \in Z - B'$. Enumerate $\tilde{z} = \{x_1, \cdots, x_q\}$ such that (6.2) holds. Define $P(z)$ by (6.3). Then

$$\|\mathfrak{v}(x_j) \wedge \mathfrak{v}(x_k)\| \leq \|\mathfrak{v}(x_j)\| \, \|\mathfrak{v}(x_k)\| \leq c^2$$

for all $(j,k) \in P(z)$. Also a pair $(j_0, k_0) \in P(z)$ exists such that $x_{j_0} \in Z_p$ and $x_{k_0} \in Z_p$ for some $p \in \mathbb{N}[1,s]$ since $s < q$. Then

$$\|\mathfrak{v}(x_{j_0}) \wedge \mathfrak{v}(x_{k_0})\| < \frac{c\sqrt{\varepsilon}}{c_0}.$$

Since $(j_0, k_0) \in P(z)$ implies $(k_0, j_0) \in P(z)$ we obtain

$$\|\delta(z) - \delta(a)\| = \|\delta(z)\| < \frac{c^2\varepsilon}{c_0^2} c^{2q(q-1)-4} = \varepsilon.$$

Therefore $\mathfrak{d}$ is continuous at a. Thus $\mathfrak{d}$ is holomorphic on U; q.e.d.

Let $\{U_\lambda\}_{\lambda\in\Lambda}$ be a covering of N-E′ by open, connected subsets $U_\lambda \neq \phi$ of N-E′ such that there is a reduced representation $\mathfrak{v}_\lambda : \tilde{U}_\lambda \longrightarrow V$ of f for each $\lambda \in \Lambda$. Let $\mathfrak{y}_\lambda : U_\lambda \longrightarrow \amalg_q V$ be the algebroid reduction of $\mathfrak{v}_\lambda$ for π. For $\lambda = (\lambda_0, \cdots, \lambda_p) \in \Lambda^{p+1}$ abbreviate

$$(6.6) \qquad U_\lambda = U_{\lambda_0 \cdots \lambda_p} = U_{\lambda_0} \cap \cdots \cap U_{\lambda_p}$$

$$(6.7) \qquad \Lambda[p] = \{\lambda \in \Lambda^{p+1} \mid U_\lambda \neq \phi\}.$$

For each $(\lambda,\mu) \in \Lambda[1]$, there exists one and only one holomorphic function $g_{\lambda\mu}: \tilde{U}_{\lambda\mu} \longrightarrow \mathbb{C}_*$ such that $\mathfrak{v}_\lambda = g_{\lambda\mu}\mathfrak{v}_\mu$ on $\tilde{U}_{\lambda\mu}$. Let $G_{\lambda\mu}: U_{\lambda\mu} \longrightarrow \mathbb{C}_*$ be the algebroid reduction of $g_{\lambda\mu}$ which is given by

$$(6.8) \qquad G_{\lambda\mu}(z) = \prod_{x\in\tilde{z}} g_{\lambda\mu}(x)^{\nu_\pi(x)} \qquad \text{for all } z \in U_\lambda,$$

Then $\mathfrak{y}_\lambda = G_{\lambda\mu}\mathfrak{y}_\mu$ on $U_{\lambda\mu}$. Hence $\{G_{\lambda\mu}\}_{(\lambda,\mu)\in\Lambda[1]}$ is a basic cocycle for the hyperplane section bundle L_g of the map g on N-E′.

<u>LEMMA 6.2</u>. Take $(\lambda,\mu) \in \Lambda[1]$. Then $\mathfrak{d}_\lambda = G_{\lambda\mu}^{2(q-1)}\mathfrak{d}_\mu$ on $U_{\lambda\mu}$.

<u>Proof</u>. Take $z \in U_{\lambda\mu}$. Enumerate $\tilde{z}$ such that (6.2) holds. Define P(z) by (6.3). There

$$\mathfrak{d}_\lambda(z) = \amalg_{(j,p)\in P(z)} \mathfrak{v}_\lambda(x_j)\wedge\mathfrak{v}_\lambda(x_p)$$

$$= \amalg_{(j,p)\in P(z)} (g_{\lambda\mu}(x_j)\mathfrak{v}_\mu(x_j))\wedge(g_{\lambda\mu}(x_p)\mathfrak{v}_\mu(x_p))$$

$$= \Big[\prod_{(j,p)\in P(z)} (g_{\lambda\mu}(x_j)g_{\lambda\mu}(x_p))\Big]\Big[\amalg_{(j,p)\in P(z)} (\mathfrak{v}_\mu(x_j)\wedge\mathfrak{v}_\mu(x_p))\Big]$$

$$= G_{\lambda\mu}(z)^{2(q-1)}\mathfrak{d}_\mu(z)$$

q.e.d.

Therefore one and only one holomorphic section D of $(N\times W)\otimes L_g^{2(q-1)}$ exists over N-E′ such that

$$D|U_\lambda = \delta_\lambda \otimes (\mathfrak{y}_\lambda^\Delta)^{2(q-1)} \tag{6.9}$$

where $\mathfrak{y}_\lambda^\Delta$ is the holomorphic frame of L_g over U_λ associated to $\mathfrak{y}_\lambda$. Because dim E′ ≤ m-2, the section D extends to a holomorphic section of $(N\times W)\otimes L_g^{2(q-1)}$ over N and is called <u>the discriminant section</u> of f.

The map f: M ⟶ ℙ(V) is said to <u>separate</u> the fibers of π if and only if there is a point $z \in N-(B' \cup I_f')$ such that $f|\tilde{z}$ is injective.

<u>LEMMA 6.3</u>. The following statements are equivalent.

(1) f separates the fibers of π.

(2) There exists $\lambda \in \Lambda$ such that $\delta_\lambda \not\equiv 0$.

(3) $\delta_\lambda \not\equiv 0$ for all $\lambda \in \Lambda$.

(4) $D \not\equiv 0$ is not the zero section.

<u>PROOF</u>. Since N-E′ is connected, (2) implies (4) which implies (3) which implies (2). Thus (2), (3) and (4) are equivalent. Take $z \in N-(B' \cup I_f')$. Thus $z \in U_\lambda$ and $\tilde{z} = \{x_1,\dots,x_q\}$ with $x_j \neq x_k$ if $j \neq k$. Then $f(x_j) \neq f(x_t)$ if and only if $\mathfrak{v}(x_j)$ and $\mathfrak{v}(x_t)$ are linearly independent. Hence $f|\tilde{z}$ is injective if and only if $\delta_\lambda(z) \neq 0$; q.e.d.

Assume that f separates the fibers of π. Then $D \not\equiv 0$ defines a non-negative divisor $\delta = \mu_D \geq 0$ called the <u>discriminant divisor</u>.

<u>THEOREM 6.4.</u>Assume that (G1)-(G5) hold. Assume that f separates the fibers of π. Let β be the branching divisor of π. Let δ be the discriminant divisor of f. Then

$$\pi_*(\beta) \leq \delta. \tag{6.10}$$

<u>REMARK</u>: In the case dim V = 2, the estimate was proved by J. Noguchi [12].

<u>Proof</u>: If supp $\pi_*(\beta) = \phi$, then $\pi_*(\beta) = 0$ and (6.10) is trivial. Assume that $C = \mathrm{supp}\, \pi_*(\beta) \neq \phi$. Then C is the union of all (m-1)-dimensional branches of B' and C-E' = B'-E'. Because $D|(B'-E') = 0$ by Lemma 6.1, we see that $C \subseteq \mathrm{supp}\, \delta$. Let H be the union of all branches of supp δ not contained in C. Then $\mathrm{supp}\, \delta = C \cup H$ and $H_0 = C \cap H$ is analytic with $\dim H_0 \leq m-2$. Taking $\nu = \beta$ define S by (4.59) and adopt the situation (S1)-(S10). Then $S_1 = C \cap (S \cup H_0)$ is analytic with $\dim S_1 \leq m-2$. Here $E' \subseteq S$ implies $B'-S_1 = C-S_1$. It suffices to prove $\pi_*(\beta) \leq \delta$ on $C-S_1$.

Take $a \in C-S_1$. Then $\tilde{a} = \{b_1,\ldots,b_s\}$ with $b_j \neq b_k$ for $j \neq k$ and $s < q$. Put $p = q - s > 0$. Define $q_j = \nu_\pi(b_j)$. Then $q_1+\ldots+q_s = q$. Also we can take the enumeration such that $q_j > 1$ if $1 \leq j \leq t$ and $q_j = 1$ if $t < j \leq s$. Since $a \in B'-E'$, we have $t \in \mathbb{N}[1,s]$.

An open, connected neighborhood Y of a exists in $N-(S \cup H) \subseteq N-(S_1 \cup H)$ such that there is a reduced representation $\mathfrak{v}: \tilde{Y} \longrightarrow V$ of f which defines the discriminant $\mathfrak{d}: Y \longrightarrow W$. By Lemma 4.20 there is an open, connected neighborhood U of a in Y and there are open, connected subsets U_j of $M-(\tilde{S}_1 \cup \tilde{H})$ with $b_j \in U_j$ such that $\tilde{U} = U_1 \cup \ldots \cup U_s$ is a disjoint union and such that $\pi: U_j \longrightarrow U$ is a proper, light, surjective, holomorphic map of sheet number q_j. If $t < j \leq s$, then $B \cap U_j = \phi$. If $1 \leq j \leq t$, then $\hat{B} \cap U_j = B \cap U_j$. There are biholomorphic maps

$$\varphi_j = (\varphi_{1j},\ldots,\varphi_{mj}): U_j \longrightarrow \mathbb{C}(1)^m \qquad \text{with } \varphi_j(b_j) = 0.$$

$$\alpha = (\alpha_1,\ldots,\alpha_m): U \longrightarrow \mathbb{C}(1)^m \qquad \text{with } \alpha(a) = 0.$$

and a holomorphic map

$$\psi_j = (\psi_{j1},\ldots,\psi_{jm}): \mathbb{C}(1)^m \longrightarrow \mathbb{C}(1)^m$$

$$\psi_j(x_1,\ldots,x_m) = (x_1,\ldots,x_{m-1},x_m^{q_j})$$

for all $x = (x_1,\ldots,x_m) \in \mathbb{C}(1)^m$ such that $\alpha\circ\pi|U_j = \psi_j\circ\varphi_j$. Here $B'\cap U = \alpha_m^{-1}(0)$. If $1 \le j \le t$, then

$$\alpha(B'\cap U) = \mathbb{C}(1)^{m-1} \times \{0\} = \varphi_j(B\cap U_j) \qquad B\cap U_j = \varphi_{mj}^{-1}(0)$$

Define $\pi_\lambda = \alpha_\lambda\circ\pi$ for $\lambda = 1,\ldots,m$. Then

$$d\pi_1\wedge\ldots\wedge d\pi_m|U_j = q_j(\varphi_{m,j})^{q_j-1}\, d\varphi_{1,j}\wedge\ldots\wedge d\varphi_{m,j}.$$

Since $d\varphi_{1,j}\wedge\ldots\wedge d\varphi_{m,j}(x) \ne 0$ for all $x \in U_j$, we obtain $\beta(x) = q_j-1$ for all $x \in B\cap U_j$ and $\beta(x) = 0$ for all $x \in U_j-B$ for $j = 1,\ldots,s$. Take $z \in B'\cap U$. Then $\alpha(z) = (x,0)$ with $x \in \mathbb{C}(1)^{m-1}$ and $\psi_j^{-1}(x,0) = \{(x,0)\}$. Define $z_j = \varphi_j^{-1}(x,0)$. Then $\{z_j\} = \tilde{Z}\cap U_j$ and $\tilde{z} = \{z_1,\ldots,z_s\}$. Theorem (4.21) implies

$$\pi_*(\beta)(z) = \sum_{j=1}^{s}\beta(z_j) = \sum_{j=1}^{s}(q_j-1) = q-s = p.$$

In particular $\pi_*(\beta)(a) = p$.

Since $U\cap \operatorname{supp}\delta = B'\cap U = \alpha_m^{-1}(0)$ with $d\alpha_m(z) \ne 0$ for all $z \in U$, there is a positive integer d such that $\delta(z) = d$ for all $z \in U\cap B'$. Moreover there is a holomorphic vector function $\mathfrak{d}_1 : U \longrightarrow W$ such that $\mathfrak{d} = \alpha_m^d\mathfrak{d}_1$ and such that $S_2 = \mathfrak{d}_1^{-1}(0)$ is an analytic set in U with $\dim S_2 \le m-2$.

Take any $z \in B'\cap U-S_2$. Then $\mathfrak{d}_1(z) \ne 0$. Also $\alpha(z) = (x,0)$ with $x \in \mathbb{C}(1)^{m-1}$ and $\tilde{z} = \{z_1,\ldots,z_s\}$ where $\{z_j\} = \tilde{z}\cap U_j$ and $\varphi_j(z_j) = (x,0)$. Take any $r \in \mathbb{R}(0,1)$. Then $(x,r) \in \mathbb{C}(1)^m$ and $z(r) = \alpha^{-1}(x,r) \in U-B'$ with $z(r) \longrightarrow z$ for $r \longrightarrow 0$. For $j \in \mathbb{N}[1,s]$ and $\lambda \in \mathbb{N}[1,q_j]$ define

$$\zeta_{j,\lambda} = e^{\frac{(2\pi i\lambda)}{q_j}} \qquad r_j = r^{\frac{1}{q_j}}.$$

Then

$$\psi_j^{-1}(\alpha(z(r)) = \psi_j^{-1}(x,r) = \{(x,\zeta_{j\lambda}r_j) \mid \lambda \in \mathbb{N}[1,q_j]\}.$$

Define $z_{j,\lambda}(r) = \varphi_j^{-1}(x,\zeta_{j\lambda} r_j) \in U_j$. Then

$$\{z_{j,\lambda}(r) \mid \lambda \in \mathbb{N}[1,q_j]\} = \varphi_j^{-1}(\psi_j^{-1}(\alpha(z(r)))) = (\pi|U_j)^{-1}(z) = \tilde{z} \cap U_j$$

$$\tilde{z} = \bigcup_{j=1}^{s} \{z_{j\lambda}(r) \mid \lambda \in \mathbb{N}[1,q_j]\}$$

where the union is disjoint. Abbreviate

$$\mathfrak{m}(r) = \coprod_{j=1}^{s} \coprod_{j\neq k=1}^{s} \coprod_{\lambda=1}^{q_j} \coprod_{\rho=1}^{q_k} \mathfrak{v}(z_{j,\lambda}(r)) \wedge \mathfrak{v}(z_{k,\lambda}(r))$$

$$\mathfrak{n}(r) = \coprod_{j=1}^{t} \coprod_{\lambda=1}^{q_j} \coprod_{\lambda\neq\rho=1}^{q_j} \mathfrak{v}(z_{j,\lambda}(r)) \wedge \mathfrak{v}(z_{j,\rho}(r)).$$

Then $\mathfrak{m}(r) \longrightarrow \mathfrak{m}(0)$ and $\mathfrak{n}(r) \longrightarrow \mathfrak{n}(0) = 0$ for $r \longrightarrow 0$. Also

$$\mathfrak{d}(z(r)) = \mathfrak{n}(r)\, \mathfrak{m}(r).$$

Define $\mathfrak{v}_j = \mathfrak{v} \circ \varphi_j^{-1} : \mathbb{C}(1)^m \longrightarrow V$. Holomorphic vector functions $\mathfrak{a}_j : \mathbb{C}(1)^{m-1} \longrightarrow V$ and $\mathfrak{b}_j : \mathbb{C}(1)^m \longrightarrow V$ exist such that

$$\mathfrak{v}_j(u,w) = \mathfrak{a}_j(u) + w\mathfrak{b}_j(u,w)$$

for all $u \in \mathbb{C}(1)^{m-1}$ and $w \in \mathbb{C}(1)$. Then

$$\mathfrak{v}(z_{j,\lambda}(r)) = \mathfrak{v}_j(x,\zeta_{j\lambda}(r)) = \mathfrak{a}_j(x) + \zeta_{j\lambda} r_j \mathfrak{b}_j(x,\zeta_{j\lambda}(r))$$

Abbreviate

$$\mathfrak{c}_{j\lambda\rho}(r) = \mathfrak{a}_j(x) \wedge (\zeta_{j\rho}\mathfrak{b}_j(x,\zeta_{j\rho} r_j) - \zeta_{j\lambda}\mathfrak{b}_j(x,\zeta_{j\lambda} r_j))$$

$$+ \zeta_{j\rho}\zeta_{j\lambda} r_j \mathfrak{b}_j(x,\zeta_{j\lambda} r_j) \wedge \mathfrak{b}_j(x,\zeta_{j\rho} r_j)$$

Then $\mathfrak{c}_{j\lambda\rho}(r) \longrightarrow \mathfrak{c}_{j\lambda\rho}(0)$ for $r \longrightarrow 0$ and

$$\mathfrak{v}(z_{j\lambda}(r)) \wedge \mathfrak{v}(z_{j\rho}(r)) = r_j \mathfrak{c}_{j\lambda\rho}(r)$$

Define

$$\mathfrak{n}_0(r) = \coprod_{j=1}^{s} \coprod_{\lambda=1}^{q_j} \coprod_{\lambda\neq\rho=1}^{q_j} \mathfrak{c}_{j\lambda\rho}(r)$$

we have

$$\prod_{j=1}^{s} \prod_{\lambda=1}^{q_j} \prod_{\lambda \neq \rho=1}^{q_j} r_j = \prod_{j=1}^{s} \prod_{\lambda=1}^{q_j} r_j^{q_j-1} = \prod_{j=1}^{s} r_j^{q_j(q_j-1)}$$

$$= \prod_{j=1}^{s} r^{q_j-1} = r^{q-s} = r^p$$

Therefore $\mathfrak{n}(r) = r^p \, \mathfrak{n}_0(r)$ and

$$r^d \mathfrak{d}_1(z(r)) = \mathfrak{d}(z(r)) = \mathfrak{n}(r)\mathfrak{m}(r) = r^p \mathfrak{n}_0(r)\mathfrak{m}(r)$$

where $\mathfrak{n}_0(r) \longrightarrow \mathfrak{n}_0(0)$ converges for $r \longrightarrow 0$. If $p > d$, then

$$0 \neq \mathfrak{d}_1(z) = \lim_{r \to 0} \mathfrak{d}_1(z(r)) = \lim_{r \to 0} r^{p-d} \mathfrak{n}_0(r)\mathfrak{m}(r) = 0.$$

Contradiction! Hence $\pi_*(\beta)(a) = p \le d = \delta(a)$. Thus $\pi_*(\beta) \le \delta$ on $C-S_1$. Because $\dim S_1 \le m-2$ we obtain $\pi_*(\beta) \le \delta$ on $C = \operatorname{supp} \pi_*(\beta)$. Thus $\pi_*(\beta) \le \delta$ on N; q.e.d.

Since $n_\beta = n_{\pi_*(\beta)}$ and $N_\beta = N_{\pi_*(\beta)}$ we obtain:

<u>COROLLARY 6.5</u>. Assume that (G1)-(G5) hold. Assume that f separates the fibers of π. If $t > 0$ and $0 < s < r$, then

(6.11) $$n_\beta(t) \le n_\delta(t) \qquad N_\beta(r,s) \le N_\delta(r,s)$$

Under the same assumptions, Lemma 6.2 and Lemma 6.3 imply that there is one and only one meromorphic map $d: N \longrightarrow \mathbb{P}(W)$ such that for each reduced representation $\mathfrak{v}: \tilde{U} \longrightarrow V$ over an open, connected subset U of N-E′ the discriminant $\mathfrak{d}: U \longrightarrow W$ is a representation of d on U. The map originally exists on N-E′ only, but continues to a meromorphic map on N since $\dim E' \le m-2$. Here d is called the <u>discriminant</u> <u>map</u> of f. Here Lemma 6.2 and (6.9) show that $L_d = (L_g)^{2q-2}$ is the hyperplane section bundle of d. Also (6.9) shows that D is a representation section of d. Let G be the reduced representation section of the

algebroid reduction map $g: N \longrightarrow \mathbb{P}(\amalg_q V)$. Define u_f by (4.27). Observe that $I_g \subseteq I'_f \cup E'$.

PROPOSITION 6.6. Assume that (G1)-(G5) holds. Let ℓ be the given hermitian metric on V. Also the induced hermitian metrics on $\amalg_q V$ and W and along the fibers of the trivial bundles $N \times \amalg_q V$ and $N \times W$ are denoted by ℓ. Let $\varkappa$ be a hermitian metric along the fibers of L_g inducing hermitian metrics $\rho = \ell \otimes \varkappa$ along the fibers of $(N \times \amalg_q V) \otimes L_g$ and $\gamma = \ell \otimes \varkappa^{2q-2}$ along the fibers of $(N \times W) \otimes L_d$ with $L_d = L_g^{2q-2}$. Thus non-negative functions $\|D\|_\gamma$ and $\|G\|_\rho$ on N are given where $\|G\|_\rho > 0$ on $N - I_g$. Then the function

$$\Delta = \frac{\|D\|_\gamma \, u_f^{2q-2}}{\|G\|_\rho} \quad \text{on } N-(I'_f \cup E') \tag{6.12}$$

does not depend on the choice of the hermitian metric $\varkappa$. Moreover

$$0 \leq \Delta \leq 1. \tag{6.13}$$

If $U \neq \phi$ is an open, connected subset of $N-E'$, if $\mathfrak{v}: \tilde{U} \longrightarrow V$ is a reduced representation of f, if $\mathfrak{y}$ is the algebroid reduction of $\mathfrak{v}$ and if $\mathfrak{d}$ is the discriminant of $\mathfrak{v}$, then

$$\Delta = \frac{\|\mathfrak{d}\| u_f^{2q-2}}{\|\mathfrak{y}\|^{2q-2}} \quad \text{on } U-(I'_f \cup E') \tag{6.14}$$

PROOF. Take $U, \mathfrak{v}, \mathfrak{y}$ and $\mathfrak{d}$ as indicated. Then

$$\|D\|_\gamma = \|\mathfrak{d} \otimes (\mathfrak{y}^\Delta)^{2q-2}\|_\gamma = \|\mathfrak{d}\| \, \|\mathfrak{y}^\Delta\|_\varkappa^{2q-2}$$

$$\|G\|_\rho = \|\mathfrak{y} \otimes \mathfrak{y}^\Delta\|_\rho = \|\mathfrak{y}\| \, \|\mathfrak{y}^\Delta\|_\varkappa^{2q-2}.$$

Thus (6.12) implies (6.14) on $U-(I'_f \cup E')$. Also (6.14) does not depend on $\varkappa$ and every point is covered by some such open set U. Thus Δ does not depend on $\varkappa$.

Take any $z \in U-(B' \cup I'_f)$. Then $\tilde{z} = \{x_1, \ldots, x_q\}$ with $x_j \neq x_k$ if $j \neq k$. We obtain

$$u_f(z) = \square f(x_1)\bullet\ldots\bullet f(x_q)\square = \frac{\|\mathfrak{v}(x_1)\ldots\mathfrak{v}(x_q)\|}{\|\mathfrak{v}(x_1)\|\ldots\|\mathfrak{v}(x_q)\|} = \frac{\|\mathfrak{y}(z)\|}{\|\mathfrak{v}(x_1)\|\ldots\|\mathfrak{v}(x_q)\|}$$

$$\|\delta(z)\| = \Big\|\coprod_{j=1}^{q}\coprod_{j\neq k=1}^{q}\mathfrak{v}(x_j)\wedge\mathfrak{v}(x_k)\Big\| \le \prod_{j=1}^{q}\prod_{j\neq k=1}^{q}\|\mathfrak{v}(x_j)\wedge\mathfrak{v}(x_k)\|$$

$$\le \prod_{j=1}^{q}\prod_{j\neq k=1}^{q}\|\mathfrak{v}(x_j)\|\,\|\mathfrak{v}(x_k)\| = \left(\prod_{j=1}^{q}\|\mathfrak{v}(x_j)\|\right)^{2q-2} = \left(\frac{\|\mathfrak{y}(z)\|}{u_f(z)}\right)^{2q-2}$$

$$\Delta(z) = \frac{\|\delta(z)\|\;u_f(z)^{2q-2}}{\|\mathfrak{y}(z)\|^{2q-2}} \le 1$$

By continuity and covering. $\Delta \le 1$ on $N-(I'_f \cup E')$, q.e.d.

If f separates the fibers of π, then $\Delta > 0$ on $N-((\mathrm{supp}\,\delta)\cup I_d \cup I'_f \cup E')$ and the integral

$$m_\Delta(r) = \int_{N\langle r\rangle} \log\frac{1}{\Delta}\,\sigma \ge 0 \tag{6.15}$$

exists for all $r \in \mathfrak{C}_\tau$. The function m_Δ is called the <u>discriminant compensation function</u>.

<u>THEOREM 6.7.</u> <u>THE FIRST MAIN THEOREM FOR DISCRIMINANTS</u>. Assume that (G1)-(G5) hold. Assume that f separates the fibers of π. Take $0 < s < r \in \mathfrak{C}_\tau$ with $s \in \mathfrak{C}_\tau$. Then

$$T_d(r,s) + N_\delta(r,s) + m_\Delta(r) - m_\Delta(s) = (2q-2)\,T_f(r,s). \tag{6.16}$$

<u>Proof</u>. Let $\varkappa$ be a hermitian metric along the fibers of L_g. Determine the hermitian metric γ and ρ as before. Then D is a representation section of d with divisor δ. Hence Theorem 2.3 and (2.56) implies

$$T_d(r,s) + N_\delta(r,s) = T(r,s,L_d,\varkappa^{2q-2}) + \int_{N\langle r\rangle}\log\|D\|_\gamma\sigma - \int_{N\langle s\rangle}\log\|D\|_\gamma\sigma \tag{6.17}$$

Since G is the reduced representation section of the reduction map g, we have

$$(6.18)\qquad T_g(r,s) = T(r,s,L_g,\varkappa) + \int_{N\langle r\rangle} \log\|G\|_\rho\sigma - \int_{N\langle s\rangle} \log\|G\|_\rho\sigma$$

Because $L_d = L_g^{2q-2}$ we have

$$(6.19)\qquad T(r,s,L_d,\varkappa^{2q-2}) = (2q-2)\,T(r,s,L_g,\varkappa)$$

Now (6.19), (6.18), (6.17) and (5.15) imply

$$\begin{aligned} &T_d(r,s) + N_\delta(r,s) - (2q-2)\,T_g(r,s) \\ &= \int_{N\langle r\rangle} \log\frac{\|D\|_\gamma}{\|G\|_\rho}\,\sigma - \int_{N\langle s\rangle} \log\frac{\|D\|_\gamma}{\|G\|_\rho}\,\sigma \\ &= -m_\Delta(r) + m_\Delta(s) + (2q-2)\left[\int_{N\langle r\rangle} \log\frac{1}{u_f}\,\sigma - \int_{N\langle s\rangle} \log\frac{1}{u_f}\,\sigma\right] \\ &= -m_\Delta(r) + m_\Delta(s) + (2q-2)\,(T_f(r,s) - T_g(r,s)) \end{aligned}$$

which gives us (6.16); q.e.d.

COROLLARY 6.8. Assume that (G1)-(G5) hold and that f separates the fibers of π. Take $0 < s < r$. Then

$$(6.20)\qquad N_\beta(r,s) \le N_\delta(r,s) + T_d(r,s) \le (2q-2)\,T_f(r,s) + m_\Delta(s),$$

If $n = 1$, then d is constant and $T_d(r,s) \equiv 0$. In this case the estimate (6.20) was proved by J. Noguchi [12].

7. THE LIFTING OF MEROMORPHIC MAPS

In this section we consider the postponed problem when is a suitable meromorphic map an algebroid reduction map. Let V be a complex vector space of dimension $n+1 > 1$. Let N be a connected, complex

manifold of dimension m. Take $q \in \mathbb{N}$ with $q \geq 2$. Let $g: N \longrightarrow \mathbb{P}(\amalg_q V)$ be a meromorphic map. Recall that

$$\mathbb{P}(V)^{(q)} = \{x_1 \ldots x_q \mid x_j \in \mathbb{P}(V)\} \tag{7.1}$$

is an irreducible, analytic subset of dimension nq of $\mathbb{P}(\amalg_q V)$. A necessary condition for g to be the algebroid reduction of some meromorphic map $f: M \longrightarrow \mathbb{P}(V)$ in respect to some proper, surjective holomorphic map $\pi: M \longrightarrow N$ of sheet number q is that g maps into $\mathbb{P}(V)^{(q)}$. Thus we assume that a meromorphic map $g: N \longrightarrow \mathbb{P}(V)^{(q)}$ is given. Such a map is said to be <u>reducible</u>, if and only if there are meromorphic maps $h: N \longrightarrow \mathbb{P}(V)^{(s)}$ and $k: N \longrightarrow \mathbb{P}(V)^{(t)}$ such that $s+t = q$ and $g = h \cdot k$. The map g is said to be <u>irreducible</u> if and only if g is not reducible. As before define a surjective holomorphic map

$$\rho: \mathbb{P}(V)^q \longrightarrow \mathbb{P}(V)^{(q)} \qquad \text{by } \rho(x_1, \ldots, x_q) = x_1 \ldots x_q. \tag{7.2}$$

The set Δ_q° of all $(x_1, \ldots, x_q) \in \mathbb{P}(V)^q$ with $x_j = x_k$ for at least one pair (j,k) with $j \neq k$ is a thin analytic subset of $\mathbb{P}(V)^q$. Hence $\Delta_q = \rho(\Delta_q^\circ)$ is a thin analytic subset of $\mathbb{P}(V)^{(q)}$. Lemma 3.4 implies

$$\rho^{-1}(\Delta_q) = \Delta_q^\circ, \tag{7.3}$$

A meromorphic map $g: N \longrightarrow \mathbb{P}(V)^{(q)}$ is said to be <u>non-degenerate</u> if and only if $g(N - I_g) \not\subseteq \Delta_q$.

<u>THEOREM 7.1</u>. Let V be a complex vector space of dimension $n+1 > 1$. Let N be a connected, complex manifold of dimension m. Take $q \in \mathbb{N}$ with $q \geq 2$. Let $g: N \longrightarrow \mathbb{P}(V)^{(q)}$ be a non-degenerate, irreducible meromorphic map. Then there exists a connected, complex manifold M of dimension m, a proper, surjective holomorphic map $\pi: M \longrightarrow N$ of sheet number q and a holomorphic map $f: M \longrightarrow \mathbb{P}(V)$ which separates the fibers of π such that g is the algebroid reduction of f for π.

The proof requires a number of preparations. Abbreviate

$$(7.4)\qquad \hat{\Omega}_q = \mathbb{P}(V)^q - \Delta_q^{\circ} \qquad \Omega_q = \mathbb{P}(V)^{(q)} - \Delta_q$$

Obviously $\hat{\Omega}_q$ is an open, connected subset of $\mathbb{P}(V)^q$ and Ω_q is an irreducible, analytic subset of $\mathbb{P}(\amalg_q V)-\Delta_q$ with dim $\Omega_q = nq$.

<u>LEMMA 7.2</u>. Ω_q is a closed, connected, complex submanifold of $\mathbb{P}(\amalg_q V)-\Delta_q$. The map $\rho\colon \hat{\Omega}_q \longrightarrow \Omega_q$ is proper, surjective and locally biholomorphic and constitutes a covering space of sheet number q!

<u>Proof</u>.(7.3) and the definitions of $\mathbb{P}(V)^{(q)}$and ρ show that $\rho\colon \hat{\Omega}_q \longrightarrow \Omega_q$ is well-defined, proper, surjective with $\#\rho^{-1}(y) = q!$ for each $y \in \Omega_q$. Take any $a \in \hat{\Omega}_q$. Let $d\rho(a)$ be the differential of the map $\rho\colon \hat{\Omega}_q \longrightarrow \mathbb{P}(\amalg_q V)$ at a.

1. <u>CLAIM</u>: <u>The linear map $d\rho(a)$ is injective</u>.

<u>Proof of the 1. Claim</u>. We have $a = (a_1,\ldots,a_q)$ with $a_j \neq a_k$ if $j \neq k$. Take $\mathfrak{a}_j \in V_*$ with $a_j = \mathbb{P}(\mathfrak{a}_j)$. Then $\mathfrak{a}_j \wedge \mathfrak{a}_k \neq 0$ if $j \neq k$. Take $\mathfrak{b}_j \in V_*^*$ such that $\langle \mathfrak{a}_j, \mathfrak{b}_j\rangle = 1$. Put $b_j = \mathbb{P}(\mathfrak{b}_j)$. Then $E[b_j]$ can be identified with the tangent space T_j of $\mathbb{P}(V)$ at a_j and $W_j = \mathbb{P}(V)-\ddot{E}[b_j]$ is an open neighborhood of a_j in $\mathbb{P}(V)$. Define

$$(7.5)\qquad D_j = \mathfrak{a}_j + E[b_j] = \{\mathfrak{x} \in V \mid \langle \mathfrak{x}, \mathfrak{b}_j\rangle = 1\}.$$

Then $T_j = E[b_j]$ is the tangent plane of the affine plane D_j at every $\mathfrak{x} \in D_j$. The restriction $\alpha_j = \mathbb{P}|D_j\colon D_j \longrightarrow W_j$ is biholomorphic. Then $W = W_1 x\ldots x W_q$ is an open, connected neighborhood of a in $\mathbb{P}(V)^q$ and $D = D_1 x\ldots x D_q$ is a connected complex manifold of dimension nq with $\mathfrak{a} = (\mathfrak{a}_1,\ldots,\mathfrak{a}_q) \in D$. The map $\alpha = (\alpha_1,\ldots,\alpha_q)\colon\ D \longrightarrow W$ is biholomorphic with $\alpha(\mathfrak{a}) = a$. Also $T = T_1\oplus\ldots\oplus T_q$ is the tangent space of D at $\mathfrak{a}$. Define $\mathfrak{a}_0 = \mathfrak{a}_1\ldots\mathfrak{a}_q$. Then $\mathbb{P}(\mathfrak{a}_0) = \rho(a) = a_0$. Take $\mathfrak{b}_0 \in V_*^*$ such that $\langle \mathfrak{a}_0, \mathfrak{b}_0^q\rangle = 1$. Put $\mathfrak{c} = \mathfrak{b}_0^q$ and $c = \mathbb{P}(\mathfrak{c})$. An affine plane is given

by

$$D_0 = \mathfrak{a}_0 + E[\mathfrak{c}] = \{\mathfrak{x} \in \amalg_q V \mid \langle \mathfrak{x}, \mathfrak{c} \rangle = 1\}. \tag{7.6}$$

Then $T_0 = E[c]$ is the tangent plane of D_0 at $\mathfrak{a}_0$ and $W_0 = \mathbb{P}(\amalg_q V) - \ddot{E}[c]$ is an open neighborhood of a_0 in $\mathbb{P}(\amalg_q V)$. The restriction $\alpha_0 = \mathbb{P}: D_0 \longrightarrow W_0$ is biholomorphic. Since $\rho(\alpha(\mathfrak{a})) = \rho(a) = a_0 \in W_0$, there is an open neighborhood U of $\mathfrak{a}$ in D such that $\rho(\alpha(U)) \subseteq W_0$. A holomorphic map $\varphi = \alpha_0^{-1} \circ \rho \circ \alpha: U \longrightarrow D_0$ is defined. Since α and α_0 are biholomorphic, it suffices to show that $d\varphi(\mathfrak{a}): T \longrightarrow T_0$ is injective. For $\mathfrak{x} = (\mathfrak{x}_1, \ldots, \mathfrak{x}_q) \in U$ define $h(\mathfrak{x}) \doteq 1/\langle \mathfrak{x}_1 \ldots \mathfrak{x}_q, \mathfrak{c} \rangle$. Then

$$\varphi(\mathfrak{x}) = h(\mathfrak{x})\mathfrak{x}_1, \ldots \mathfrak{x}_q. \tag{7.7}$$

Take $\mathfrak{v} = (\mathfrak{v}_1, \ldots, \mathfrak{v}_q) \in T$ with $d\varphi(\mathfrak{a}, \mathfrak{v}) = 0$. Then

$$0 = d\varphi(\mathfrak{a}, \mathfrak{v}) = dh(\mathfrak{a}, \mathfrak{v})\mathfrak{a}_1 \ldots \mathfrak{a}_q + h(\mathfrak{a}) \sum_{j=1}^{q} \mathfrak{a}_1 \ldots \mathfrak{a}_{j-1}\, \mathfrak{v}_j\, \mathfrak{a}_{j+1} \ldots \mathfrak{a}_q. \tag{7.8}$$

Take $k \in \mathbb{N}[1,q]$. Then $\mathfrak{v}_k \in T_k = E[b_k]$ implies $\langle \mathfrak{v}_k, \mathfrak{b}_k \rangle = 0$. For every $j \in \mathbb{N}[1,q]$ with $j \neq k$, the vectors $\mathfrak{a}_j$ and $\mathfrak{a}_k$ are linearly independent. Assume that $\mathfrak{v}_k \neq 0$. If $\lambda \in \mathbb{C}$ and $\mu \in \mathbb{C}$ are given with $\lambda \mathfrak{v}_k + \mu \mathfrak{a}_k = 0$. Then $0 = \langle \lambda \mathfrak{v}_k + \mu \mathfrak{a}_k, \mathfrak{b}_k \rangle = \mu$. Hence $\lambda = 0$. Thus $\mathfrak{v}_k$ and $\mathfrak{a}_k$ are linearly independent. Hence $\mathfrak{n} \in V_*^*$ exists such that $\langle \mathfrak{a}_k, \mathfrak{n} \rangle = 0$ and $\langle \mathfrak{v}_k, \mathfrak{n} \rangle \neq 0$ and $\langle \mathfrak{a}_j, \mathfrak{n} \rangle \neq 0$ for all $j \in \mathbb{N}[1,q]$ with $j \neq k$. Now (7.8) implies

$$0 = \langle d\varphi(\mathfrak{a}, \mathfrak{v})\mathfrak{n}^q \rangle = h(\mathfrak{a}) \langle \mathfrak{v}_k, \mathfrak{n} \rangle \prod_{k \neq j=1}^{q} \langle \mathfrak{a}_j, \mathfrak{n} \rangle \neq 0$$

The contradiction shows that $\mathfrak{v}_k = 0$ for all $k \in \mathbb{N}[1,q]$. Hence $\mathfrak{v} = 0$ the 1. Claim is proved.

This shows that $\rho: \hat{\Omega}_q \longrightarrow \mathbb{P}(\amalg_q V)$ is an immersion.

<u>2. CLAIM:</u> Ω_q <u>is a submanifold and</u> $\rho: \hat{\Omega}_q \longrightarrow \Omega_q$ <u>is locally biholomorphic</u>.

<u>PROOF OF THE 2. CLAIM</u>. The symmetric group $\mathfrak{S}_q$ operates on $\mathbb{P}(V)^q$. If $\pi \in \mathfrak{S}_q$, then a biholomorphic map $\hat{\pi}: \mathbb{P}(V)^q \longrightarrow \mathbb{P}(V)^q$ is defined by $\hat{\pi}(x_1,\ldots,x_q) = (x_{\psi(1)},\ldots,x_{\psi(q)})$ where $\psi = \pi^{-1}$. Then $\rho\circ\hat{\pi} = \rho$ and $\hat{\pi}(\hat{\Omega}_q) = \hat{\Omega}_q$. Take $a_0 \in \Omega_q$. Take $a \in \hat{\Omega}_q$ with $\rho(a) = a_0$. Then $\hat{\pi}(a) \neq \hat{\chi}(a)$ if $\pi \in \mathfrak{S}_q$ and $\chi \in \mathfrak{S}_q$ with $\pi \neq \chi$. Hence there exists an open neighborhood Y of a in $\hat{\Omega}_q$ such that $\hat{\pi}(Y) \cap \hat{\chi}(Y) = \phi$ if $\pi \in \mathfrak{S}_q$ and $\chi \in \mathfrak{S}_q$ with $\pi \neq \chi$. Also by the 1. Claim and the injective mapping theorem, there exist open, connected neighborhoods W of a in Y and Z of a_0 in $\mathbb{P}(\amalg_q V)$ such that $W' = \rho(W)$ is a smooth, connected, closed submanifold of Z and such that $\rho: W \longrightarrow W'$ is biholomorphic. Then $\hat{\pi}(W)$ is an open, connected neighborhood of $\hat{\pi}(a)$ in $\hat{\Omega}_q$ such that $\rho: \hat{\pi}(W) \longrightarrow W'$ is biholomorphic.Since $W \subseteq Y$ we have $\hat{\pi}(W) \cap \hat{\chi}(W) = \phi$ if $\pi \in \mathfrak{S}_q$ and $\chi \in \mathfrak{S}_q$ with $\pi \neq \chi$. Since $\mathbb{P}(V)^{(q)} = \mathbb{P}(V)^q/\mathfrak{S}_q$ carries the quotient topology,

$$\rho^{-1}(W') = \bigcup_{\pi\in\mathfrak{S}_q} \hat{\pi}(W) \tag{7.9}$$

is a disjoint union of open sets and W' is open in $\mathbb{P}(V)^{(q)}$ with $W' \subseteq \Omega_q$. Thus Ω_q is a connected complex manifold, $\rho: \hat{\Omega}_q \longrightarrow \Omega_q$ is locally biholomorphic and a covering space of sheet number q!, q.e.d.

Let N be a connected, complex manifold of dimension m. Let V be a complex vector space of dimension n+1 > 1. A holomorphic map $\mathbb{P}: N\times V_* \longrightarrow N\times\mathbb{P}(V)$ is defined by $\mathbb{P}(z,\mathfrak{x}) = (z,\mathbb{P}(\mathfrak{x}))$ for all $z \in N$ and $\mathfrak{x} \in V_*$. If $S \subseteq N\times V$ abbreviate $\mathbb{P}(S) = \mathbb{P}(S \cap (N\times V_*))$. A subset S of $N\times V$ is said to be <u>$\mathbb{P}$-saturated</u> if and only if $(z,\mathfrak{x}) \in S$ and $0 \neq \zeta \in \mathbb{C}$ imply $(z,\zeta\mathfrak{x}) \in S$.

<u>LEMMA 7.3.</u> Let A be a $\mathbb{P}$-saturated subset of $N\times V_*$. Let $\bar{A}$ be the closure of A in $N\times V$. Then the following statements are equivalent

(1) $\bar{A}$ is analytic in $N \times V$.

(2) A is analytic in $N \times V_*$.

(3) $B = \mathbb{P}(A)$ is analytic in $N \times \mathbb{P}(V)$.

<u>PROOF</u>. Trivially, (1) implies (2). Assume (2). Take $a \in \mathbb{P}(V)$. Take $\mathfrak{a} \in V_*$ and $\mathfrak{b} \in V_*^*$ with $\langle \mathfrak{a}, \mathfrak{b} \rangle = 1$ and $\mathbb{P}(\mathfrak{a}) = a$. Put $b = \mathbb{P}(\mathfrak{b})$. Then $W = \mathbb{P}(V) - \ddot{E}[b]$ is an open neighborhood of a and $D = \mathfrak{a} + E[b]$ is an affine plane with $\mathfrak{a} \in D$ and the restriction $\alpha = \mathbb{P}: N \times D \longrightarrow N \times W$ is biholomorphic. Hence $A' = \alpha(A \cap (N \times D))$ is an analytic subset of $N \times W$. Take $(z,x) \in B \cap (N \times W)$. Then $(z,\mathfrak{x}) \in A$ with $x = \mathbb{P}(\mathfrak{x})$ exists. Then $\mathfrak{x} \in V - E[b]$ and $\langle \mathfrak{x}, \mathfrak{b} \rangle \neq 0$. Because A is $\mathbb{P}$-saturated

$$\left(z, \frac{\mathfrak{x}}{\langle \mathfrak{x}, \mathfrak{b} \rangle}\right) \in A \cap (N \times D) \text{ and } (z,x) = \mathbb{P}(z,\mathfrak{x}) = \mathbb{P}\left(z, \frac{\mathfrak{x}}{\langle \mathfrak{x}, \mathfrak{b} \rangle}\right) \in A'.$$

Also $A' = \alpha(A \cap (N \times D)) = \mathbb{P}(A \cap (N \times D)) \subseteq \mathbb{P}(A) = B$. Hence $A' \subseteq B \cap (N \cap N)$. Hence $B \cap (N \times W) = A'$ is analytic in $N \times W$. Thus B is analytic in $N \times \mathbb{P}(V)$. Thus (2) implies (3).

Assume (3). Let $Y = \mathcal{O}(-1) = \{(x,\mathfrak{x}) \in \mathbb{P}(V) \times V \mid \mathfrak{x} \in E[x]\}$ be the universal bundle which is a smooth, closed complex submanifold of $\mathbb{P}(V) \times V$. Let $\pi: N \times Y \longrightarrow N \times \mathbb{P}(V)$ and $\sigma: N \times Y \longrightarrow N \times V$ be the projections. Here σ is proper. Hence $\hat{B} = \pi^{-1}(B)$ is analytic in $N \times Y$ and $C = \sigma(\hat{B})$ is analytic in $N \times V$. Take $(z,\mathfrak{x}) \in C$. Then $x \in \mathbb{P}(V)$ exists such that $(z,x,\mathfrak{x}) \in \hat{B}$. Thus $(z,x) \in B$. Then $\mathfrak{g} \in V_*$ exists such that $(z,x) = \mathbb{P}(z,\mathfrak{g})$ and $(z,\mathfrak{g}) \in A$. Then $x = \mathbb{P}(\mathfrak{g})$. If $\mathfrak{x} \neq 0$, then $\mathbb{P}(\mathfrak{x}) = x = \mathbb{P}(\mathfrak{g})$. A number $\zeta \in \mathbb{C}_*$ exists such that $\mathfrak{x} = \zeta\mathfrak{g}$. Since A is $\mathbb{P}$-saturated, $(z,\mathfrak{x}) \in A$. Therefore $C \subseteq \bar{A}$. Take $(z,\mathfrak{x}) \in A$. Then $\mathfrak{x} \neq 0$ and $(z,\mathbb{P}(\mathfrak{x}),\mathfrak{x}) \in N \times Y$ with $(z,\mathbb{P}(\mathfrak{x})) = \mathbb{P}(z,\mathfrak{x}) \in \mathbb{P}(A) = B$. Hence $(z,\mathbb{P}(\mathfrak{x}),\mathfrak{x}) \in \hat{B}$ and $(z,\mathfrak{x}) \in C$. Thus $A \subseteq C$ which implies $\bar{A} \subseteq C$. Therefore $\bar{A} = C$ is analytic; q.e.d.

Let B be an analytic subset of $\mathbb{P}(V)$. Then $A = \mathbb{P}^{-1}(B) \cup \{0\}$ is analytic and $\mathbb{P}$-saturated in V. Let $f: N \longrightarrow \mathbb{P}(V)$ be a meromorphic map.

The restriction $f_0 = f\colon N-I_f \longrightarrow \mathbb{P}(V)$ is holomorphic. The closure F of the graph F_0 of f_0 in $N\times\mathbb{P}(V)$ is analytic in $N\times\mathbb{P}(V)$. Let $\pi\colon F \longrightarrow N$ and $\varphi\colon F \longrightarrow \mathbb{P}(V)$ be the projections. Here π is proper. Then $f^{-1}(B) = \pi(\varphi^{-1}(B))$ is analytic in N with

$$f_0^{-1}(B) = f^{-1}(B) \cap (N-I_f). \tag{7.10}$$

Here $C = f_0^{-1}(B)$ is analytic in $N-I_f$ and $\bar{C}$ is analytic in N with

$$\bar{C} \subseteq f^{-1}(B) \subseteq C \cup I_f. \tag{7.11}$$

If U is an open, connected subset of N and if $\mathfrak{v}\colon U \longrightarrow V$ is a reduced representation of f, we easily obtain

$$\mathfrak{v}^{-1}(A) = U \cap (f^{-1}(B) \cup I_f) \tag{7.12}$$

which implies

<u>LEMMA 7.4</u>. Under these assumption the following statements are equivalent

(1) $f(M-I_f) \not\subseteq B$

(2) $f^{-1}(B)$ is thin analytic in N.

(3) $\mathfrak{v}(U) \not\subseteq A$ for some reduced representation $\mathfrak{v}\colon U \longrightarrow V$ of f.

(4) $\mathfrak{v}(U) \not\subseteq A$ for all reduced representations $\mathfrak{v}\colon U \longrightarrow V$ of f.

<u>PROOF OF THEOREM 7.1</u>. Let V be a complex vector space of dimension $n + 1 > 1$. Let N be a connected complex manifold of dimension m. Take $q \in \mathbb{N}$ with $q \geq 2$. Let $g\colon N \longrightarrow \mathbb{P}(V)^{(q)}$ be a non-degenerate, irreducible meromorphic map. We consider g as a map into $\mathbb{P}(\amalg_q V)$ also. Then $S = g^{-1}(\Delta_q) \cup I_f$ is a thin analytic subset of N. Let $\{\mathfrak{y}_\lambda\}_{\lambda\in\Lambda}$ be a family of reduced representations $\mathfrak{y}_\lambda\colon N_\lambda \longrightarrow \amalg_q V$ of g where $N_\lambda \neq \phi$ is open and connected and where $N = \bigcup_{\lambda\in\Lambda} N_\lambda$. Use the notations defined by (2.66) and (2.67). For each $(\lambda,\mu) \in \Lambda[1]$, there is a holomorphic

function $h_{\lambda\mu}: N_{\lambda\mu} \longrightarrow \mathbb{C}_*$ without zeroes such that $\mathfrak{v}_\lambda = h_{\lambda\mu}\mathfrak{v}_\mu$ on $N_{\lambda\mu}$. The set $\tilde{\Delta}_q = \mathbb{P}^{-1}(\Delta_q) \cup \{0\}$ is analytic and $\mathbb{P}$-saturated in $\sqcup_q V$ and $S \cap N_\lambda = \mathfrak{v}_\lambda^{-1}(\tilde{\Delta}_q)$ for all $\lambda \in \Lambda$.

Take $0 \neq e \in \bigwedge_{n+1} V^*$. A holomorphic map

$$\psi_\lambda: N_\lambda \times V \longrightarrow \sqcup_q \bigwedge_{n-1} V^* \tag{7.13}$$

is defined by

$$\psi_\lambda(z,\mathfrak{x}) = (e \llcorner \mathfrak{x})^q \llcorner \mathfrak{v}_\lambda(z) \qquad \text{for } (z,\mathfrak{x}) \in N_\lambda \times V. \tag{7.14}$$

If $(\lambda,\mu) \in \Lambda[1]$, then $\psi_\lambda(z,\mathfrak{x}) = h_{\lambda\mu}(z)\psi_\mu(z,\mathfrak{x})$ for $(z,\mathfrak{x}) \in N_{\lambda\mu} \times V$. Hence there exists one and only one analytic subset Y of $N\times V$ such that

$$Y \cap (N_\lambda \times V) = Y_\lambda = \psi_\lambda^{-1}(0)$$

$$= \{(z,\mathfrak{x}) \in N_\lambda \times V \mid (e \llcorner \mathfrak{x})^q \llcorner \mathfrak{v}_\lambda(z) = 0\}. \tag{7.15}$$

The set Y is $\mathbb{P}$-saturated. Hence $Q = \mathbb{P}(Y)$ is an analytic subset of $N\times\mathbb{P}(V)$ where $\mathbb{P}: N\times V_* \longrightarrow N\times\mathbb{P}(V)$ is defined by $\mathbb{P}(z,\mathfrak{x}) = (z,\mathbb{P}(\mathfrak{x}))$. Let $f_0: Q \longrightarrow \mathbb{P}(V)$ and $\pi_0: Q \longrightarrow N$ be the projections. Here π_0 is proper. Then $S_0 = \pi_0^{-1}(S) = Q \cap (S\times\mathbb{P}(V))$ is analytic in Q.

<u>1. CLAIM</u>: Take $a \in N-S$. Then there exists an open, connected neighborhood U of a and holomorphic maps $g_j: U \longrightarrow \mathbb{P}(V)$ and $\mathfrak{v}_j: U \longrightarrow V_*$ for $j = 1,\ldots,q$ such that the following properties are satisfied:

(P1): $g|U = g_1\cdots g_q$.

(P2): $U_j = \{(z,g_j(z)) \mid z \in U\}$ is an open connected subset of Q.

(P3): $\pi_0^{-1}(U) = U_1\cup\ldots\cup U_q = Q \cap (U\times\mathbb{P}(V))$ is a disjoint union.

(P4): $f_0|U_j = g_j\circ\pi_0|U_j$ for $j = 1,\ldots,q$.

(P5): $\mathfrak{v}_j: U_j \longrightarrow V_*$ is a reduced representation of f_0 for $j = 1,\ldots,q$.

(P6): $\gamma_j: U \longrightarrow U_j$ is biholomorphic with $\gamma_j(z) = (z, g_j(z))$ for all $z \in U_j$.

(P7): $\pi_0: U_j \longrightarrow U$ is biholomorphic with $(\pi_0|U_j)^{-1} = \gamma_j$ for $j = 1,\ldots,q$.

PROOF OF THE 1. CLAIM: Since $I_g \subseteq S$ and $a \in N-S$, the map g is holomorphic at a and $b = g(a) \in \Omega_q$ is defined. Take $\hat{b} = (b_1,\ldots,b_q) \in \hat{\Omega}_q$ such that $\rho(\hat{b}) = b$. Then $b = b_1 \ldots b_q$ and $b_j \neq b_k$ for $j \neq k$. By Lemma 7.2 there are open, connected neighborhoods W of b and W_j and b_j in $\mathbb{P}(V)$ with $W_j \cap W_k = \phi$ for $j \neq k$ such that $\tilde{W} = W_1 x \ldots x W_q$ is an open, connected neighborhood of $\hat{b}$ in $\hat{\Omega}_q$ and such that $\rho_a = \rho: \tilde{W} \longrightarrow W$ is biholomorphic. Here $W \subseteq \Omega_q$. There is an open neighborhood U of a in N-S such that $g(U) \subseteq W$ and such that $U \subseteq N_\lambda$ for some $\lambda \in \Lambda$. Also we can assume that U is biholomorphically equivalent to a ball in $\mathbb{C}^m$. A holomorphic map

$$\tilde{g} = \rho_a^{-1} \circ g = (g_1,\ldots,g_q): U \longrightarrow \tilde{W} \subseteq \mathbb{P}(V)^q \tag{7.16}$$

is defined where $g_j: U \longrightarrow \mathbb{P}(V)$ is holomorphic with $g_j(U) \subseteq W_j$ for $j = 1,\ldots,q$. Hence $g_j(U) \cap g_k(U) = \phi$ if $j \neq k$. then $g_1 \ldots g_q = \rho(g_1,\ldots,g_q) = g|U$ which proves (P1). Define U_j as in (P2) and γ_j as in (P6). Then $\gamma_j: U \longrightarrow U_j$ is biholomorphic and (P6) is established. Also $\tilde{U} = U_1 \cup \ldots \cup U_q$ is a disjoint union and a pure m-dimensional, closed, complex submanifold of $U \times \mathbb{P}(V)$. Because U is biholomorphically equivalent to a ball in $\mathbb{C}^m$, a reduced representation $\mathfrak{w}_j: U \longrightarrow V_*$ of g_j on U exists for $j = 1,\ldots,q$. Then

$$\mathfrak{w} = \mathfrak{w}_1 \ldots \mathfrak{w}_q: U \longrightarrow (\amalg_q V)_* \tag{7.17}$$

is a reduced representation of $g|U = g_1 \ldots g_q$ on U. A holomorphic function $h: U \longrightarrow \mathbb{C}_*$ exists such that $h\mathfrak{w} = \mathfrak{v}_\lambda|U$.

Take $j \in \mathbb{N}[1,q]$ and $z \in U$. Then

$$(e \llcorner \mathfrak{w}_j(z))^q \llcorner \mathfrak{y}_\lambda(z) = h(z)(e \llcorner \mathfrak{w}_j(z))^q \llcorner \mathfrak{w}_1(z) \ldots \mathfrak{w}_q(z)$$

$$= h(z) \coprod_{k=1}^{q} (e \llcorner \mathfrak{w}_j(z)) \llcorner \mathfrak{w}_k(z)$$

$$= h(z) \coprod_{k=1}^{q} e \llcorner (\mathfrak{w}_j(z) \wedge \mathfrak{w}_k(z)) = 0.$$

Therefore $(z, \mathfrak{w}_j(z)) \in Y$ and $(z, g_j(z)) = \mathbb{P}(z, \mathfrak{w}_j(z)) \in Q$. Thus $U_j \subseteq Q$ and $\tilde{U} \subseteq Q \cap (U \times \mathbb{P}(V))$.

Take $(z,x) \in Q \cap (U \times \mathbb{P}(V))$. Then $\mathfrak{x} \in V_*$ exists such that $(z, \mathfrak{x}) \in Y$ and $\mathbb{P}(\mathfrak{x}) = x$. Since $z \in U \subseteq U_\lambda$, we have $(z, \mathfrak{x}) \in Y_\lambda$. Therefore

$$0 = (e \llcorner \mathfrak{x})^q \llcorner \mathfrak{y}_\lambda(z) = h(z)(e \llcorner \mathfrak{x})^q \llcorner \mathfrak{w}_1(z) \ldots \mathfrak{w}_q(z)$$

$$= h(z) \coprod_{j=1}^{q} (e \llcorner \mathfrak{x}) \llcorner \mathfrak{w}_j(z)$$

$$= h(z) \coprod_{j=1}^{q} e \llcorner (\mathfrak{x} \wedge \mathfrak{w}_j(z)).$$

Since $h(z) \neq 0$, an index $j \in \mathbb{N}[1,q]$ exists such that $e \llcorner (\mathfrak{x} \wedge \mathfrak{w}_j(z)) = 0$. Thus $\mathfrak{x} \wedge \mathfrak{w}_j(z) = 0$ which implies $x = \mathbb{P}(\mathfrak{x}) = \mathbb{P}(\mathfrak{w}_j(z)) = g_j(z)$. Therefore $(z,x) = (z, g_j(z)) \in U_j \subseteq \tilde{U}$. Thus (P3) is proved which implies (P2) and (P7).

Take $(z,x) \in U_j$, then $x = g_j(z)$ and $z = \pi_0(z,x)$ and

$$f_0(z,x) = x = g_j(z) = g_j(\pi_0(z,x)).$$

Hence (P4) holds. Also $\mathfrak{v}_j = \mathfrak{w}_j \circ \pi_0 \colon U_j \longrightarrow V_*$ is holomorphic with

$$f_0(z,x) = g_j(z) = \mathbb{P}(\mathfrak{w}_j(z)) = \mathbb{P}(\mathfrak{v}_j(z,x)).$$

Thus $\mathfrak{v}_j \colon U_j \longrightarrow V_*$ is a reduced representation of f_0. The 1. Claim is proved.

Consequently, $Q - S_0$ is a pure m-dimensional, complex manifold with $Q - S_0 = \pi_0^{-1}(N-S)$. The map $\pi \colon Q - S_0 \longrightarrow N - S$ is surjective, proper,

locally biholomorphic and constitutes a q-sheeted covering space.

Let $\mathfrak{B}$ be the set of branches of Q. Define $\mathfrak{B}_0 = \{B \in \mathfrak{B} | B \subseteq Q\}$. Then dim B = m for all $B \in \mathfrak{B}_0$. Also $\mathfrak{B}_0 \neq \phi$. Hence $P = \bigcup_{B \in \mathfrak{B}_0}$ is a pure m-dimen sional analytic subset of $N \times \mathbb{P}(V)$. Also $P \subseteq Q$ and P is the closure of $Q-S_0$ in Q (or in $N \times \mathbb{P}(V)$). The set $S_1 = P \cap S_0$ is thin analytic in P and $P-S_1 = Q-S_0$ is a pure m-dimensional, complex submanifold of $(N-S) \times \mathbb{P}(V)$. Hence the singular set $\Sigma(P)$ of P is contained in S_1. By Hironaka's desingularization theorem, there exists a pure m-dimensional, complex manifold M, a proper, surjective holomorphic map $\eta: M \longrightarrow P$ such that $T = \eta^{-1}(\Sigma(P))$ is thin analytic in M and such that $\eta: M-T \longrightarrow P-\Sigma(P)$ is biholomorphic. The set $\tilde{S} = \eta^{-1}(S_1)$ is thin analytic in M with $T \subseteq \tilde{S}$. Hence $\eta: M-S \longrightarrow P-\tilde{S}$ is biholomorphic. Then $\pi = \pi_1 \circ \eta: M \longrightarrow N$ is a surjective, proper, holomorphic map and $\pi: M-\tilde{S} \longrightarrow N-S$ is proper, surjective, locally biholomorphic and constitutes a q-sheeted covering space. Thus $\pi: M \longrightarrow N$ has sheet number q. The map $f = f_1 \circ \eta: M \longrightarrow \mathbb{P}(V)$ is holomorphic.

Still we have to prove that M is connected and that g is the algebroid reduction of f. Let $\{M_\alpha\}_{\alpha \in A}$ be the family of connectivity components of M such that $M_\alpha \cap M_\beta = \phi$ if $\alpha \neq \beta$. Each M_α is open and closed in M. Hence M_α is a connected complex manifold of dimension m and $\pi: M_\alpha \longrightarrow N$ is a proper holomorphic map which is light on $M_\alpha - \tilde{S} \neq \phi$. Hence $\pi: M_\alpha \longrightarrow N$ is surjective and has finite sheet number q_α. Because $M = \bigcup_{\alpha \in A} M_\alpha$ is a disjoint union, $\#A \leq q$ and $\sum_{\alpha \in A} q_\alpha = q$. Let h_α be the algebroid reduction of $f|M_\alpha$.

Take $a \in N-S$. Because $\eta: M-\tilde{S} \longrightarrow P-S_1 = Q-S_0$ is biholomorphic we can identify $M-\tilde{S} = Q-S_0$ such that η becomes the identity on this subset and such that $\pi = \pi_1 = \pi_0$ on this subset. The 1. Claim holds. Define

$$I_\alpha = \{j \in \mathbb{N}[1,q] \mid U_j \subseteq M_\alpha\}$$

Then $\mathbb{N}[1,q] = \bigcup_{\alpha\in A} I_\alpha$ is a disjoint union. Take $z \in U$. Then

$$h_\alpha(z) = \amalg_{j\in I_\alpha} f(\gamma_j(z)) = \amalg_{j\in I_\alpha} g_j(\pi(\gamma_j(z))) = \amalg_{j\in I_\alpha} g_j(z)$$

$$\amalg_{\alpha\in A} h_\alpha(z) = \amalg_{\alpha\in A} \amalg_{j\in I_\alpha} g_j(z) = g_1(z)\ldots g_q(z) = g(z).$$

Hence $g = \amalg_{\alpha\in A} h_\alpha$ on U and consequently on N. Because g is irreducible #A = 1 and M is connected. Also g is the algebroid reduction of f; q.e.d.

REFERENCES

[1] L. Ahlfors. The theory of meromorphic curves. Acta Soc. Sci. Fenn, Nova Ser. A 3 (4) (1941) 171-173.

[2] A. Andreotti and W. Stoll. Analytic and algebraic dependence of meromorphic functions. Lecture Notes in Mathematics 234 (1971) 390 pp. Springer-Verlag.

[3] D. Barlet. Espace analytique reduit des cycles analytiques complexes compacts d'un espace analytique complexe de dimension finie. Lecture Notes in Mathematics 482 Fonctions des Plusieurs variables complexes II (1975) 1-158. Springer-Verlag.

[4] H. Cartan. Sur les systèmes de fonctions holomorphes à varieties linéaires et leurs applications. Ann Sci. École Norm. Sup. (3) 45 (1928) 255-346.

[5] H. Cartan. Sur les zéros des combinaisons linéaires de p fonctions holomorphes données. Mathematica(Cluj) 7 (1933) 5-31.

[6] W. H. Greub: Multilinear Algebra. Die Grundl. d. Math. Wiss. 136 (1967) pp. 224. Springer-Verlag.

[7] W.K. Hayman: Meromorphic functions. Oxford Math. Monographs. (1964) pp. 191 Calderon Press, Oxford.

[8] R. Nevanlinna: Le Théorème de Picard-Borel et la Théorie des Fonctions Meromorphes. Gauthiers-Villars, Paris (1929) reprint Chelsea Publ.Co. New York (1974) pp. 171.

[9] R. Nevanlinna: Eindeutige analytische Funktionen. 2nd ed. Die Grundl. d. Math. Wiss. 46 (1953) pp. 379. Springer-Verlag.

[10] K. Niino and M. Ozawa: Deficiencies of an entire algebroid function. Kōdai Math. Sem. Report 22 (1970) 98-113.

[11] J. Noguchi. On the deficiencies and the existence of Picard's exceptional values of entire algebroid functions. Kōdai. Math. Sem. Report 26 (1974) 29-35.

[12] J. Noguchi: Meromorphic Mappings of a Covering Space over $\mathbb{C}^m$ into a Projective Variety and Defect Relations. Hiroshima Math. J. 6 (1976) 265-280.

[13] J. Noguchi: On value distribution of meromorphic mappings of covering spaces over $\mathbb{C}^m$ into algebroid varieties. Preprint pp. 35.

[14] V.P. Petrenko: Growth and distribution of values of algebroid functions. Mat. Zametki 26 (1979) 513-522, 653. Engl. Math. Notes 26 (1979) 3-4, 746-751 (1980).

[15] R. Remmert: Holomorphe und meromorphe Abbildungen komplexer Räume. Math. Annalen 133 (1957) 328-370.

[16] R. Remmert and K. Stein: Über die wesentlichen Singularitäten analytischer Mengen. Math. Annalen 126 (1953), 263-306.

[17] H.L. Selberg: Algebroide Funktionen und Umkehrfunktionen Abelscher Integrale Avh. Norske Vid. Acad. Oslo 8 (1934)1-72.

[18] B.V. Shabat: Distribution of Values of Holomorphic Mappings. Transl. of Math. Monog. 61 (1985) pp. 225.

[19] N. Steinmetz: Eine Verallgemeinerung des zweiten Nevanlinnaschen Hauptsatzes. Preprint 1986 pp. 11.

[20] W. Stoll: Die beiden Hauptsätze der Wertverteilungstheorie bei Funktionen mehrerer komplexer Veränderlichen. I Acta Math. 90 (1953) 1-115 II Acta Math. 92 (1954) 55-169.

[21] W. Stoll: The multiplicity of a holomorphic map. Invent. Math. 2 (1966) 15-58.

[22] W. Stoll: Value distribution of holomorphic maps into compact complex manifolds. Lecture Notes in Mathematics 135 (1970) pp. 267. Springer-Verlag.

[23] W. Stoll: Deficit and Bezout estimates. Value Distribution Theory. Part B. (ed. by R.O. Kujala and A.L. Vitter III). Pure and Appl. Math. 25 Marcell Dekker, New York, (1973) pp. 271.

[24] W. Stoll: Value distribution on parabolic spaces. Lecture Notes in Mathematics 600 (1977), pp. 216. Springer-Verlag.

[25] W. Stoll: The characterization of strictly parabolic manifolds. Annali di Pisa 7 (1980) 87-154.

[26] W. Stoll: The characterization of strictly parabolic spaces. Compositio Mathematics 44 (1981) 305-373.

[27] W. Stoll: Introduction to value distribution theory of meromorphic maps. Lecture Notes in Mathematics 950 (1982) 210-359. Springer-Verlag.

[28] W. Stoll: Value distribution and the lemma of the logarithmic derivative on poly-discs. Internat. J. Math. Sci. 6 (1983) No. 4. 617-669.

[29] W. Stoll: The Ahlfors-Weyl theory of meromorphic maps on parabolic manifolds. Lecture Notes in Mathematics 981 (1983) 101-219. Springer-Verlag.

[30] W. Stoll: Value distribution theory of meromorphic maps. Aspects of Mathematics E7 1985 pp. 347. Vieweg-Verlag.

[31] N. Toda: Sur la croissance de fonctions algebroides a valeurs deficientes. Kōdai Math. Sem. Rep. 22 (1970) 324-337.

[32] N. Toda: Sur lex combinaisons exceptionnelles de fonctions holomorphes, applications aux fonctions algebroides. Tôhoku Math. Journ. 22 (1970) 290-319.

[33] N. Toda: Sur le nombre de combinaissons exceptionnelles: applications aux fonctions algebroides. Tôhoku Math. Journ. 22 (1970) 480-491.

[34] Ch.Ch.Tung: The first main theorem of value distribution theory on complex spaces. Atti della Acc. Naz. d. Lincei Serie VIII 15 (1979), 93-262.

[35] G. Valiron: Sur la derivé de fonctions algebroides. Bull Soc. Math. France 59 (1931) 17-29.

[36] H. Weyl and J. Weyl. Meromorphic functions and analytic curves. Annals of Math. Studies 12 Princeton Univ. Pr.(1943) pp. 269.

[37] H. Wu: The equidistribution theory of holomorphic curves. Annals of Mathematics Studies 64 Princeton Univ. Pr (1970) pp. 219.

Author's address

University of Notre Dame
Department of Mathematics
Notre Dame, Indiana 46556

AN INFINITE-DIMENSIONAL GENERALIZATION OF THE SHILOV BOUNDARY AND INFINITE DIMENSIONAL ANALYTIC STRUCTURES IN THE SPECTRUM OF A UNIFORM ALGEBRA

Toma V. Tonev

We introduce a generalization of the classical Shilov boundary of a commutative Banach algebra that is suitable for the investigation of infinite-dimensional analytic structures living in the maximal ideal space (the spectrum) of a uniform algebra. The one-dimensional case, created (for the boundary) by Shilov [1], and (for analytic structures in the spectrum) by Bishop [2], was carried over to n-dimensions by Sibony [5] and Basener [6] (for the boundary), and for n-dimensional analytic structures in the spectrum by Basener [6], Sibony [5], Kramm [7] and others. Another, simpler definition as well as a detailed investigation of Sibony-Basener's generalization of the Shilov boundary, based on the class of all nonvanishing continuous mappings from the spectrum into $\mathbb{C}^n$ was given in [8]. In [9] and [10] this definition was carried out to the case of some classes of continuous mappings from the spectrum to a normal space. Here we interprete these results for the case of the Banach space ℓ^∞. In section 3 an ∞-dimensional generalization of Bishop's [2] and Basener's [6] results about existence of analytic structures in the spectrum of a uniform algebra are given.

1. Classical boundaries and Shilov boundary.

Let A be a commutative Banach algebra over $\mathbb{C}$ with unit. By SpA will be denoted as usual the spectrum of A, i.e. the maximal ideal space of A, $\hat{f}$ will denote the Gelfand transform of $f \in A$ (by definition $\hat{f}(m)=m(f)$, $m \in A$) and by $\sigma(f)$ will be denoted the spectrum of $f \in A$, i.e. $\sigma(f)=\hat{f}(SpA)$. Recall that a boundary of A is said to be any subset E of SpA on which all Gelfand transforms $\hat{f}(f \in A)$ assume the maximum of their modulus $|\hat{f}(m)|$.

One of the important properties of any closed boundary E of A is that

$$(1) \qquad \hat{f}(E) \supset b\sigma(f) \text{ for any } f \in A,$$

where $b\sigma(f)$ denotes the topological boundary of $\sigma(f)$ (e.g.[4]). It is easy to see on the other hand, that for a given set $E \subset SpA$ the fulfillment of (1) implies that E is a boundary of A. Indeed since $b\sigma(f) \subset \hat{f}(E) \subset \hat{f}(SpA) = \sigma(f)$ it follows:

$$\max_{b\sigma(f)} |\hat{f}(m)| \le \max_{E} |\hat{f}(m)| \le \max_{SpA} |\hat{f}(m)| = \max_{\sigma(f)} |z| =$$

$$= \max_{b\sigma(f)} |z| = \max_{b\sigma(f)} |\hat{f}(m)|$$

Hence $\max_{E} |\hat{f}(m)| = \max_{SpA} |\hat{f}(m)|$, i.e. E is a boundary.

So we have the following

<u>Proposition 1</u> A closed set $E \subset SpA$ is a boundary of A iff (1) is fulfilled for any $f \in A$.

Equivalently,

<u>Proposition 2</u>. A closed subset E of SpA is a boundary of A if and only if

$$(2) \qquad \min |f(m)| \le \min\{|z| : z \in b\sigma(f)\} \quad \text{for any } f \in A$$

<u>Proof</u>: If E is a closed boundary then, according to (1), $\hat{f}(E) \supset b\sigma(f)$ and hence

$$\min_{E} |\hat{f}(m)| = \min_{\hat{f}(E)} |z| \le \min_{b\sigma(f)} |z| .$$

If on the other hand we suppose that a closed Set E satisfying (2) is not a boundary, then according to Proposition 1 there exists a $f \in A$, such that $\hat{f}(E) \not\supset b\sigma(f)$. If $m_o \in SpA$ is such that $\hat{f}(m_o) \in b\sigma(f) \setminus \hat{f}(E)$, then for $h = \hat{f} - f(m_o) \in A$ we have: $\hat{h}(m_o) = 0 \in b\sigma(h)$, but $\hat{h}(E) \not\ni 0$, so that $0 = \min_{b\sigma(h)} |z| < \min_{E} |h(m)|$ and we are done.

It is easy to see that condition (2) can be written as:

(2') $\Delta(\min_E |\hat{f}(m)|)$ is contained either entirely inside or entirely outside $\sigma(f)$,

where $\Delta(r)$ is the disk with radius r, centered at 0.

By A^{-1} in the sequel will be denoted the set of all invertible elements, by A_o, will be denoted the set of all elements $f \in A$, for which $0 \in b\sigma(f)$, by A_o' will be denoted the set of all elements $f \in A$, for which $0 \in \sigma(\partial A)$, and by A_o'' will be denoted the set of all generalized divisors of zero of A (Recall that $f \in A$ is a generalized divisor of zero of A iff there exists a sequence $\{g_n\} \subset A$ with: a) $\inf \|g_n\| > 0$; b) $\lim_{n \to \infty} \|fg_n\| = 0$).

The following inclusions take place: $A \setminus A^{-1} \supset A_o'' \supset A_o$, $A \setminus A^{-1} \supset A_o' \supset A_o$ and $A_o' = A_o''$ for uniform algebras (see [1]). Because of the invariance of condition (1) under translations, to be a boundary of A, a closed set $E \subset SpA$ needs to satisfy (1) only for those $f \in A$ that belong to one of the sets: $A^{-1}, A_o, A \setminus A^{-1}, A_o', A_o''$. Now we have the following

<u>Proposition 3</u>. A closed set $E \subset SpA$ is a boundary of A iff one of the following equivalent conditions holds true:

(3_1) $\min |\hat{f}(m)| = \min_{spA} |\hat{f}(m)|$ for any $f \in A^{-1}$;

(3_2) $\min_E |\hat{f}(m)| = 0$ for any $f \in A_o$;

(3_3) $\Delta(\min_E |f(m)|) \subset \sigma(f)$ for any $f \in A \setminus A^{-1}$;

(3_4) Condition (3_3) for any $f \in A_o'$ only;

(3_5) Condition (3_3) for any $f \in A_o''$ only.

Indeed conditions (3_1)-(3_5) are the explicit expression of (1) (or (2), (2')) for the corresponding classes.

<u>REMARK</u>. Condition (3_5) implies that $\hat{f}$ vanishes at ∂A for any generalized divizor of zero. As shown by Shilov (see [1]) this is also sufficient for E to be a boundary.

Recall that the <u>Shilov boundary</u> ∂A of A is the (nonempty) intersection of all closed boundaries of A. As a corollary from Proposition 3 we get the following characterization of Shilov boundary:

<u>THEOREM 1.</u> The Shilov boundary ∂A of a commutative Banach algebra A coincides with the smallest among all closed subsets E of A, that satisfy one of the equivalent conditions (1), (2), (2') and (3_1)-(3_5).

2. <u>∞-boundaries and minimal ∞-boundary.</u>

Let $\ell^{\infty}A$ be the class of all ℓ^{∞}-sequences $F=(f_1,\dots,f_j,\dots)$ over A, i.e. $\ell^{\infty}A=\{F=(f_1,\dots,f_j,\dots);\ \|F\|_{\infty}=\sup_j\|f_j\|<\infty\}$. By $\hat{F}$ we will denote the (not necessarily continuous) open map: $\hat{F}\colon SpA \to \ell^{\infty}=\ell^{\infty}(\mathbb{C})$, $\hat{F}(m)=(\hat{f}_1(m),\dots,\hat{f}_j(m),\dots)$ and $\sigma(F)$ will denote the spectrum of F, i.e. $\sigma(F)=\hat{F}(SpA)$.

<u>Definition 1</u>. A subset E of SpA is said to be an <u>∞-boundary</u> of A if

(4) $[\hat{F}(E)]\supset b\sigma(F)$ for any $F\in\ell^{\infty}A$.

<u>Proposition 4</u>. A subset E of SpA is an ∞-boundary of A iff

(5) $$\inf_E\|\hat{F}(m)\|_{\infty} \le \inf\{\|Y\|_{\infty}\colon Y\in b\sigma(F)\} \text{ for any } F\in\ell^{\infty}A.$$

<u>Proof.</u> If E is a boundary, then according to (4), $[\hat{F}(E)]\supset b\sigma(F)$. Hence

$$\inf_E\|\hat{F}(m)\|_{\infty} = \inf_{\hat{F}(E)}\|Y\|_{\infty} \le \inf_{b\sigma(F)}\|Y\|_{\infty}$$

Suppose on the other hand that E is not an ∞-boundary. Then there will exist a $F\in\ell^{\infty}A$, such that $[\hat{F}(E)]\not\supset b\sigma(F)$. If $m_o\in SpA$ is such that $\hat{F}(m_o)\in b\sigma(F)\setminus[\hat{F}(E)]$, then for the element $H=F-\hat{F}(m_o)\in\ell^{\infty}A$ we have: $\hat{H}(m_o)=0\in b\sigma(H)$, but $[\hat{H}(E)]\not\ni 0$ and therefore $0=\inf_{b\sigma(H)}\|Y\|_{\infty} < \inf_E\|\hat{H}(m)\|_{\infty}$.

Condition (5) can be rewritten in an equivalent way as:

(5') For any $F\in\ell^{\infty}A$, $B_{\infty}(\inf_E\|\hat{F}(m)\|_{\infty})$ is either contained entirely in $\sigma(F)$ or in $\ell^{\infty}\setminus\sigma(F)$,

where $B_\infty(r)$ is the open ball with radius r in ℓ^∞, centered at 0.

Because of the invariance of condition (5) under translations in ℓ^∞, an ∞-closed set $E \subset \mathrm{SpA}$ will be a ∞-boundary of A if it satisfies (5) for f belonging to one of the following classes only:

$\ell^\infty_* A = \{F \in \ell^\infty A: \sigma(F) \not\ni 0\}$ (<u>regular</u>-elements of $\ell^\infty A$)

$\ell^\infty_o A = \{F \in \ell^\infty A: b\sigma(F) \ni 0\}$;

$\ell^\infty A \setminus \ell^\infty_* A = \{F \in \ell^\infty A: \sigma(F) \ni 0\}$ (non-<u>regular</u>-elements $\ell^\infty A$).

Now we have the following:

<u>Proposition 5.</u> A set $E \subset \mathrm{SpA}$ is an ∞-boundary of A iff one of the following equivalent conditions holds true:

(6_1) $\inf_E \|\hat{F}(m)\|_\infty = \inf_{\mathrm{SpA}} \|\hat{F}(m)\|_\infty$ for any $F \in \ell^\infty_* A$;

(6_2) $\inf_E \|\hat{F}(m)\|_\infty = 0$ for any $F \in \ell^\infty_o A$;

(6_3) $B_\infty(\inf_E \|\hat{F}(m)\|_\infty) \subset \sigma(F)$ for any $F \in \ell^\infty A \setminus \ell^\infty_* A$.

Indeed conditions (6_1)-(6_3) are the explicit expression of (1) (or (5), (5')) for the corresponding classes.

It is obvious that for each ∞-boundary E of A the following condition also holds true:

(7) $\sup_E \|\hat{F}(m)\|_\infty = \sup_{b\sigma(F)} \|Y\|_\infty = \sup_{\mathrm{SpA}} \|\hat{F}(m)\|_\infty$ for any $F \in \ell^\infty A$,

but unlike the situation in the classical case, as we shall see below, (7) is not sufficient to guarantee that E is an ∞-boundary.

As we mentioned at the beginning of the paragraph, any $\hat{F} \in \ell^\infty A$ is an open mapping but not necessarily continuous with respect to the weak* topology on SpA. That is why we shall consider SpA as a bitopological space, providing it except with the weak* topology also with the weakest topology τ_∞ with respect to which all the mappings $\hat{F} \in \hat{\ell}^\infty A$ are continuous. Let us give some notations.

Let $X=(x_1,\ldots, x_{j-1}, x_j, x_{j+1},\ldots)$ is a point of ℓ^∞. By X^j will be denoted the point $X^j=(x_1,\ldots,x_{j+1},0,x_{j+1}\ldots)$. The complex line in

$\mathbb{C}^\infty$ through X parallel to the j-th coordinate axis will be denoted by $\mathbb{C}(X,j)$, i.e. $\mathbb{C}(X,j) = \{(x_1,\dots, x_{j-1}, z, x_{j+1},\dots): z\in\mathbb{C}\}$ and the complex plane through X and Y, namely $\{Z=X+zY: z\in\mathbb{C}\}$ will be denoted by $\mathbb{C}(X,Y)$. Obviously $\mathbb{C}(X,j)=\mathbb{C}(X^j,e_j)$ where $e_j = (0,\dots, 0,\underset{-j-}{1},0,\dots 0)$. Given a closed subset V of SpA by A_V will be denoted the closure in C(V) of restrictions of all elements of $\hat{A}$.

An important role in what follows will be played by the following set:

$$(8)\quad \partial_\infty A = [\cup\{\partial A_{\hat{G}^{-1}(\mathbb{C}(X,j))};\ G\in\ell^\infty,\ X\in\ell^\infty,\ j\in\mathbb{Z}_+\}]_\infty,$$

where the closure is taken with respect to topology τ_∞. Obviously, $\partial_\infty A$ is a boundary of A. Indeed, by taking $G=X^j$ we get: $\hat{G}^{-1}(\mathbb{C}(x,j)) =$ SpA, hence $A_{\hat{G}^{-1}(\mathbb{C}(X,j))} = \hat{A}$ and $\partial A_{\hat{G}^{-1}(\mathbb{C}(X.j))} = \partial\hat{A} = \partial A$. Therefore $\partial_\infty A\supset\partial A$ and consequently $\partial_\infty A$ is a boundary.

<u>THEOREM 2</u>. The set $\partial_\infty A$ is the smallest ∞-closed ∞-boundary of A.

<u>Proof:</u> Let $F=(f_i,\dots,f_j,\dots)\in\ell^\infty A$. For a given $X\in\ell^\infty$ and $j\in\mathbb{Z}_+$ define the set

$$Z(X,j) = \hat{F}^{-1}(\mathbb{C}(X,j))$$

that obviously contains X. According to Proposition 2

$$\min_{\partial A_{Z(X,j)}} |\hat{f}_j(m)| \le \min_{b\sigma(\hat{f}_{j|Z(X,j)})} |z|$$

and hence

$$\inf_{\partial_\infty A}\|\hat{F}(m)\|_\infty \le \inf_{\partial A_{Z(X,j)}}\|(m)\|_\infty = \min_{\partial A_{Z(X,j)}} \max(\|x^j\|_\infty, |(\hat{f}_j(m)|) \le$$

$$\le \min_{b\sigma(\hat{f}_j|Z(X,j))} \max(\|x^j\|_\infty, |z|) = \min_{b\sigma(\hat{f}_{j|Z(X,j)})} \|x^j+ze_j\|_\infty$$

$$= \min_{b\sigma(\hat{F}_{|Z(X,j)})} \|x^j+ze_j\|_\infty = \min_{b(\sigma(F)\cap\mathbb{C}(X,j))} \|Y\|$$

Since $\sigma(F)=\cup[\alpha(F)\cap\mathbb{C}(X,j): X\in\ell^\infty, j\in\mathbb{Z}_+]$; and

$b\sigma(F) = \cup[b(\sigma(F)\cap\mathbb{C}(X,j)): X\in\ell^\infty, j\in\mathbb{Z}_+\}$ we have:

$$\inf_{\partial_\infty A}\|\hat F(m)\|_\infty \le \inf\{\|Y\|_\infty : Y\in b(\sigma(F)\cap\mathbb{C}(X,j));\ X\in\ell^\infty,\ j\in\mathbb{Z}_+\}$$

$$= \inf_{b\sigma(F)}\|Y\|_\infty .$$

Because $\inf_{b\sigma(F)}\|Y\|_\infty \le \inf_{\partial A}\|\hat F(m)\|_\infty$ for any regular $F\in\ell^\infty_* A$, we get $\inf_{b\sigma(F)}\|Y\|_\infty = \inf_{\partial_\infty A}\|F(m)\|_\infty$ for any $F\in\ell^\infty_* A$. Now (6_1) gives us that $\partial_\infty A$ is an ∞-boundary of A. Let now E be an ∞-closed ∞-boundary of A. Then $\inf_E\|\hat F(m)\|_\infty = \inf_{SpA}\|\hat F(m))\|_\infty$ for any $F\in\ell^\infty_* A$. Let $\mathbb{C}(X,j)$, $X\in\ell^\infty$, $j\in\mathbb{Z}_+$, be a fixed complex line in ℓ^∞. Suppose that there exists a $G\in\ell^\infty A$, such that:

$$\inf_{E\cap\hat G^{-1}(\mathbb{C}(X,j))}|\hat f(m)| > \inf_{\hat G^{-1}(\mathbb{C}(X,j))}|\hat f(m)|,$$

for some $f\in A^{-1}$, $\|X^j\|_\infty \le \min_{SpA}|\hat f(m)|$ and let $F=X^j+f$. New $F^{-1}(\mathbb{C}(X,j)) = SpA$ and:

$$\inf\|\hat F(m)\|_\infty = \inf_{E\cap\hat G^{-1}(\mathbb{C}(X,j))}\|\hat F(m))\|_\infty = \inf_{E\cap\hat G^{-1}(\mathbb{C}(X,j))}\max(\|X^j\|_\infty,|\hat f(m)|)$$

$$> \inf_{\hat G^{-1}(\mathbb{C}(X,j))}\max(\|X^j\|_\infty,|\hat f(m)|) = \inf_{\hat G^{-1}(\mathbb{C}(X,j))}\|\hat F(m)\|_\infty = \inf_{SpA}\|\hat F(m)\|_\infty,$$

a contradiction, because E is an ∞-boundary. Hence for any $G\in\ell^\infty$ the equality

$$\inf_{E\cap\hat G^{-1}(\mathbb{C}(X,j))}|\hat f(m)| = \inf_{\hat G^{-1}(\mathbb{C}(X,j))}|f(m)|$$ holds true for any $f\in A^{-1}$.

Consequently $E\cap\hat G^{-1}(\mathbb{C}(X;j)) \supset \partial A\hat G^{-1}{}_{(\mathbb{C}(X,j))}$ for any $G\in\ell^\infty A$. Now $E \supset \bigcup_{G,X,j}(E\cap\hat G^{-1}(\mathbb{C}(X,j)) \supset \bigcup_{G,X,j}\partial A_{\hat G^{-1}}(\mathbb{C}(X,j))$ and therefore $E\supset\partial_\infty A$ by taking ∞-closure. The theorem is proved.

Because of Theorem 2, $\partial_\infty A$ is said to be the <u>minimal ∞-boundary</u> of A.

As an immediate corollary from Proposition 5 we get the following.

<u>Corollary 1.</u> The minimal ∞-boundary $\partial_\infty A$ of a commutative Banach algebra A is the smallest among all ∞-closed subsets E of SpA,

satisfying one of the equivalent conditions: (4),(5),(5^1), (6_1)-(6_3),(9).

<u>Remark.</u> Following the same line of reasoning in the setting of Sibony-Basener boundary (see [5], [6], we can get the following:

<u>Corollary 2.</u> The Sibony-Basener boundary $\partial_{n-1}A$ is the smallest among all closed subsets E of maximal ideal space SpA of a commutative Banach algebra A, satisfying one of the equivalent conditions:

(10_1) $\hat{F}(E) \supset b\sigma(F)$ for any $F \in A^n$;

(10_2) $\min_E \|\hat{F}(m)\| \leq \min \{\|z\|,\ z \in b\sigma(F)\}$, for any $F \in A^n$,

where $\|z\| = (\sum_{j=1}^{n} |z_j|^2)^{1/2}$;

(10_3) $B_\infty(\min_E \|\hat{F}(m)\|)$ is contained entirely either in $\sigma(F)$ or in $\mathbb{C}^n \backslash \sigma(F)$ for any $F \in A^n$;

(10_5) $\min_E \|\hat{F}(m)\| = \min_{SpA} \|\hat{F}(m)\|$ for any $F \in A^n, \sigma(F) \ni 0$;

(10_6) $\min_E \|\hat{F}(m)\| = 0$ for any $F \in A^n$ with $b\sigma(F) \not\ni 0$;

(10_7) $B_\infty(\min_E \|\hat{F}(m)\|) \subset \sigma(F)$ for any $F \in A^n$, $\sigma(F) \ni 0$.

Actually the case (10_5) (and partly the case (10_1)) from corollary 2 were proved in [8]. Proceeding analogously with the corresponding property for the Shilov boundary (e.g.[1]) we get the following local characterization of the minimal ∞-boundary $\partial_8 A$.

<u>Corollary 3.</u> A point $m \in SpA$ belongs to $\partial_\infty A$ iff for any ∞-neighborhood U of m there exists:

a) some element $F \in \ell^\infty_* A$, so that $\inf_{SpA \backslash U} \|F(m)\|_\infty > \inf_{b\sigma(F)} \|Y\|_\infty$;

b) some element $F \in \ell^\infty_o A$, so that $\inf_{SpA \backslash U} \|\hat{F}(m)\|_\infty > 0$;

c) some element $F \in \mathcal{E}^\infty A \backslash \ell^\infty_* A$, so that $B_\infty(\inf_{SpA \backslash U} \|\hat{F}(m)\|_\infty) \subset \sigma(F)$.

<u>Example.</u> Let X be the infinitely dimensional polydisc $\bar{\Delta}^{\infty}=\{Y\in\mathbb{C}^{\infty} \mid \|Y\|_{\infty} \leq \leq 1\}$ provided with Tichonov weak topology, under which it is a compact set. Let $A=P(\bar{\Delta}^{\infty})$ be the algebra of all continuous functions on $\bar{\Delta}^{\infty}$, uniformly approximable on $\bar{\Delta}^{\infty}$ by polynomials, depending on finitely many arguments. It is easy to see (e.g.[2]) that $SpA=\bar{\Delta}^{\infty}$ and $\partial A=T^{\infty}=\{Y\in\mathbb{C}^{\infty} \mid |y_j|=1 \text{ for any } j\in\mathbb{Z}_+\}$. By definition

$$\partial_{\infty}A=[\cup\{\partial A_{\hat{G}^{-1}(\mathbb{C}(X,j))};\ G\in\ell^{\infty}A,\ X\in\ell^{\infty},\ j\in\mathbb{Z}_+\}]\subset\bar{\Delta}^{\infty}$$

Since $\partial_{\infty}A$ is an ∞-boundary, $[\hat{F}(\partial_{\infty}A)]_{\infty}\supset b\sigma(F)$ for any $F\in\ell^{\infty}A$. In particular, if $F=id_{\Delta^{\infty}}$, we get $[\partial_{\infty}A]\supset b_{\infty}\bar{\Delta}^{\infty}$, and therefore $\partial_{\infty}A\supset b_{\infty}\bar{\Delta}^{\infty}$because of the ∞-closedness of $\partial_{\infty}A$. We see that $\bar{\Delta}^{\infty}\supset\partial_{\infty}A\supset b_{\infty}\bar{\Delta}^{\infty} = = \{Y\in\ell^{\infty} \mid \|Y\|=1\}$. Suppose that $\partial_{\infty}A$ coinsides with the smallest among all ∞-closed subsets E of SpA, satisfying condition (7). An analogue to corollary 3 will hold true, namely a point in SpA will belong to $\partial_{\infty}A$ iff for any ∞-neighborhood U of m there exists some $F\in\ell^{\infty}A$, so that $\sup_{SpA\setminus U}\|\hat{F}(m)\|_{\infty}.<\sup_{SpA}\|\hat{F}(m)\|_{\infty}$. If $\|\hat{F}(m)\|_{\infty}' = \sup_j|\hat{f}_j(m)|$ then $\sup_{SpA}\|\hat{F}(m)\|_{\infty}' < \sup_{SpA}\|\hat{F}(m)\|_{\infty}'$. But $\|\hat{F}(m)\|_{\infty}'$ is a function, subharmonic on any complex line in $\bar{\Delta}^{\infty}$, and according to the maximum modulusprinciple for subharmonic functions, $\sup\|\hat{F}(m)\|_{\infty}'$ (and hence $\sup\|\hat{F}(m)\|_{\infty}$) is assumed within T^{∞}. Hence $[T^{\infty}]_{\infty}=T^{\infty}=\partial_{\infty}A$, a contradiction. Here T^{∞} satisfies (7) without being a ∞-boundary. On the other hand it is obvious that $T^{\infty}=\partial A$ is the smallest ∞-closed (even weak closed) subject of SpA satisfying condition (7).

3. <u>∞-dimensional analytic structure in the spectrum of a uniform algebra</u>.

The contents of this paragraph in general follows Basener's article [6], replacing $\partial_{n-1}A$ with $\partial_{\infty}A$.

<u>Lemma 1</u>. Let K be a closed subset of SpA and let $F\in\ell^{\infty}A$. Suppose that $K\cap\hat{F}^{-1}(\mathbb{C}(F(x),j))\neq\phi$ for some $x\in K$. Then for any $g\in A/_{\hat{F}^{-1}(\mathbb{C}(F(x),j))}$ the following holds true:

$$(11)\quad |g(x)| \leq \max\{|g(y)|:\ y\in[b_{\infty}K\cap\hat{F}^{-1}(\mathbb{C}(F(x),j)]\cup [\partial_{\infty}A\cap\hat{F}^{-1}(\mathbb{C}(F(x),j))\cap K]\}.$$

Proof: Denote by Z the set $Z=\hat{F}^{-1}\mathbb{C}(F(x),j)$. Applying to algebra A_Z the local maximum modulus principle (e.g.[4]) gives:

$$|g(x)| \le \max\{|g(y)|: y\in b_Z(K\cap Z)\cup[\partial A_Z\cap K\cap Z]\}.$$

But $\partial A_Z\subset\partial_\infty A$ and $b_Z(K\cap Z) \subset [b_\infty K\cap Z]_Z$ and therefore (11) holds.

Proposition 6. Let K be a compact A-convex subset of $SpA\backslash\partial_\infty A$. Then $\partial_\infty A_K\subset b_\infty K$. (A - convexity here means that $SpA_K=K$).

Proof (of [6]): Suppose that $\partial_\infty A_K\not\subset b_\infty K$. Then $\partial_\infty A_K\cap \operatorname{int} K\neq\phi$ and hence there exists some $F\in\ell^\infty A_K$, $X\in\ell^\infty$ and $j\in\mathbb{Z}_+$, such that $\operatorname{int} K\cap\hat{F}^{-1}(\mathbb{C}(X,j))\neq\phi$ and therefore $\operatorname{int} K\cap\partial(A_{K\cap\hat{F}^{-1}(\mathbb{C}(X,j))})\neq\phi$. Consequently there exists a $h\in A$ and a $m\in\operatorname{int} K\cap\hat{F}^{-1}(\mathbb{C}(X,j))$, so that $|h(m)| > 1$ but $|h(x)|<1$ on $bK\cap\hat{F}^{-1}(\mathbb{C}(X,j))$. Let $\varepsilon>0$ be such that $|h(x)|< 1$ on the set $bK\cap\{x\in K: |\hat{f}_n(x)-x_n|<\varepsilon$ for $n\neq j\}$. Let $G\in\ell^\infty A$ is such that $\hat{g}_n(m)=x_n$ for $n\neq j$, while $|g_n-f_n| < \varepsilon$ on K for $n\neq j$. Then $m\in\hat{G}^{-1}(\mathbb{C}(X,j))$ and $|h(x)| <1$ on $b_\infty K\cap\hat{G}^{-1}(\mathbb{C}(X,j))$, but $|\hat{h}(m)|>1$ in contradiction with Lemma 1.

Lemma 2. Let $F\in\ell^\infty A$ be such that $\sigma(F)$ is closed in ℓ^∞ and let W be a connected component of the set $\ell^\infty\backslash[\hat{F}(\partial_\infty A)]_\infty$. If in addition $\sigma(F)\cap W\neq\phi$, then $W\subset\sigma(F)$.

Proof (of [10]): Let W_1 be the (relatively) open set: $W_1=E\backslash\sigma(F)$. We shall show that at the same time W_1 is a closed set. Let $Y\in W\backslash W_1=W\cap\sigma(F)$ and let $\{Y_n\}\subset W_1$ is a sequence tending to Y. Then there exists $m\in SpA$ with $\hat{F}(m)=Y$, since $Y\in W\backslash W_1 \subset\sigma(F)$. Then

$$\inf_{x\in\partial_\infty A}\|\hat{F}(x)-Y_n\|_\infty\in\|\hat{F}(m)-Y_n\|_\infty=\|Y-Y_n\|_\infty \to \text{ as } n\to\infty.$$

In other words $\operatorname{dist}(Y_n, \hat{F}(\partial_\infty A)) \to 0$ as $n\to\infty$ and therefore $Y\in[\hat{F}((\partial_\infty A)]$ in contradiction with $Y\in W\subset\ell^\infty\backslash[\hat{F}(\partial_\infty A)]$. Hence W_1 is a open set in W, and so is $W\backslash W_1$, which is nonempty by assumption. Hence $W\cap\sigma(F)=W\backslash W_1=W$ and therefore $W\subset\sigma(F)$.

Theorem 3. (maximality theorem). Let K be a bounded polynomially convex set in $\mathbb{C}^\infty$ with connected ∞-interior. Let B be a closed subalgebra of C(K) containing the polynomials and such that $\partial_\infty B=b_\infty K$, then the mapping: $\Pi:SpB \to \ell^\infty$: $\Pi(m)=(\hat{z}_1(m),\dots, \hat{z}_j(m), \dots)$ is

injective, and either $f \circ \Pi^{-1} \in P(\Pi(K))$ for any $f \in B$ with $\Pi(SpB)=K$, or $\Pi(SpB)=b_\infty K$.

Proof (cf [5]). Since $\mathrm{int}_\infty K$ is connected, according to Lemma 2 either $\Pi(SpB)=K$ or $\Pi(SpB) = b_\infty K$. In the first case let $Y \in K$ and

$$Z(Y,j) = SpB \cap \mathbb{C}(Y,j),\ S(Y,j) = b_\infty K \cap \mathbb{C}(Y,j),\ \tilde{S}(K,j) = K \cap \mathbb{C}(Y,j).$$

Since $\partial_\infty K = b_\infty K$ we see that $\partial B_{Z(Y,j)} = S(Y,j)$

The algebra B restricted to $S(Y,j)$ contains z_j and its spectrum projects on $\tilde{S}(Y,j)$, so by [5, Lemma 9] $Z(Y,j) \cong \hat{S}(Y,j)$ and Π is injective. Moreover, since Π is open, then Π^{-1} is continuous and $f \circ \Pi^{-1} \in (K)$ for any $f \in B$.

Theorem 4. Let $F \in \ell^\infty A$ and suppose that $\|\hat{F}(x)\|_\infty = 1$ on $\partial_\infty A$, that $0 \in \sigma(F)$ and that $\sigma(F)$ is closed in ℓ^∞. Assume there exists a subset S' of the unit sphere $S_\infty(1)$ of ℓ^∞, such that:

(i) for any $Y \in S^1$ there exists a unique $q \in \partial_\infty A$ with $\hat{F}(q)=Y$;

(ii) for any $Z \in B_\infty(1)$ there exists some complex line $\mathbb{C}(Z,j)$ in ℓ^∞, such that $\mathrm{meas}(S^1 \cap \mathbb{C}(z,j)) > 0$. Then for any $Y \in B_\infty(1)$ there exists a unique $x \in SpA$, such that $\hat{F}(x)=Y$ and for any $f \in A$ the function $\hat{f} \circ \hat{F}^{-1}$ is ∞-holomorphic on $B_\infty(1)$.

By an ∞-holomorphic function on a subset N of ℓ^∞ we mean any continuous function on N, such that $g_{|\mathbb{C}(X,j)}$ is holomorphic in $\mathrm{int}(N \cap \mathbb{C}(X,j))$ for any $X \in \ell^\infty, j \in \mathbb{Z}_+$.

Proof (cf [6]): Let $Y \in B_\infty(1)$. Without loss of generality we can assume that $Y=0$ (since $0 \in \sigma(F)$, $\sigma(F) \supset B_\infty(1)$ according to Lemma 2). Let $\mathbb{C}(0,j)$ be a complex line through 0 such that $\mathrm{meas}(S_\infty(1) \cap C(0,j)) > 0$. Then because the Wermer's theorem holds for $\hat{F}|\mathbb{C}(0,j) = \hat{f}_j \in A_{\mathbb{C}(0,j)}$, for any $z \in B_\infty(1) \cap \mathbb{C}(0,j)$ there exists a unique $x \in \hat{F}^{-1}(\mathbb{C}(0,j))$, such that $\hat{f}_j(x) = \hat{F}(x) = z$. Hence for any $Y \in B_\infty(1)$ there exists a unique $x \in SpA$, such that $\hat{F}(x)=Y$.

Let, for a fixed $\rho (0 < \rho < 1)$, K_ρ be the compact set

$$K_\rho = \{x \in SpA : \|\hat{F}(x)\|_\infty \le \rho\}$$

According to Proposition 6, $\partial_\infty A_{K_\rho} \subset bK_\rho = \{x \in SpA : \|\hat{F}(x)\|_\infty = \rho\}$. Because $\hat{F}$ is one-to-one on K, Wermer's theorem says that for any $f \in A$ the function $\hat{f} o \hat{F}^{-1}$ is holomorphic on the intersection of any line $\mathbb{C}(X,j)$ with the open ρ-ball in ℓ^∞. Because of the arbitrariness of ρ it holds that for any $f \in A$ $\hat{f} o \hat{F}^{-1}$ is a ∞-holomorphic function on $B_\infty(1)$. Q.E.D.

Let D be a domain in ℓ^∞ and Λ be a subset in D. We call Λ a <u>negligible set</u> if Λ is nowhere dense and if for any subdomain $D' \subset D$ every function of, ∞-holomorphic on $D' \setminus \Lambda$ and locally bounded in D^1 admits a unique ∞-holomorphic extension on the whole of D^1.

<u>Definition 2</u>. <u>∞-analytic semi-cover</u> It is any triple (X, Π, U), for which:

1) X is a locally compact Hansdorff space;
2) U is a domain in ℓ^∞;
3) Π is an open mapping of X onto U, for which the set $\Pi^{-1}(Y)$ is discrete for any $Y \in U$;
4) there exists a negligible set $\Lambda \subset U$ and an integer k, so that Π is a k-sheeted semi-covering mapping of $X \setminus \Pi^{-1}(\Lambda)$ onto $U \setminus \Lambda$;
5) the set $X \setminus \Lambda^{-1}(\Lambda)$ is dense in X.

Here Π is said to be a k-sheeted semi-covering mapping, if for any $x \in U \setminus \Lambda$ there exists a neighborhood $N \ni x$, such that $\Pi^{-1}(N) = \{U_1, U_2, \ldots U_k\}$, $U_1 \cap U_j = \phi (i \neq j)$, and Π / U_j is a bijective map onto N for any $j = 1, \ldots, k$.

A continuous complex valued function f defined on an open subset V of a ∞-analytic semi-cover X is said to be <u>∞-analytic</u> V if for any open subset $V' \subset X \setminus \Pi^{-1}(\Lambda)$ on which Π is bijective, the function $(f/V') o \Pi^{-1}$ is ∞-analytic on $\Pi(V') \subset \ell^\infty$.

As in the classical one-dimensional and in n-dimensional situation, Theorem 4 and preeceding Lemmas enalbe us to prove the following;

<u>Theorem 5.</u> Let $F \in \ell^\infty A$ with $\sigma(F)$ closed and W be a component of $\mathbb{C}^\infty \setminus \hat{F}(\partial_\infty A)$. Suppose that $\sigma(F) \cap W \neq \phi$ and let there exists some subset $W' \subset W$, such that too any $Y \in W'$ meas$(W' \cap \mathbb{C}(Y,j)) > 0$ for any complex line

$\mathbb{C}(Y,j)$ in ℓ^{∞}, and that for any $Y \in W'$, $\#\hat{F}(Y)$ is finite. Let $W_{\ell}=\{Y \in W : \#\hat{F}^{-1}(Y)=l\}$ for $l=1,2,\ldots,$

Then there exists a positive number k, such that:

i) $W=\bigcup_{j=1}^{k} W_j$,

ii) $(\hat{F}^{-1}(W), \hat{F}, W)$ is a (branched) ∞-analytic semicover;

iii) any $f \in A$ is ∞-analytic on $F^{-1}(W)$.

Actually a careful reading of the argument used by Wermer [11] in the investigation of 1-dimensional analytic structures, and afterwards used again by Basener [6] in the investigation of n-dimensional analytic structures in the algebra spectrum shows that independently of the very different (∞-analytic) environment here; the same argument is applicable now to prove theorem 5.

REFERENCES

1. I. Gelfand, D. Raikov, G. Shilov, Kommutatinye normirovannye koljea, Moscow, 1960 (Russian)

2. E.L. Stout, the theory of uniform algebras, Bogden & Qugely, Tarrytown-on-Hudson, N.Y., 1971.

3. E. Bishop, Holomorphic completions, analytic continuation, and the interpolation of semi-norm, Ann. of Math. (2) 78 (1963), 468-500.

4. T. Gamelin, Uniform Algebras, Prentice-Hall Inc., Englewood Cliffs, N.J., 1969.

5. N. Sibony, Multidimensional analytic structure in the spectrum of a uniform algebra. In: Spaces of Analytic Functions, Kristiansand, Norway 1975. Lecture Notes in Mathematics V.512. Springer, Berlin-Heidelberg, New York, 139-175.

6. R. Basener, A generalized shilov boundary and analytic structure, Proc. Amer. Math. Soc. 47(1975), 98-104.

7. B. Kramm, Nuclearity (resp. Schwartzity) help to embed holomorphic structure into spectra - a survey, to appear in Rickart Conf. Proc. at Yale University, AMS Contemproary Math. Series.

8. T. Tonev, New relations between Sibony-Basener boundaries, these proceedings.

9. T. Tonev, Minimal boundaries of commutative Banach algebras, containing the Shilov boundary. In: Complex Analysis and Applications, Varna '85, Sofia (to appear).

10. T. Tonev, Minimal boundaries of commutative Banach algebras, commuted with normal spaces, In: Proc. Int. Symp. on Aspects of Positivity in Functional analysis, Tubingen '85 (to appear).

11. T. Wermer, Banach algebras and several complex variables, Springer-Verlog, New York-Heidelberg-Berlin, 1976.

Institute of Mathematics
Bulgarian Academy of Sciences
BG-1090 Sofia, P.O.B.373, Bulgaria

NEW RELATIONS BETWEEN SIBONY-BASENER BOUNDARIES

T. V. Tonev

The most advanced progress in the investigation of analytic structures in the maximal ideal space of uniform algebras during the last ten years is related to the discovering of n-dimensional generalizations of the Shilov boundary by Sibony [1] and Basener [2]. The classical Shilov boundary is the intersection of all closed boundaries, i.e. of these subsets of the maximal ideal space sp A, on which any element f from the algebra A takes the maximum of absolute value of its Gelfand transform $\hat{f}$. The Shilov boundary plays an important role in many branches of mathematics, in particular in commutative Banach algebra theory and in Complex analysis. Only in the last ten years it was realized that the Shilov boundary is too small for some purposes; for instance for instance for investigation of higher dimensional analytic structures in the maximal ideal space. The Sibony-Basener boundary $\partial_{n-1}A$ turns out to be the natural and exact fence for any n-dimensional analytic structure in sp A (see [1], [2]). Afterwards this boundary was used by B. Kramm, D. Kumagai, P. Jakóbczak and others to find higher-dimensional analytic structures in the algebra spectrum.

Here we give some new and simple characterizations of the Sibony-Basener boundaries and establish new relations between them.

Let A be a commutative Banach agebra over $\mathbb{C}$ with unit. As usual sp A denotes the maximal ideal space, ∂A denotes its Shilov boundary and $\hat{f}$ denotes the Gelfand transform of the element f from A (i.e. $\hat{f}(m) = m(f)$ for any $m \in$ sp A). Recall that a <u>boundary</u> of A is a subset of sp A on which all Gelfand transforms $\hat{f}$, $f \in A$, take the maximum of their absolute values $\hat{f}$. The <u>Shilov boundary</u> is the intersection of all closed boundaries of A.

In 1975 Sibony and Basener introduced independently n-dimensional generalisations of the Shilov boundary, known also as Sibony-Basener

boundaries. Denote by $V(f_1,\dots,f_n)$ the vanishing set of the n-tuple $(\hat{f}_1,\dots,\hat{f}_n)$ over $\hat{A}$, i.e.

$$V(f_1,\dots,f_n) = \{m \in \text{sp } A \mid \hat{f}_1(m) = \hat{f}_2(m) = \dots = \hat{f}_n(m) = 0\},$$

and by A_V the closure on the closed set V of the restrictions of elements of $\hat{A}$ on V.

Definition. (n-1)-st Sibony-Basener boundary of a commutative Banach algebra A is the following subset $\partial_{n-1}A$ of sp A:

$$\partial_{n-1}A = [\cup\{\partial A_V(f_1,\dots,f_{n-1}) \mid (f_1,\dots,f_{n-1}) \in A^{n-1}\}].$$

Note that $\partial_o A$ is exactly the usual Shilov boundary ∂A of A.

The starting point of our investigation is the simple fact that boundaries are those subsets of sp A on which Gelfand transforms of all invertible elements of A together with the maximums take also the minimums of their absolute values.

Recall that a n-tuple $(f_1,\dots,f_n)$ over A is called regular if the elements $\{f_j\}_1^n$ do not belong to one maximal ideal, or equivalently, if the functions $\{f_j\}_1^n$ do not have common zeroes on sp A, i.e. if $V(f_1,\dots,f_n) = \phi$. The spectral mapping $\sigma_{f_1,\dots,f_n}$ of the n-tuple $(f_1,\dots,f_n)$ is the map $\sigma_{f_1,\dots,f_n}$: sp $A \to \mathbb{C}^n$, defined as follows: $\sigma_{f_1,\dots,f_n}(m) = (\hat{f}_1(m),\dots,\hat{f}_n(m))$ and the joint spectrum $\sigma(f_1,\dots,f_n) \subset \mathbb{C}^n$ of $(f_1,\dots,f_n) \in A^n$ is the image of sp A through the spectral mapping, i.e. $\sigma(f_1,\dots,f_n) = \sigma_{f_1,\dots,f_n}(\text{sp } A)$ $= \{z \in \mathbb{C}^n \mid \exists m \in \text{sp } A: z_j = \hat{f}_j(m), j=1,\dots,n\}$.

Theorem 1. The (n-1)-th Sibony-Basener boundary $\partial_{n-1}A$ coincides with the smallest closed subset of sp A, on which the minimum of the expression

$$\|\sigma_{f_1,\dots,f_n}(m)\| = \left(\sum_{j=1}^{n} |\hat{f}_j(m)|^2\right)^{1/2}$$

is taken for any regular n-tuple $(f_1,f_2,\dots,f_n)$ over A.

Proof: Let $(f_1,\dots,f_n)$ be a regular n-tuple over A and m_o be an arbitrary element of sp A. Without loss of generality (applying, if

necessary, an orthogonal transformation in $\mathbb{C}^n$) we can assume that $\hat{f}_j(m_o) = 0$ for all $j>1$, so that $\|\sigma_{f_1,\dots,f_n}(m_o)\| = |\hat{f}_1(m_o)|$. Consider the set

$$Z_{m_o} = V(f_2,\dots,f_n) = \{m\in \text{sp } A \,|\, \hat{f}_j(m) = 0 \text{ for } j>1\}.$$

The element $\hat{f}_1|Z_{m_o}$ is invertible in the algebra $A' = A_{Z_{m_o}}$ and according to our above observation, $|\hat{f}_1(m_o)| \geq \min_{\partial A'} |\hat{f}_1(m)|$. It follows that:

$$\|\sigma_{f_1,\dots,f_n}(m_o)\| = (\sum_j^n |\hat{f}_j(m_o)|^2)^{1/2} = |f_1(m_o)|$$

$$\geq \min_{\partial A'} |\hat{f}_1(m)| = \min_{\partial A'} (\sum_1^n |f_j(m)|^2)^{1/2}$$

$$\geq \min \{(\sum_1^n |f_j(m)|^2)^{1/2} | m\in \cup\, \partial A_{V(g_1,\dots,g_{n-1})},\ (g_1,\dots,g_{n-1})\in A^{n-1}\}$$

$$= \min_{\partial_{n-1}A} \|\sigma_{f_1,\dots,f_n}(m)\|.$$

Hence the expression $\|\sigma_{f_1,\dots,f_n}(m)\|$ takes its minimum on $\partial_{n-1}A$ for any regular n-tuple $(f_1,\dots,f_n)$ over A.

Let now E be a closed subset of sp A, on which the minimum of the expression $\|\sigma_{f_1,\dots,f_n}(m)\| = (\sum_{i=1}^n |\hat{f}_j(m)|^2)^{1/2}$ is taken for any regular n-tuple over A. Take an arbitrary (n-1)-tuple $(g_1,\dots,g_{n-1})$ over A and suppose that for an element f from A the restriction $\hat{f}|_{V(g_1,\dots,g_{n-1})}$ is invertible in the algebra $\hat{A}_{V(g_1,\dots,g_{n-1})}$ and satisfies the inequality $|\hat{f}(m)| \geq r>0$ on the set $V(g_1,\dots,g_{n-1})\cap E$. For each ε, $0<\varepsilon<r$, there exists a neighborhood V_ε of the set $V(g_1,\dots,g_{n-1})\cap E$ on which $|\hat{f}(m)| \geq r-\varepsilon$. Hence on E we have

$$(C_\varepsilon^2 \sum_{j=1}^{n-1} |\hat{g}_j(m)|^2 + |\hat{f}(m)|^2)^{1/2} \geq r-\varepsilon \tag{1}$$

for some large enough positive constant C_ε. But according to our choice of E the minimum of the expression $\|\sigma(C_\varepsilon g_1,\dots,C_\varepsilon g_{n-1},f)(m)\|$ is taken on E since the n-tuple $(C_\varepsilon g_1,\dots,C_\varepsilon g_{n-1},f)$ is regular. Consequently (1) holds on the whole sp A. In particular on

$V(g_1,\dots,g_{n-1})$ we have: $|\hat{f}(m)| \geq r-\varepsilon$ and hence $|\hat{f}(m)| \geq r$ on $V(g_1,\dots,g_{n-1})$ because we can choose ε small enough. We obtained that the set $V(g_1,\dots,g_{n-1})\cap E$ is a boundary for the algebra $A_{V(g_1,\dots,g_{n-1})}$ and consequently $V(g_1,\dots,g_{n-1})\cap E \supset \partial A_{V(g_1,\dots,g_{n-1})}$ for any $(n-1)$-tuple $(g_1,\dots,g_{n-1})$ over A. Now

$$E \supset \cup\{V(g_1,\dots,g_{n-1}) \cap E \mid (g_1,\dots,g_{n-1})\in A^{n-1}\} \supset$$

$$\cup\{A_{V(g_1,\dots,g_{n-1})} \mid (g_1,\dots,g_{n-1})\in A^{n-1}\}$$

and hence $E\supset\partial_{n-1}A$ after taking the closure. The theorem is proved. □

Following the same line of reasoning as for the corresponding property of the Shilov boundary, now we can get the following local characterization of Sibony-Basener boundaries:

<u>Corollary 1</u>. A point $m \in$ sp A belongs to $\partial_{n-1}A$ iff for any neighborhood $U\ni m$ there exists a regular n-tuple $(f_1,\dots,f_n)\in A^n$, such that the set where the expression $\|\sigma_{f_1,\dots,f_n}\|$ takes its minimum is contained in U.

<u>Theorem 2</u>. Let $(f_1,\dots,f_n)\in A^n$ be a n-tuple over A. The image $\sigma_{f_1,\dots,f_n}(\partial_{n-1}A)$ of the (n-1)-th Sibony-Basener boundary contains the topological boundary of the joint spectrum $\sigma(f_1,\dots,f_n)$, i.e.

$$\sigma_{f_1,\dots,f_n}(\partial_{n-1}A) \supset b\sigma(f_1,\dots,f_n).$$

<u>Proof</u>: Supposing that there exists a $m_o \in$ sp A, such that $\sigma_{f_1,\dots,f_n}(m_o)\in b\,\sigma(f_1,\dots,f_n)\setminus\sigma_{f_1,\dots,f_n}(\partial_{n-1}A)$, let $\delta = \inf_{\partial_{n-1}A} \|\sigma_{f_1,\dots,f_n}(m_o) - \sigma_{f_1,\dots,f_n}(\varphi)\| > 0$. Fix a z from $\mathbb{C}^n\setminus\sigma(f_1,\dots, f_n)$ with $\|\sigma_{f_1,\dots,f_n}(m_o)\| < \delta/2$ and consider the continuous function $\|z - \sigma_{f_1,\dots,f_n}(m_o)\|$ on sp A. For $x\in\partial_{n-1}A$ we have:

$$\|\sigma_{f_1,\dots,f_n}(x) - z\| \geq$$

$$|\ \|\sigma_{f_1,\dots,f_n}(x) - \sigma_{f_1,\dots,f_n}(m_o)\| - \|\sigma_{f_1,\dots,f_n}(m_o) - z\|\ | \geq$$

$$\delta/2.$$

Because $\sigma_{f_1,\dots,f_n}(x) - z = \sigma_{f_1-z_1,\dots,f_n-z_n}(x)$, $(f_1-z_1,\dots, f_n-z_n)$ being a regular n-tuple, it follows that $\|\sigma_{f_1,\dots,f_n}(x) - z\| \geq \delta/2$ on the whole spectrum sp A in contradiction with $\|z - \sigma_{f_1,\dots,f_n}(m_o) \leq \delta/2$. Consequently $b\sigma_{f_1,\dots,f_n}(m)\backslash\sigma_{f_1,\dots,f_n}(\partial_{n-1}A) = \phi$. □

The following theorem gives another characterization of the Sibony-Basener boundaries, which, as far as we know, is new even for the case of the Shilov boundary.

<u>Theorem 3</u>. The Sibony-Basener boundary $\partial_{n-1}A$ coincides with the intersection of all closed subsets K of sp A, such that

$$\sigma_{f_1,\dots,f_n}(K) \supset b\sigma(f_1,\dots,f_n)$$

for any regular n-tuple $(f_1,\dots,f_n)$ over A, i.e. $\partial_{n-1}A = \cap\{K \subset \text{sp } A,\ K$ is closed $|\sigma_{f_1,\dots,f_n}(K) \supset b\sigma(f_1,\dots,f_n)$ for any regular $(f_1,\dots,f_n) \in A^n\}$.

<u>Proof</u>: Theorem 2 shows that $\partial_{n-1}A$ contains the inter-section on the right. Supposing that $\sigma_{f_1,\dots,f_n}(K) \supset b\sigma(f_1,\dots,f_n)$ for some regular n-tuple $(f_1,\dots,f_n) \in A^n$ and for some closed subset K of sp A, we can find a $m_o \in b\sigma(f_1,\dots,f_n)$ on which the expression $\|\sigma_{f_1,\dots,f_n}(m)\|$ takes its minimum, i.e. for which $\|\sigma_{f_1,\dots,f_n}(m_o)\| = \min_K \|\sigma_{f_1,\dots,f_n}(m)\| = \min_{spA} \|\sigma_{f_1,\dots,f_n}(m)\|$. Hence for any regular n-tuple $(f_1,\dots,f_n)\in A^n$ the minimum of $\|\sigma_{f_1,\dots,f_n}(m)\|$ is taken on K. Since $\partial_{n-1}A$ is the smallest set on which $\|\sigma_{f_1,\dots,f_n}(m)\|$ takes its minimum on all regular n-tuples over A, $\partial_{n-1}A$ is contained in K, which completes the proof. □

<u>Theorem 4</u>. Let k be a fixed integer, $1 \leq k < n$. Then

$$\partial_{n-1}A = \left[\cup \left\{\partial_{k-1}A_{V(g_1,\dots,g_{n-k})} \mid (g_1,\dots,g_{n-k}) \in A^{n-k}\right\} \right].$$

<u>Proof</u>: The proof follows the same line of argument as the proof of Theorem 1. Let $(f_1,\dots,f_n)$ be a regular n-tuple over A and m_o be an arbitrary element of sp A. Again without loss of generality we

shall assume that $\hat{f}_j(m_o) = 0$ for all $j > k$. Consider the set $Z_o = V(\{f_j\}, j > k)$. For the regular k-tuple $(\hat{f}_1|Z_o,\dots,\hat{f}_k|Z_o)$ over the algebra $A' = A_{Z_o}$ it holds that $\|\sigma_{\hat{f}_1|Z_o,\dots,\hat{f}_k|Z_o}(m_o)\| \geq \min_{\partial_{k-1}A'} \|\sigma_{\hat{f}_1|Z_o,\dots,\hat{f}_k|Z_o}(m)\|$ and consequently $\|\sigma_{f_1,\dots,f_n}(m_o)\| = \|\sigma_{f_1,\dots,f_k}(m_o)\| \geq \min_{\partial_{k-1}A'} \|\sigma_{f_1,\dots,f_k}(m)\| = \min_{\partial_{k-1}A'} \|\sigma_{f_1,\dots,f_n}(m)\| \geq \min \{\|\sigma_{f_1,\dots,f_n}(m)\| \mid m \in \cup\, \partial_{k-1}A_{V(g_1,\dots,g_{n-k})};\ (g_1,\dots,g_{n-k}) \in A^{n-k}\}$. Hence the set $[\cup\{\partial_{k-1}A_{V(g_1,\dots,g_{n-k})};\ (g_1,\dots,g_{n-k}) \in A^{n-k}\}]$ contains $\partial_{n-1}A$, since according to Theorem 1 the last set is the smallest one on which $\min \|\sigma_{f_1,\dots,f_n}(m)\|$ is taken for all $(f_1,\dots,f_n)$. Let now $(g_1,\dots,g_{n-k}) \in A^{n-k}$ and suppose that $\|\sigma_{f_1,\dots,f_k}(m)\| \geq r > 0$ on $V(g_1,\dots,g_{n-k}) \cap \partial_{n-1}A$ for some regular k-tuple $(f_1,\dots,f_k)$ over A. For each ε: $0 < \varepsilon < r$ we can find a neighborhood V_ε of the set $V(g_1,\dots,g_{n-k}) \cap \partial_{n-1}A$, such that $\|\sigma_{f_1,\dots,f_k}(m)\| \geq r-\varepsilon$ on V_ε. Hence on $\partial_{n-1}A$ we have:

$$C_\varepsilon^2 \|\sigma_{g_1,\dots,g_{n-k}}(m)\|^2 + \|\sigma_{f_1,\dots,f_k}(m)\|^2 \geq (r-\varepsilon)^2. \tag{2}$$

Consequently $\|\sigma_{C_\varepsilon g_1,\dots,C_\varepsilon g_{n-k}, f_1,\dots,f_k}(m)\| \geq r - \varepsilon$ on sp A, because the n-tuple $(C_\varepsilon g_1,\dots,C_\varepsilon g_{n-k}, f_1,\dots,f_k)$ is regular. As in Theorem 1 we see that $\|\sigma_{f_1,\dots,f_k}(m)\| \geq r$ on $V(g_1,\dots,g_{n-k})$. Hence $V(g_1,\dots,g_{n-k}) \cap \partial_{n-1}A \supset \partial_{k-1}A_{V(g_1,\dots,g_{n-k})}$ for any (n-k)-tuple $(g_1,\dots,g_{n-k})$ over A, because the latter is the smallest such subset of sp $A_{V(g_1,\dots,g_{n-k})}$. Now $\partial_{n-1}A \supset \cup\{V(g_1,\dots,g_{n-k}) \cap \partial_{n-1}A \mid (g_1,\dots,g_{n-k}) \in A^{n-k}\} \supset \cup\{\partial_{k-1}A_{V(g_1,\dots,g_{n-k})} \mid (g_1,\dots,g_{n-k}) \in A^{n-k}\}$, and consequently $\partial_{n-1}A \supset [\cup\{\partial_{k-1}A_{V(g_1,\dots,g_{n-k})} \mid (g_1,\dots,g_{n-k}) \in A^{n-k}\}]$. The theorem is proved.

REFERENCES

1. N. Sibony. Multidimensional analytic structure in the spectrum of a uniform algebra. In: Spaces of analytic functions, Kristiansand, Norway 1975. Lecture Notes in Mathematics 512. Springer, Berlin-Heidelberg-New York, 139-175.

2. R. Basener. A generalized Shilov boundary and analytic structure, Proc. Amer. Math. Soc. 47(1975), 98-104.

3. T. Tonev. Minimal boundaries of commutative Banach algebras, containing the Shilov boundary. In: Complex analysis and Applications, Varna '85, Sofia (to appear).

Institute of Mathematics
P.O.B. 373, BG-1090 Sofia, Bulgaria.

Picard-Fuchs Differential Equations for the Quadratic Periods of Abelian Integrals of the First Kind

Marvin Tretkoff*

1. Introduction

The primary purpose of the present paper is to show that the quadratic periods of abelian integrals of the first kind belonging to certain one-parameter families of Riemann surfaces satisfy homogeneous linear ordinary differential equations of Fuchsian class. Thus, we have an analogue of the Picard-Fuchs differential equations for the classical periods of abelian integrals that are well known to algebraic geometers. However, the motivation for our work comes from another source: the desire to identify an example of the equations of Fuchsian class whose existence was proved in [21]. Namely, we hope to find an equation of Fuchsian class whose solutions cannot be expressed as rational combinations of hypergeometric functions, possibly involving several values for the parameters, even though the singularities of the equation are at $z=0,1$, and ∞. (The precise meaning of this statement is given in [21]). Unfortunately, we have not yet been able to determine whether the equation discussed herein serves this purpose. The apparent reason for this unsatisfactory state of affairs is that we have not been able to establish that the monodromy group of our equation satisfies a group-theoretic condition proved in [21] to be characteristic of equations of the desired type. However, although we have not included this material here, it is possible to determine generating matrices for the monodromy groups arising from families of cyclic coverings of the Riemann sphere. Thus, the present paper may be viewed as the first steps in a search for a differential equation with three regular singular points that does not belong to the seemingly ubiquitous family of hypergeometric functions. If this program ultimately succeeds, then the quadratic periods will provide integral representations for the solution of the differential equation.

The term quadratic period refers to an interated integral of a pair of holomorphic 1-forms on a compact Riemann surface, V, of genus $g>1$. The precise definition will be given later; here we

---

* Supported in part by National Science Foundation Grant MCS-8103453.

merely note that, in order to make this notion meaningful, a base point, p_0, must be selected and the integration carried out along loops based at p_0. In this paper we will consider a family, V_x, $x \varepsilon X$, of Riemann surfaces and we will view the quadratic periods as functions of the parameter space X. In this context, it is traditional to take for X a complex manifold and to expect the quadratic periods to be holomorphic functions on it. Here, in view of the intended application, X must be the Riemann sphere with the points $x = 0,1$ and ∞ deleted.

The simplest families of Riemann surfaces that meet our requirements are families of cyclic coverings, V_x, of the z-sphere that are branched over the four distinct points $z = 0,1,x$ and ∞. The variable branch point $z=x$ serves as the complex parameter of the families and the unique point on V_x above $z=\infty$ is an expedient choice for the base point of its fundamental group. The latter permits us to make a locally constant choice of the fundamental group. That is, for each $x_0 \varepsilon X$ there is a set of paths on the z-sphere that lift to yield generators for the fundamental group of V_x for all x in a sufficiently small neighborhood of x_0. Moreover, there are simple expressions for the abelian differentials of first kind in which the parameter x appears rationally. Since the path of integration is fixed, we easily conclude that *locally* the quadratic periods are holomorphic functions on X. In fact, these turn out to be *multi-valued* functions because, as x traverses a loop in X that starts at x_0 and is not null-homotopic, the original set of generators for the fundamental group of V_{x_0} is replaced by another set. In other words, there is a non-trivial monodromy action on the fundamental group and this induces a non-trivial representation on the complex vector space of germs of holomorphic functions spanned by the quadratic periods at x_0.

Our main result about these functions is contained in the following

Theorem : Suppose that V_x is the Riemann surface defined by the equation

$$w^q=(z-x)^{a_0}\Pi(z-x_j)^{a_j}; \quad 1 \leq j \leq m,\ x_j \varepsilon C,$$

where $q>0$ is an odd prime that does not divide any of the a_j or their sum. Then, the quadratic periods of the abelian integrals of the first kind on V_x satisfy a homogeneous linear differential equation of Fuchsian class. The singular points of this equation are $x=x_j$, $j=1,\ldots,m$.

Since the proof of our Theorem is virtually independent of the choice of branch points and the corresponding exponents in the equation defining V_x, we have stated it in some generality. However, we have only presented the proof for the special family $w^5=z(z-1)(z-x)$. Hopefully, the resulting simplification of notation will serve both to clarify the details of the exposition and to bring the main lines of the proof into clearer focus. Moreover, the technique employed herein actually applies more generally. These generalizations are mentioned very briefly at the end of the paper; we hope to return to them in a future publication.

Throughout, we have adopted the viewpoint of the classical theory of functions and Riemann surfaces. We want to mention two ramifications of this decision. First, we have utilized growth estimates to prove the regularity of the singularities of our differential equation. However, Professor B. Dwork has kindly shown us that a purely arithmetic proof is possible. Next, we note that because we work exclusively with abelian differentials of the *first kind,* we have given a transcendental proof, based on analytic continuation, for the existence of the differential equation satisfied by the quadratic periods (see, Theorem 1, below). The usual algebro-geometric treatments involve the vector spaces of differentials of the second kind modulo exact differentials. Roughly speaking, these spaces are shown to be closed under differentiation with respect to the parameter, x, so some linear combination of a differential and its derivatives up to order 2g is exact. Integrating both sides of this equation yields the classical Picard-Fuchs equations for the periods. We note that our method applies to ordinary as well as quadratic periods. In the former case, we also obtain a differential equation of order 2g because the fundamental group requires 2g generators.

Investigation of the periods of abelian integrals and their special case, elliptic integrals, has played a major role in the development of function theory and its relationship to other branches of mathematics. Therefore, it would be futile to attempt a sketch of the subject here. However, it might be useful to illustrate the importance of this topic in the theory of Riemann surfaces by paraphrasing Torelli's theorem: the periods of the abelian integrals of the first kind uniquely determine the complex structure of a Riemann surface (See, for example, Gunning, [8], for a pre-

cise statement of Torelli's theorem and the allied theory of Jacobian varieties). Of course, the quadratic periods represent a far more limited topic, and a brief sketch of their development seems appropriate here.

Certain combinations of quadratic periods play a role in the classical theory of the Riemann theta function. In particular, a vector known as the "vector of Riemann constants" whose entries consist of such expressions appears in the famous "Riemann Vanishing Theorem" (see, for example, Siegel [20], Volume 2, Theorem 2, page 167). The explicit investigation of quadratic periods was initiated by R. C. Gunning in [7], where additional motivation from the theory of Riemann surfaces is provided. The main result contained in [7] is the calculation of the quadratic periods in case V is hyperelliptic; they are shown to equal certain quadratic expressions involving the ordinary periods. The explicit determination of the periods of abelian integrals is known to be a formidable task. Obviously, the case of the quadratic periods is no easier. Besides the hyperelliptic case, that is, 2-sheeted cyclic coverings of the sphere, the only published examples known to us are some special cyclic coverings: the Klein-Hurwitz curve and the Fermat curves. These may be found in [22] and, for the Fermat quartic curve, in [11]. In the course of proving Theorem 2 of the present paper, we essentially determine the quadratic periods for an arbitrary cyclic covering of the Riemann sphere. Despite the difficulty of determining them, the quadratic periods have played a role in a number of significant investigations.

First, we note an important paper by C. J. Earle, [4], in which the universal Teichmuller curve is embedded in the corresponding family of Jacobian varieties. In view of Hubbard's theorem [23], one of the key difficulties here is to pick a section of the latter family; essentially, the vector of Riemann constants provides the desired section. Next, the work of E. Jablow, [14], provides information about the natural domain for the quadratic periods qua functions of the family of all Riemann surfaces. Here, too, the vector of Riemann constants makes an important appearance. We note that Jablow works with pointed curves, that is, Riemann surfaces together with a point on them, and the Teichmuller space that parametrizes them. Presumably, this is necessitated by difficulties related to those that confronted Earle. The work of B. Harris, [10]

and [11], uses the quadratic periods to investigate an outstanding problem in algebraic geometry: does homological equivalence imply algebraic equivalence? In [11] he shows that the Fermat quartic curve and its negative, viewed as 1-cycles in the Jacobian variety of the former, provide a striking illustration that the answer is "no." Finally, we mention the work of R. Hain (see, for example, [9]) on iterated integrals. Among other things, the homotopy theoretic applications of these pioneered by K. T. Chen (see, [3], for a survey) are developed and applied to yield a mixed Hodge structure on the homotopy groups of algebraic varieties. Using this theory, together with the classical Torelli theorem, Hain and Pulte, [17], have obtained an analogue of Torelli's theorem for pointed Riemann surfaces Other interesting studies involving quadratic periods have been carried out by Hwang-Ma, [13] and Ramakrishnan, [18].

2. Basic Notions

We begin by recalling the definition of a quadratic period. Thus, let V denote a compact Riemann surface of genus $g>1$ and let $du_1,\dots,du_g$ be a basis for the complex vector space of abelian differentials of the first kind, that is, the vector space of holomorphic 1-forms on V. Next, let us fix a base point, p_0, on V and a parametrized loop $\gamma=\gamma(t)$, $0\leq t\leq 1$, that begins and ends at p_0. Then, the expression

$$Q_{jk}(\gamma)=\int_{\gamma}\Big(\int_{p_0}^{p} du_j\Big)du_k,\quad 1\leq j,k\leq q,$$

will be called a *quadratic period.* Here, the outer integral sign denotes a line integral along the loop γ. The integrand is the product of the holomorphic 1-form du_k and the "abelian integral of the first kind" $\int_{p_0}^{p} du_j$. The latter is a multi-valued holomorphic function on V whose branches at p_0 differ from one another by a period of du_j, that is, by the integral of du_j along a loop on V. In order to avoid any ambiguity, we require that: (1) as the parameter t runs over the unit interval, the upper limit $p=p(t)=\gamma(t)$ traverses γ, and (2) the initial value of our abelian integral is zero.

It is a routine matter to verify that a quadratic period is unchanged if we replace the loop γ by a homotopic loop on V. Therefore, for fixed j and k, $Q_{jk}(\gamma)$ may be viewed as a function defined

on the fundamental group $\pi_1(V,p_0)$. The following formulas, due to Gunning [7], express the behavior of this function with respect to the group operations:

$$\begin{aligned} Q_{jk}(\delta\gamma) &= Q_{jk}(\delta)+Q_{jk}(\gamma)-\pi_j(\delta)\pi_k(\gamma) \\ 0 &= Q_{jk}(\gamma^{-1})+Q_{jk}(\gamma)+\pi_j(\gamma)\pi_k(\gamma). \end{aligned} \tag{G}$$

Here, $\pi_j(\gamma)$ denotes the integral of du_j along γ. Of course, $\pi_j(\delta\gamma)=\pi_j(\delta)+\pi_j(\gamma)$ and $\pi_j(\gamma^{-1})=-\pi_j(\gamma)$, so the classical periods are homomorphisms from the abelianized fundamental group to the additive group of complex numbers. The quadratic periods do *not* share this property and it is hoped that these transcendental invariants actually reflect the more subtle, non-abelian, nature of the fundamental group.

Before continuing, we wish to point out that Gunning's definition of the quadratic periods is different from ours, but equivalent to it. Namely, he works on the universal covering, $\bar{V}$, of V and views $\pi_1((V,p_0)$ as a discontinuous group, Γ , acting on $\bar{V}$ with V as its quotient. The result is a somewhat more streamlined presentation than ours; for example, the complications arising from the multi-valued nature of the abelian integral disappear. Nevertheless, the "classical" spirit of the definition given herein seems better suited to the discussion of algebraically defined families that is our goal. When we turn to Teichmuller space and Bers fibre space over it, we shall adopt Gunning's approach.

Suppose that V is the Riemann surface defined by the equation

$$w^q=\Pi(z-x_j)^{a_j}, \quad 1\leq j\leq m+1, \; x_j \varepsilon C,$$

where $q>0$ is an odd prime that does not divide any of the a_j or their sum. Then, V is a q-sheeted cyclic covering of the z-sphere that is totally ramified over each of the x_j and the point $z=\infty$. Using the Riemann-Hurwitz formula (see, for example, [20]):

$$2 - 2g = 2q - r,$$

where r is the sum of the ramification indices, we may compute the genus, g, of V. In our case, we have $m+2$ ramification points, each with ramification index q-1, so we find that

$$g = \frac{1}{2}m(q-1).$$

Here is is important to realize the extent to which the branching behavior depends on the exponents. There would be no branch point at $z=\infty$ if the sum of the a_j were divisible by q. Moreover, the ramification index at $z=x_j$ would decrease if q and a_j were not relatively prime.

Abelian differentials of the first kind on V are of the form (see, for example, [15])

$$\frac{p(z,w)dz}{w^n},$$

where $n=1,\ldots,q\text{-}1$ and where p(z,w) is a polynomial whose degree, d, satisfies the equation:

$$1+d = \left[\sum \frac{na_j}{q}\right] - \sum \left[\frac{na_j}{q}\right], \quad j=1,\ldots,m+1.$$

Here, [s] denotes the greatest integer less than or equal to s. If, for each admissable value of n, we pick a basis for the polynomials of the corresponding degree d, then we shall arrive at a basis for the space of holomorphic 1-forms on V.

We now consider the family of Riemann surfaces V_x defined by the equation

$$w^5 = z(z-1)(z-x), \quad x \neq 0,1,\infty$$

Applying the preceding discussion to the present situation, we find that the genus, $g = 4$, and that a basis for the abelian differentials of the first kind on V_x is given by

$$du_1 = \frac{dz}{w^2}, \quad du_2 = \frac{dz}{w^3}, \quad du_3 = \frac{dz}{w^4}, \quad du_4 = \frac{zdz}{w^3}.$$

Next, we turn our attention to the topology of cyclic coverings of the Riemann sphere. In particular, we will give an *explicit* set of generators for the fundamental group. These generators will be described in terms of lifts of specific paths joining the branch points on the z-sphere. Once again, we will mainly discuss the Riemann surfaces defined by

$$w^5 = z(z-1)(z-x), \quad x \neq 0,1,\infty.$$

We begin our work on the z-sphere. Let e_j denote an oriented path that starts at $z=\infty$ and

ends at z=j, where j = 0,1, or x. In addition, we shall suppose, as indeed we may, that the three paths have been selected so that: (1) the e_j are simple paths, that is, they have no self-intersections, (2) each pair of distinct paths e_j and e_k meet at the point $z=\infty$ and nowhere else, (3) there is an open set U containing $z=\infty$ such that every circle C centered at $z=\infty$ and belonging to U meets each e_j in precisely one point, (4) the paths e_j are encountered in the order e_0, e_1, e_x, or a cyclic permutation thereof, when C is traversed in the positive (counterclockwise) direction.

It is obvious that the complement, D, of the union of the three paths e_0, e_1, and e_x on the z-sphere is homeomorphic to the interior E, of the unit disk $\bar{E}$. In fact, the Riemann Mapping Theorem asserts the existence of a conformal map, f, of E onto D; moreover f admits a continuous extension, $\bar{f}$, that carries $\bar{E}$ onto $\hat{C}$. Thus, we are led to the following cell decomposition of $\hat{C}$:

0−cells	1−cells	2−cells
0, 1, x, ∞	e_0, e_1, e_x	$\bar{E}$

The attaching maps are indicated by the following diagrams:

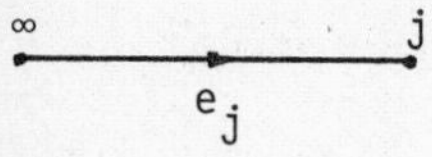

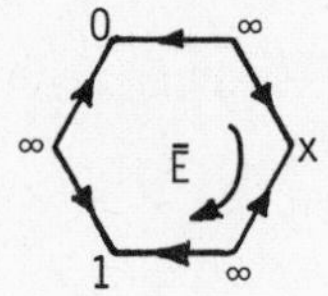

Although there are other cell decompositions of the sphere that require fewer cells, the one we have given is useful in the present situation because it lifts to a cell decomposition of V_x that is invariant under the cyclic group of covering transformations.

Lifting each cell in our decomposition of the sphere to V_x, we obtain the following:

0−cells	1−cells	2−cells	
$\bar{0}, \bar{1}, \bar{x}, \bar{\infty}$	$\bar{e}_j(k)$	$\bar{E}_k$	$j=0,1,x,\infty,\ k=1,\ldots,5.$

Here, our notation is intended to indicate that $\bar{j}$ lies over j, that each $\bar{e}_j(k)$, $k=1,\ldots,5$, projects onto e_j, and that $\bar{E}_k$ projects onto $\bar{E}$. The attaching maps are indicated by the following diagrams:
(See FIGURE on page preceding REFERENCES.)

Of course, the index k is taken modulo 5.

In order to describe a set of generators for the fundamental group of V_x, we first note the existence of the following loops:

$$\gamma_j(k)=\bar{e}_j(1)-\bar{e}_j(k+1),\quad k=1,\dots,4,\quad j=0,1,x.$$

Here $\gamma_j(k)$ is supposed to be parametrized so that we begin at $\bar{\infty}$ and traverse $\bar{e}_j(1)$ until we reach $\bar{j}$, then we return to $\bar{\infty}$ along the path $\bar{e}_j(k+1)$.

<u>Proposition</u> <u>1:</u> The eight loops $\gamma_0(k)$ and $\gamma_1(k)$, $k = 1,\dots,4$, represent a set of generators for the fundamental group $\pi_1(V_x,\bar{\infty})$.

<u>Sketch</u> <u>of</u> <u>Proof:</u> The 1-skeleton, $V_x^{(1)}$, of V_x consists of the union of all the 1-cells $\bar{e}_j(k)$, so its fundamental group is a free group whose generators are in one to one correspondence with the edges comprising the complement of a maximal tree. In particular, such a tree is formed by the union of $\bar{e}_0(1)$, $\bar{e}_1(1)$ and $\bar{e}_x(1)$. Therefore, the twelve loops $\gamma_j(k)$, $j=0,1,x$, $k = 1,\dots,4$, represent a free set of generators for $\pi_1(V_x^{(1)},\bar{\infty})$.

Attaching each of the 2-cells $\bar{E}_k$ introduces a single relation among these generators. It is easily checked that as k runs from 1 to 4 the relations introduced by the $\bar{E}_k$ allow us to express $\gamma_\infty(k)$ in terms of the $\gamma_0(m)$ and $\gamma_1(n)$. Finally, the relation introduced by $\bar{E}_5$ turns out to be a reduced word in the $\gamma_0(m)$ and $\gamma_1(n)$, so we obtain a single relation among these eight generators. Thus we have obtained a presentation for $\pi_1(V_x,\bar{\infty})$ that has the eight required generators and a single relation. This completes the proof.

<u>Remarks:</u>

1. Although we have obtained a presentation of the fundamental group as a 1- relator group, we have not obtained the "standard" presentation because the intersection matrix for the homology classes of our generators is not the standard form

$$J = \begin{bmatrix} 0 & E \\ -E & 0 \end{bmatrix},$$

E the identity matrix.

2. A similar presentation can be given for any cyclic covering of the sphere. If $z=a_1, \ldots, a_{m+1}$ and $z = \infty$ are the branch points, then the generators are given in terms of paths joining $z=\infty$ to $z=a_j$, $j=1,\ldots,m$.

For more details, see [22].

3. The Differential Equation

We now view the quadratic periods as multi-valued functions defined on the parameter space of our family of Riemann surfaces. To indicate this more clearly, we augment our notation and write $Q_{jk}(\gamma,x)$ and $\pi_j(\gamma,x)$ to indicate, respectively, the quadratic and ordinary periods on V_x. For brevity, we will also refer to products such as $\pi_j(\gamma)\pi_k(\delta)$ as quadratic periods.

Our goal is to prove the following

Theorem 1: The quadratic periods defined by the family $w^5=z(z-1)(z-x)$ satisfy a homogeneous linear differential equation,

$$L(u)=p_0(x)\frac{d^n u}{dx^n} + \ldots + p_n(x)u = 0 ,$$

whose coefficients $p_j(x)$ are single valued holomorphic functions with a finite set of singularities on the Riemann sphere.

Proof : First, because the fundamental group of V_x is finitely generated, any loop γ can be expressed as a product of a fixed finite set of elements: the generators $\gamma_0(k)$, $\gamma_1(k)$, $k=1,\ldots,4$, and their inverses. Therefore, by repeated application of (G), we see that each quadratic period is a linear combination, with complex coefficients, of the following finite collection:

$$Q_{jk}(\gamma_r(s),x) \text{ and } \pi_j(\gamma_r(s),x)\pi_k(\gamma_m(n),x), \qquad \text{(QP)}$$
$$1\le j,\ k\le g;\quad 0\le r,\ ,m\le 1;\quad 1\le s,\ n\le 4.$$

Next, letting X denote the x-sphere punctured at $x=0,1,\infty$, we observe that each of the expressions (QP) defines a single valued holomorphic function in the neighborhood of each point $x \varepsilon X$. This is true because each of the integrands that occurs is a holomorphic function of the parameter and because we may differentiate under the integral sign. Needless to say, the new

integrand need not be of the same type; for example, a differential of the second kind may occur, but this is of no matter here. Now, if we pick a base point $x_0 \varepsilon X$, it follows that each of the expressions (QP) can be continued analytically along any path in X that begins at x_0. Of course, continuation along a loop in X need not end with the original expression; that is, the expressions (QP) may well define *multi-valued* functions on X. However, because we must end with a quadratic period, we will arrive at a linear combination of the original expressions.

From a more modern viewpoint, the preceding discussion means that the quadratic periods together with the products of ordinary periods span a complex vector bundle, E, over X. Namely, for each $x \varepsilon X$, the fibre E_x of E is the complex vector space spanned by the expressions (QP). Although we shall not use it, the fact is that E is a topologically trivial bundle because the base space, X, has the homotopy type of a finite graph. Moreover, non-compact Riemann surfaces are Stein manifolds (see, for example, [5]), and a famous theorem of Grauert [6] asserts that topologically trivial complex vector bundles defined over Stein manifolds are also analytically trivial; thus E is analytically trivial.

In order to complete the proof of Theorem 1, we modify an argument given without reference by Poole, [16], pages 46-47, under the title "Riemann's Converse Theorems". Conceivably, this argument is due to Riemann, but we have not investigated the matter. In any event, we have

Proposition 2: Suppose that X is a Zariski open subset of C and that $x_0 \varepsilon X$. Moreover, suppose that we are given a finite collection $w_1,\dots,w_t$ of functions that are holomorphic in an open disk centered at x_0 and belonging to X. In addition, suppose that: (1) each w_j admits analytic continuation along any path in X beginning at x_0, and (2) analytic continuation of w_j along any loop in X based at x_0 ends with a linear combination of $w_1,\dots,w_t$. Then there is a homogeneous linear ordinary differential equation with single valued holomorphic coefficients defined on X that is satisfied by each of the w_j, $j=1,\dots,n$.

Proof: We may suppose that $w_1, \dots, w_t$ are linearly independent; otherwise, we simply work with a maximal linearly independent subset of them. Now, let u denote an unknown function of

x and let

$$W = W(u) = W(u, w_1, \ldots, w_n) = \begin{vmatrix} u & u' \ldots & u^{(n)} \\ w_1 & w_1' \ldots & w_1^{(n)} \\ \vdots & \vdots & \vdots \\ w_n & w'_n \ldots & w_n^{(n)} \end{vmatrix}$$

denote the Wronskian. Clearly, $W(u)$ vanishes when u is replaced by any of the w_j. Moreover, if we expand $W(u)$ with respect to the first row, then we obtain a differential equation whose coefficients $a_j = a_j(w_1,\ldots,w_n)$ are holomorphic functions of x.

Unfortunately, these coefficients need not be single valued on X. For example, analytic continuation along a loop σ may transform the column vector

$$\vec{w} = \vec{w}(x) = \begin{bmatrix} \vdots \\ w_j(x) \\ \vdots \end{bmatrix}$$

to $A_\sigma \vec{w}$, where A_σ is a non-singular complex matrix of degree n. It follows that $W(w_1,\ldots,w_n)$ is multiplied by the determinant of A_σ upon continuation along σ. Moreover, the determinant of each of the n by n minors formed from the last n rows of $W(u)$ is easily seen to be multiplied by this same factor after continuation along σ. Setting

$$b_j = b_j(w_1,\ldots,w_n) = \frac{a_j(w_1,\ldots,w_n)}{W(w_1, \ldots, w_n)},$$

we obtain single-valued functions on X that are holomorphic except, perhaps, at the points where the denominator vanishes. Of course, W is not identically zero because $w_1,\ldots,w_n$ are linearly independent.

Although $W(w_1,\ldots,w_n)$ may be multi-valued on X, the orders of its zeros are well defined. Namely, at any point of X, any one branch of the function may be obtained from any of its other branches by multiplication by a non-zero constant. It follows that the zeros of $W(w_1,\ldots,w_n)$ form

a discrete subset of X and a positive integer, its order of vanishing, is associated with each of these points. Now, X is a Stein manifold, so there is a single valued holomorphic function, $g(x)$, on X whose zeros, as well as their orders, coincide with those of $W(w_1,...,w_n)$. Setting

$$p_j = p_j(w_1, \ldots, w_n) = g(x)b_j,$$

we obtain single-valued holomorphic functions on X. Moreover, each of the functions $w_j(x)$ satisfies the differential equation

$$L(u) = p_0(x)u^{(n)} + \ldots + p_n(x)u = 0$$

This completes the proof of Proposition 2.

Finally, substituting the expressions (QP) for the functions w_j in Proposition 2, we see that the proof of Theorem 1 is complete.

Remark : Poole, [16], apparently does not consider the possibility that the Wronskian vanishes on X. Since he does not know apriori that the w_j satisfy a linear differential equation, his assumption seems unwarrented. Of course, he may have tacitly intended additional hypothesis.

4. Regularity

We now turn our attention to the singularities of our differential equation. In particular, we have

Theorem 2: The differential equation $L(u) = 0$ satisfied by the quadratic periods is of Fuchsian class. That is, its three singular points, $x=0,1,\infty$, are each regular singular points.

Proof: Since the proof is virtually the same for each of the expressions (QP), we will restrict our discussion to one example: $Q_{13}(\gamma_1(1),x)$. In order to simplify notation, we will set $Q=Q_{13}$, $\gamma=\gamma_1(1)$, $e=e_1$, $\bar{e}_1=e_1(1)$, $\bar{e}_2=e_1(2)$, $du=du_1$ and $dv=du_3$. In addition, we will indicate the multi-valued abelian integrals by writing

$$u(\bar{z}) = \int_{\infty}^{\bar{z}} du \text{ and } v(\bar{z}) = \int_{\infty}^{\bar{z}} dv.$$

Moreover, we will only consider the singularity at $x=1$; the remaining cases are treated

similarly. None of the essential ideas are lost by making these assumptions.

Thus, we wish to prove that the isolated singularity $x=1$ of the homogeneous linear equation $L(u) = 0$ is a regular singular point. We shall accomplish this by showing that as x tends to 1, none of its solutions tend to infinity faster than a fixed power of x-1 in any sector with $x=1$ as its vertex. It is well-known (see, for example, [12]) that this growth condition is equivalent to the more algebraic definition of a regular singularity that is often found in textbooks. Therefore, we wish to prove that there is a positive integer N such that

$$|x-1|^N|Q(\gamma,x)|$$

is bounded as x aproaches 1 in any sector with $x=1$ as its vertex. In order to accomplish this, we will first express the quadratic period $Q(\gamma,x)$ in terms of integrals in the z-plane and then we will show that these integrals satisfy an estimate of the desired type.

We will now evaluate the quadratic period $Q(\gamma)$. Thus, we have

$$Q(\gamma)=\int_{\bar{e}_1} u\,dv - \int_{\bar{e}_2} u\,dv,$$

with the understanding that the appropriate branch of $u(\bar{z})$ must be used along the paths $\bar{e}_1$ and $\bar{e}_2$. We will also suppose that the path $e=e_1$ is that part of the positive real axis joining $z=\infty$ to $z=1$. Implicit in this assumption is the fact that x is not on e. Our simplifications are merely for illustrative purposes; they are not essential to the technique employed. Finally, we indicate the classical period of du along γ by writing

$$\pi(\gamma)=\int_{\gamma} du.$$

Recall that in order to evaluate a line integral along a path, we first cover the path with a finite set of coordinate neighborhoods and then we write the path as a sum of subpaths, each belonging to one of the coordinate neighborhoods. Expressing the integrand in terms of the appropriate local parameter, we may then carry out the integration along each subpath. Summation of the results of integration along these subpaths then yields the value of the desired line integral. In the present situation, we shall see that only four coordinate charts are required.

Suppose that $\bar{z}(t)$ belongs to γ and let z(t) denote its projection on the Riemann sphere. If $z(t) \neq 1$ or ∞, then we may use the ordinary complex variable z as a parameter in a neighborhood of $\bar{z}(t)$. More precisely, for some positive number, δ, depending on z(t), the disk $|z-z(t)| < \delta$ serves as a coordinate neighborhood of $\bar{z}(t)$. Now, suppose that U is a simply connected open subset of the Riemann sphere that contains all points of e, except $z=1$ and $z=\infty$. Then, V_x contains two disjoint open subsets, U_1 and U_2, that each project homeomorphically onto U and contain, respectively, all the points on $\bar{e}_1$ and $\bar{e}_2$ other than the end points $\bar{1}$ and $\bar{\infty}$. Of course, the complete preimage of U on V_x consists of five disjoint copies of U.

In order to introduce local coordinates at the point above $z=1$, we let W denote the disk $|z-1| < r < 1$, where r is an arbitrarily selected positive number less than 1. Then, corresponding to W, a disk $\overline{W}$, is defined in the τ-plane by the equation $\tau^5=z-1$. This disk may be mapped conformally onto the totally ramified neighborhood of $\bar{1}$ in V_x that lies above W. Thus, τ may be viewed as a local parameter on V_x in a neighborhood of $\bar{1}$ and $\overline{W}$ is its domain of definition. In fact, we will abuse notation and use the same symbol, $\overline{W}$, for the disk in the τ-plane and its image on V_x

Local coordinates are introduced in the same fashion at the point $\bar{\infty}$ of V_x. Here, we pick a positive number $R > 1$ and we let W_∞ denote the disk centered at $z=\infty$ that is defined by $|z| > R$. Then, the equation $z=\tau^{-5}$ allows us to use τ as a local uniformizing parameter in the disk, $\overline{W}_\infty$, on V_x lying over W_∞.

The loop γ will be divided into six paths, each lying entirely in one of the four coordinate neighborhoods introduced in the preceding discussion. This is accomplished by simply subdividing e into three subintervals and lifting them to the appropriate coordinate neighborhoods on V_x. Thus, we pick positive numbers a and b that satisfy the inequalities $1 < a < r < R < b$, and we write e as the union of the intervals [1,a], [a,b], [b,∞]. Of course, [a,b] has two disjoint lifts, one in U_1 and the other in U_2, whereas [1,a] has two preimages in $\overline{W}_1$ with the common end point $\bar{1}$. If α is the real number defined by $\alpha^5+1=a$, then $0 \leq \tau \leq \alpha$ is one of these preimages of [1,a]. The other preimage is given by the segment $\eta\tau$, where $\eta=e^{2\pi i/5}$ and $0 \leq \tau \leq \alpha$ Similarly, the interval

$[b,\infty]$ has two preimages on $\overline{W}_\infty$. These are given by the segments $0 \le \tau \le \beta$ and $\eta\tau$, $0 \le \tau \le \beta$, where $b = \beta^{-5}$ defines the real number β.

Next, we wish to express the integrand udv in terms of the appropriate local coordinates. We begin by selecting a point z_0 in U and a power series $w_1(z-z_0)$ that satisfies the equation $w^5 = z(z-1)(z-x)$ and converges in a neighborhood of z_0. This series admits analytic continuation throughout the z-sphere with $z = 0,1$, and ∞ deleted; in particular, it defines a single-valued holomorphic function on the simply connected domain D obtained from $\hat{C}$ by deleting the paths e_0, e_1 and e_∞. We indicate this function by writing

$$w_1 = w_1(z) = z^{1/5}(z-1)^{1/5}(z-x)^{1/5}.$$

Identifying D with E_1, the interior of $\overline{E}_1$, we may view w_1 as a holomorphic function on E_1. This function admits analytic continuation to yield a single-valued meromorphic function on V_x; we wish to examine the relation between its expansions at pairs of points that lie, respectively, on $\overline{e}_1$ and $\overline{e}_2$ and project to the same point on e.

If $z(t) \ne 1$ or ∞ lies on e, then there are two points $\overline{z}(t_1)$ and $\overline{z}(t_2)$ on $\overline{E}_1$ that project to $z(t)$. Suppose that $\overline{\lambda}_1$ and $\overline{\lambda}_2$ are paths on $\overline{E}_1$ joining $\overline{z}_0$ to $\overline{z}(t_1)$ and $\overline{z}(t_2)$ as illustrated in the Figure, and let λ_1 and λ_2 denote their projections to the Riemann sphere. Then, analytic continuation of $w_1(z-z_0)$ along λ_1 and along λ_2 yields two power series, $w_1(z-z(t))$ and $w_2(z-z(t))$, that converge in a neighborhood of $z = z(t)$. Since $w_1(z-z_0)$ is multiplied by $\eta = e^{\frac{2\pi i}{5}}$ upon continuation along the loop $\lambda_1^{-1}\lambda_2$ in the z-plane, we see that

$$w_2(z-z(t)) = \eta w_1(z-z(t)).$$

Therefore, if we write $z^{1/5}(z-1)^{1/5}(z-x)^{1/5}$ for the preceding continuation of w_1 from E_1 to a point on $\overline{e}_1$ whose local coordinate is z, then its continuation to the corresponding point on $\overline{e}_2$ is given by $\eta z^{1/5}(z-1)^{1/5}(z-x)^{1/5}$.

We may now evaluate the period

$$\pi(\gamma) = \int_\gamma du = \int_{\overline{e}_1} du - \int_{\overline{e}_2} du.$$

Here, we must break up the paths $\bar{e}_1$ and $\bar{e}_2$ comprising γ , as indicated earlier, into subpaths belonging to the coordinate neighborhoods $\overline{W}_1$, U_1, U_2, and $\overline{W}_\infty$. Substituting the appropriate local coordinates, we find that

$$\int_{\bar{e}_1} du = 5\int_{\alpha}^{0} \frac{\tau^2 d\tau}{(\tau^5+1)^{2/5}(\tau^5+1-x)^{2/5}} + \int_{b}^{a} \frac{dz}{z^{2/5}(z-1)^{2/5}(z-x)^{2/5}} - 5\int_{0}^{\beta} \frac{d\tau}{(1-\tau^5)^{2/5}(1-x\tau^5)^{2/5}} .$$

Since the integrands represent abelian differentials of the first kind, they are holomorphic in each coordinate chart. It follows that, as α tends to zero, the integral along the path $0 \le \tau \le \alpha$ also tends to zero. Similarly, the integral along the subpath belonging to $\overline{W}_\infty$ tends to 0 as β approaches 0. It follows that

$$\int_{\bar{e}_1} du = \int_{\infty}^{1} \frac{dz}{z^{2/5}(z-1)^{2/5}(z-x)^{2/5}} .$$

The integral of du along $\bar{e}_2$ can be treated in the same fashion. Taking into account the relation between the local expressions for $du = \frac{dz}{w^2}$ along $\bar{e}_1$ and $\bar{e}_2$, we find that

$$\int_{\bar{e}_2} du = \eta^3 \int_{\bar{e}_1} du.$$

Therefore, we have

$$\int_{\gamma} du = (1-\eta^3)\int_{\infty}^{1} \frac{dz}{z^{2/5}(z-1)^{2/5}(z-x)^{2/5}} .$$

Applying the substitution $z = \zeta^{-1}$, we obtain

$$\int_{\gamma} du = (\eta^3-1)\int_{0}^{1} \frac{d\zeta}{\zeta^{4/5}(1-\zeta)^{2/5}(1-x\zeta)^{2/5}} ,$$

so

$$\int_{\gamma} du = (e^{6\pi i/5}-1)\frac{\Gamma(1/5)\Gamma(3/5)}{\Gamma(4/5)} F(2/5,1/5,4/5;x),$$

where, as usual, F(a,b,c; x) denotes the hypergeometric function.

We now turn to the quadratic period $Q(\gamma)$. Once again, we will compare the power series expressions for the integrand at pairs of points, $\bar{z}(t_1)$ and $\bar{z}(t_2)$, on $\bar{e}_1$ and $\bar{e}_2$ that project to the

same point, $z(t)$, on the path e on the z-sphere. This time we face the additional complication of comparing the branches of the abelian integral $u(z)$ at corresponding points on $\bar{e}_1$ and $\bar{e}_2$.

We begin by writing u_0 for the branch of u that vanishes at the point $\bar{\infty}$ on V_x. Of course, we must use the appropriate local parameter in the series expansion of this branch. Next, we let $u_1 = u_1(z - z(t))$ and $u_2 = u_2(z - z(t))$ denote the branches of u that are obtained from u_0 by analytic continuation along γ to $\bar{z}(t_1)$ and $\bar{z}(t_2)$ respectively. We may suppose that both of these branches are given as power series that converge in a disk Δ given by $|z\text{-}z(t))| < \delta$, for some positive number δ. If z belongs to Δ, then we have

$$u_2(z) = \int_{\gamma(0)}^{\gamma(t_2)} du + \int_{z(t)}^{z} du,$$

where the first integral denotes the line integral of du along γ from $\bar{\infty} = \gamma(0)$ to $\bar{z}(t_2) = \gamma(t_2)$ and where the second integral denotes the integral of du from $z(t)$ to z along any path in Δ. Clearly, we also have

$$\int_{\gamma(0)}^{\gamma(t_2)} = \int_{\gamma} du + \int_{\gamma^{-1}(0)}^{\gamma^{-1}(1-t_2)} du,$$

where the first integral equals the period $\pi(\gamma)$ and where the second integral is the line integral of du along the path γ^{-1}, starting $\bar{\infty} = \gamma^{-1}(0)$ and ending at $\gamma^{-1}(1-t_2) = \gamma(t_2)$. Taking into account the expression for du along $\bar{e}_2$ in terms of its expression along $\bar{e}_1$, we see that the latter integral equals

$$\eta^3 \int_{\gamma(0)}^{\gamma(t_1)} du = \eta^3 u_1(z).$$

Now, it is easy to evaluate $Q(\gamma)$. First, note that the local expression for dv along $\bar{e}_2$ is obtained from the expression for dv along $\bar{e}_1$ by multiplication by η because $dv = \frac{dz}{w^4}$. Then, we see that

$$
\begin{aligned}
Q(\gamma) &= \int_{\bar{e}_1} u\,dv - \int_{\bar{e}_1} \Big[\eta^3 u + \pi(\gamma)\Big]\eta\,dv \\
&= (1-\eta^4)\int_{\bar{e}_1} u\,dv - \eta\pi(\gamma)\int_{\bar{e}_1} dv \\
&= (1-\eta^4)\int_{\infty}^{1}\left[\left[\int_{\infty}^{\zeta}\frac{dz}{z^{2/5}(z-1)^{2/5}(z-x)^{2/5}}\right]\left[\frac{1}{\zeta^{4/5}(\zeta-1)^{4/5}(\zeta-x)^{4/5}}\right]\right]d\zeta \\
&\quad -(1-\eta^4)\eta\pi(\gamma)\int_{\infty}^{1}\frac{dz}{z^{4/5}(z-1)^{4/5}(z-x)^{4/5}}\,.
\end{aligned}
$$

Here, as before, the Eulerian integrals might be expressed in terms of known special functions, but this is not necessary for our purposes. Here, we have also omitted the analogue of the previous discussion of the subdivision of γ into six subpaths because, as in that case, the integrals along the paths in $\overline{W}_1$ and $\overline{W}_\infty$ tend to zero as α and β approach zero.

Now, we are ready to make the estimates that prove Theorem 2. In order to indicate its dependence upon x, we once again write $Q(\gamma,x)$ for our quadratic period. At first, it will be supposed that the sector in which x approaches 1 lies in the left half plane $\mathrm{Re}(z) < 1$. A simple modification of the paths of integration will subsequently allow us to infer the validity of Theorem 2 in the remaining cases. Since we have expressed $Q(\gamma,x)$ as a linear combination of two integrals, it will suffice to estimate the growth of these integrals as x approaches 1. Our discussion will be limited to the more complicated of the two integrals because, as will become apparent, our reasoning applies to the simpler integral as well.

Thus, we wish to estimate the growth of

$$
\int_{\infty}^{1}\left[\left[\int_{\infty}^{\zeta}\frac{dz}{z^{2/5}(z-1)^{2/5}(z-x)^{2/5}}\right]\left[\frac{1}{\zeta^{4/5}(\zeta-1)^{4/5}(\zeta-x)^{4/5}}\right]\right]d\zeta
$$

as $x\to 1$ in the half plane $\mathrm{Re}(z) < 1$. It will be convenient to break up the path of integration and use local coordinates as in our discussion of the integral $\int_{\bar{e}_1} du$. Therefore, we let

$$
\int_{\infty}^{1} u\,dv = \int_{a}^{1} u\,dv + \int_{b}^{a} u\,dv + \int_{\infty}^{b} u\,dv.
$$

Setting $z=\tau^5+1$ and $\zeta=\sigma^5+1$, we obtain

$$J=\int_a^1\left(\int_a^z \frac{dz}{w^2}\right)\frac{dz}{w^4}$$

$$=25\int_\alpha^0\left[\left[\int_\alpha^\sigma \frac{\tau^2 d\tau}{(\tau^5+1)^{2/5}(\tau^5+1-x)^{2/5}}\right]\left[\frac{1}{(\sigma^5+1)^{4/5}(\sigma^5+1-x)^{4/5}}\right]\right]d\sigma.$$

Thus, we find that

$$|J|\leq 25\alpha^4 \max_{0\leq\sigma\leq\alpha}\left\{\left[\frac{1}{|\sigma^5+1-x|^{4/5}}\right]\left[\max_{\sigma\leq\tau\leq\alpha}\frac{1}{|\tau^5+1-x|^{2/5}}\right]\right\}.$$

Since Re(x) < 1 and both σ^5+1 and τ^5+1 are on the real interval [1,a], we have

$$|\sigma^5+1-x|\geq|x-1|$$

and

$$|\tau^5+1-x|\geq|x-1|.$$

It follows that

$$|J|\leq\frac{25\alpha^4}{|x-1|^{6/5}}\ ;$$

therefore, $|(x-1)^2J|$ is bounded by a constant for all values of x in the left half plane Re(z) < 1 that also satisfy, for example, $|x\text{-}1| < \frac{\alpha^5}{2}$.

It is easy to see that the preceding discussion also applies to the integrals

$$\int_a^b u\,dv \text{ and } \int_\infty^b u\,dv.$$

In fact, the argument is even simpler because, in each case, there is a positive lower bound for the distances between points on the interval of integration and admissible values of x. This completes our proof in case Re(x) < 1.

If the sector in which x approaches 1 does not lie in the half plane Re(z) < 1, then we must modify our choice of paths of integration. Suppose our sector is defined by

$$\varphi\leq \mathrm{Arg}(x-1)\leq\theta.$$

Then, we replace e by the path consisting of the following: (1) the positive real axis from $z=\infty$ to

$z=a$, (2) an arc of the circle $|z-1| = a-1$ joining $z=a$ to the point, P, in the half plane $Re(z) < 1$ where the circle $|z-1| = a-1$ meets the line $Arg(z-1) = \varphi+\frac{(\theta-\varphi)}{2}$, and (3) the radius of $|z-1| = a-1$ that joins P to $z = 1$. Moreover, in step (2) we must choose an arc that does not cut the sector $\varphi \leq Arg(z-1) \leq \theta$ into two pieces. Otherwise, it would be impossible to select a path e_x from $z=\infty$ to $z=x$ that does not meet $e=e_1$ except at $e=\infty$.

With these modifications to our path of integration, it is easy to see that the integral of udv along the radius defined in step (3) behaves like the integral J in the case $Re(x) < 1$. As in that case, integration along the remaining segments does not contribute to the growth of $Q(\gamma,x)$ as x approaches 1. Therefore, we find that the quadratic period does not grow faster than a fixed power of $|x-1|$ as $x \rightarrow 1$, so Theorem 2 is proved.

Remark: It follows from Theorem 2 that the coefficients, $p_j(x)$, of the differential equation $L(u) = 0$ are polynomials in x. Thus, the Wronskian, W, appearing in the proof of Proposition 2 has finitely many zeros, but we have not shown that none of them are on X. In other words, we have not excluded the possibility that our differential equation has *apparent singularities.*

5. Generalizations

It is clearly desirable to extend our results to more general complex analytic families of Riemann surfaces, V_x, $x \epsilon X$. In particular, we would hope that the quadratic periods give rise to interesting complex analytic vector bundles on X. Two problems immediately present themselves: (1) we must select holomorphic 1-forms on V_x that vary holomorphically with x, and (2) we must select holomorphically varying fundamental groups, $\Gamma_x = \pi_1(V_x, p_x)$.

There are several ways to deal with problem (1). First, we may work with algebraically defined families of Riemann surfaces. For example, we might define V_x by the vanishing of a polynomial $f(z,w,x)$, where z and w are complex variables and x denotes the coordinates of the ambient space of the algebraic variety that parametrizes our family. In the simplest case, x might be an additional complex variable and V_x the curve cut out on the surface $f(z,w,x) = 0$ by the plane, $x =$ constant. In any event, the holomorphic 1-forms are given by well-known

formulas. These are of the form $\frac{p(z,w,x)dz}{f_w}$ where p is a polynomial that satisfies certain conditions depending on the degree and the nature of the singularities of f, and where f_w denotes the partial derivative of f with respect to w. Since x appears rationally in these expressions, they will only be valid on a Zariski open subset, X, of the variety that parametrizes our family. For example, for certain values of x the equation $f(z,w,x) = 0$ may define a singular curve whose normalization has genus less than g. In any event, the denominator f_w may be replaced by a polynomial whose factors are $(z-z_j)$, where the z_j are the branch points of the Riemann surface V_x. To see this, we recall that: (1) $D=Af+Bf_w$, where D is the discriminant of f and where A and B are polynomials in z, and (2) D is a polynomial in z that vanishes at the common roots of f and f_w. The roots occuring in (2) include the branch points; any other roots that occur are absorbed in the numerator because of the conditions imposed on p(z,w,x). Finally, we note that this reduction leads to estimates for the growth of the periods along the lines of those presented in the proof of Theorem 2.

A second way to approach the difficulty presented by (1) is to let X be a submanifold of *Teichmuller space*, T_g, the space that parametrizes all Riemann surfaces of geneus g that are marked by a choice of canonical homology basis. If V_x denotes the Riemann surface represented by the point x, then the resulting family is called the *universal Teichmuller curve* and the corresponding family of universal coverings, suitably represented, forms *Bers fibre space* over Teichmuller space. A theorem of Bers states that there exist g holomorphic functions on the Bers fibre space whose restrictions to each fibre descend to yield a basis for the holomorphic 1-forms on the corresponding Riemann surface, V_x, in the universal curve. Therefore, if X is any submanifold of Teichmuller space, these l-forms provide a solution to problem (1). We refer to Earle's article, [4], for a convenient summary of the material concerning Teichmuller space and fibre spaces over it.

The difficulties associated with problem (2) appear to be far more serious than those posed by problem (1). For example, according to a celebrated theorem of J. Hubbard [23], it is impossible to select a holomorphic section of the universal curve over Teichmuller space. Therefore, if X

is the whole Teichmuller space we cannot solve problem (2) because we cannot select a holomorphically varying base point for the fundamental group of V_x , $x \epsilon X$. However, we can make these choices on Bers fibre space of genus g, which is well-known to be isomorphic to the Teichmuller space of pointed Riemann surfaces, that is, the space of marked Riemann surfaces with a distinguished point chosen on each of them. In this connection, the quadratic periods provide us with a vector bundle over Bers fibre space, as in Theorem 1, and we pose the

Problem: Identify the vector bundle of quadratic periods over Bers fibre space in terms of "known" vector bundles. Of course, the same question can be asked with r-fold iterated integrals, $r>2$.

Next, suppose that X is the subspace of Teichmuller space consisting of those points, x, for which: (1) V_x does not admit any non-trivial conformal self-mapping, and (2) V_x contains only normal Weierstrass points $p_j = p_j(x)$, $j=1,...,g(g-1)(g+1)$. Now, by modifying the construction given in the proof of Theorem 1, we can construct a complex analytic vector bundle, E, on X. Namely, we augment the expressions (QP) in two ways. First, we form all possible quadratic periods using *each* of the base points $p_j(x)$. Next, we introduce all possible expressions of the form $\pi_{jk}(x)w_{lmn}(x)$, where $\pi_{jk}(x)$ denotes the period of $du_j(x)$ along a l-cycle γ_k on V_x and where $w_{lmn}(x)$ is the integral of $du_l(x)$ from $p_m(x)$ to $p_n(x)$. Of course, $du_j(x)$ and γ_k belong, respectively, to bases for the space of abelian differentials of the first kind on V_x and for $H_1(V_x)$. Since Weierstrass points vary holomorphically with moduli (see, for example, [19]), each member of this finite collection is, locally, a holomorphic function on X. Here, we must work on Bers fibre space and use Gunning's formulation of the quadratic periods because the holomorphic functions that provide us with the $du_j(x)$ have Bers fibre space as their domain. Our expressions are multi-valued on X because the Weierstrass points $p_j(x_0)$ may undergo a permutation as x traverses a loop in X that is not null-homotopic. A simple formula, due to Gunning, [7], that expresses the relationship between quadratic periods calculated at distinct base points forces us to introduce the terms $\pi_{jk}w_{lmn}$. We have not observed these in previous investigations of Riemann surfaces, but we suspect they can be fitted into the framework of Carlson's theory of extensions of mixed Hodge

structures, [2]. In any event, we arrive at a vector bundle, E, on X and the

Question: Does E extend to an analytic vector bundle on a larger subset of Teichmuller space than X? In particular, does it extend to the complement of its branch locus over the moduli space? Finally, does E admit an interpretation in terms of "known" vector bundles.

Next, consider the subvariety of the *moduli space* of Riemann surfaces of genus g defined by Riemann surfaces that carry only normal Weierstrass points and that admit a representation as a g-sheeted covering of the z-sphere with one totally ramified point over, for example, $z=\infty$, and with $3g-1$ simple branch points. If, in addition, we exclude points representing Riemann surfaces with non-trivial automorphisms, then we obtain an algebraically defined family of Riemann surfaces. In fact, this family is in the image of the Weierstrass-Hurwitz space introduced in [1]. Therefore, the algebraically defined holomorphic differentials described above are meaningful on a Zariski open subset, X. Moreover, by considering the quadratic periods based at each of the Weierstrass points, we obtain an analytic vector bundle E over X and we may pose the same question as in the preceding case. In addition, if we take 1-parameter subfamilies, we may repeat the investigation carried out for cyclic families and obtain differential equations of Fuchsian class. We close with the

Question: Do the differential equations obtained by the above process serve as examples of any of the equations whose existence is proved in [21]?

Acknowledgment: The plan for the present paper, as well as many of its details, originated during the 1978-79 academic year when the author was a visitor in the Mathematics Department of Princeton University. The author is grateful to Professors B. Dwork and R. C. Gunning for their suggestions and encouragement. He also wishes to thank the following mathematicians for interesting conversations related to this paper: L. Ehrenpreis, E. Jablow, R. Hain, B. Harris, J. Ries, and M.F. Singer. Finally, he wishes to thank the Mathematics Department of the University of Maryland, especially Professor C. Berenstein, for the opportunity to participate in their Special Year in Complex Analysis.

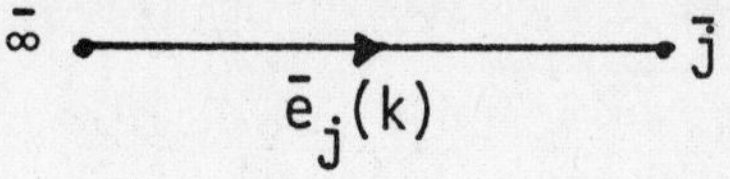

$\bar{\infty}$
$\bar{j}$
$\bar{e}_j(k)$

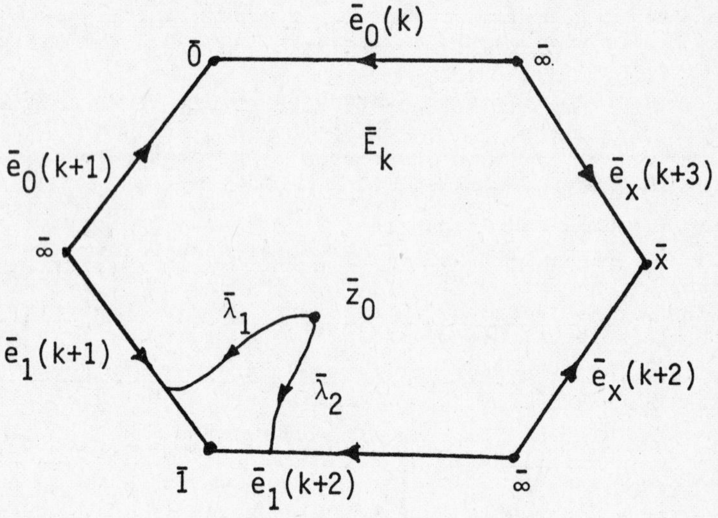

$\bar{e}_0(k)$
$\bar{0}$
$\bar{\infty}$
$\bar{E}_k$
$\bar{e}_0(k+1)$
$\bar{e}_x(k+3)$
$\bar{\infty}$
$\bar{x}$
$\bar{\lambda}_1$
$\bar{z}_0$
$\bar{e}_1(k+1)$
$\bar{\lambda}_2$
$\bar{e}_x(k+2)$
$\bar{1}$
$\bar{e}_1(k+2)$
$\bar{\infty}$

FIGURE

REFERENCES

1. Arbarello, E., On subvarieties of the moduli space of curves of genus g defined in terms of weierstrass points, Atti Della Accademia Nazionale Dei Lincei, XV (1978), 1-20.

2. Carlson, J., Extensions of mixed Hodge structures, Journées de Geometrie Algebraic d'Angers, Sijthoff and Noordhoff (1980).

3. Chen, K. T., Iterated path integrals, Bull. A.M.S., 83 (1977), 831-879.

4. Earle, C. J., Families of Riemann Surfaces and Jacobi Varieties, Annals of Math., 107 (1978), 255-286.

5. Forster, O., Lectures on Riemann Surfaces, Graduate Texts in Mathematics, Vol. 81, Springer-Verlag, New York (1981).

6. Grauert, H., Analytische Faserungen über holomorph-vollständigen Räumen, Math. Ann. 135 (1958), 263-273.

7. Gunning, R. C., Quadratic periods of Hyperelliptic Abelian Integrals, in Problems in Analysis, Princeton University Press (1970), 239-247.

8. Gunning, R.C., Lectures on Riemann Surfaces; Jacobi Varieties, Princeton University Press (1972).

9. Hain, R. M., Mixed Hodge Structures as Homotopy Groups, Bull. A.M.S., 14 (1986), 111-114.

10. Harris, B., Harmonic Volumes, Acta Math, 150 (1983), 91-123.

11. Harris, B., Homological versus Algebraic Equivalence in a Jacobian, Proc. Nat. Acad. Sci., U.S.A., 80 (1983), 1157-1158.

12. Henrici, P., Applied and Computational Complex Analysis, Vol. 2, Wiley-Interscience, New York (1977).

13. Hwang-Ma, S.-Y., Periods of iterated integrals of holomorphic forms on a compact Riemann surface, Trans. A.M.S., 264 (1981), 295-300.

14. Jablow, E., Quadratic vector classes on Riemann surfaces, Duke Math. J., 53 (1986), 221-232.

15. Lefschetz, S., On certain numerical invariants of algebraic varieties with application to abelian varieties, Trans. A.M.S. 22 (1921), 327-482.

16. Poole, E.G.C., Introduction to the Theory of Linear Differential Equations, Dover reprint, New York (1960).

17. Pulte, M., Thesis, University of Utah (1985).

18. Ramakrishnan, D., On the monodromy of higher logarithms, Proc. A.M.S., 85 (1982), 596-599.

19. Rauch, H. E., Weierstrass points, branch points and moduli of Riemann surfaces, Comm. on Pure and App. Math, 12 (1959), 543-560.

20. Siegel, C. L., Topics in Complex Function Theory, Vol. I (1969), Vol. II (1971), Vol. III (1973), Wiley-Interscience, New York.

21. Singer, M. F. and Tretkoff, M. D., Some applications of linear groups to differential equations, Am. J. Math. 107 (1985), 1111-1121.

22. Tretkoff, C. L. and Tretkoff, M. D., Combinatorial group theory, Riemann surfaces and differential equations, Contemporary Math, 33 (1984), 467-519

23. J. H. Hubbard, Sur les sections analytique de la courbe universelle de Teichmuller, Memoirs Amer. Math. Soc., 4, No 166 (1976), 1-137

Mathematics Department
Stevens Institute of Technology
Hoboken, New Jersey 07030

and

Cornell University
Department of Mathematics
White Hall
Ithaca, New York 14853-7901

HOLOMORPHIC FAMILIES OF HOLOMORPHIC ISOMETRIES

Edoardo Vesentini

Several years ago, while investigating with Aldo Andreotti deformations of quotients of bounded homogeneous domains by discrete groups of holomorphic automorphisms, the question arose whether there exist non-trivial complex analytic families of complex manifolds which are uniformized on a fixed bounded domain D in $\mathbb{C}^n$.

In the process of providing an answer to that question we came across the following result [1]. Let Δ be the open unit disc in $\mathbb{C}$, let $\mathrm{Aut}\, D$ be the group of all holomorphic automorphisms of D and let $\mathrm{Hol}(\Delta\times D, D)$ be the set of all holomorphic maps of $\Delta\times D$ into D.

Theorem. If $g \in \mathrm{Hol}(\Delta\times D, D)$ is such that $g(\zeta,\cdot) \in \mathrm{Aut}\, D$ for every $\zeta \in \Delta$, then g is independent of ζ.

This theorem raised further questions: Does the conclusion still hold if the domain D is replaced by a complex manifold? Can the condition whereby $g(\zeta,\cdot) \in \mathrm{Aut}\, D$ for every $\zeta \in \Delta$ be weakened?

Answers to those questions seemed to be out of our reach at that time due to the fact that our proof of the theorem relied heavily on properties of the Bergman kernel function of the domain D.

A few years later the theory of what is now called Kobayashi metric was developed by S. Kobayashi. Using these new tools, Kobayashi proved that there are no non-trivial complex Lie groups acting holomorphically on a hyperbolic domain or, more in general, on a hyperbolic manifold.

Later on, the theory of Kobayashi's (and Carathéodory's) metric was generalized to domains in a complex Banach space and to complex connected Banach manifolds.

Using these tools the following theorem can be proved.

Theorem. Let Z be a connected hyperbolic complex Banach manifold, and let $g \in \mathrm{Hol}(\Delta\times Z, Z)$. If there is $\zeta_0 \in \Delta$ such that $g(\zeta_0,\cdot) \in \mathrm{Aut}\, Z$, then g is independent of ζ.

Recall that the connected Banach manifold Z is called hyperbolic if the Kobayashi pseudodistance on Z is a distance defining Z the natural topology. For $x \in Z$, let Z_x be the complex tangent space to Z at $x \in Z$. For $v \in Z_x$, $\varkappa_Z(x;v)$ will denote the length of the vector v for the Kobayashi differential metric $\varkappa_Z$.

The proof of the latter theorem depends [1] on the following

<u>Lemma</u>. Let Z by hyperbolic. For every $x \in Z$ and every v, $v \in Z_x \setminus \{0\}$, $\varkappa_Z(x;v) > 0$.

In the infinite dimensional case, holomorphic maps exist which are isometries for the Kobayashi (or for the Carathéodory) differential metric without being locally surjective. The question arises therefore whether the above theorems hold more in general for holomorphic isometries.

This paper is devoted to investigating the latter question for holomorphic isometries of a bounded convex domain D in a complex Banach space. It will be shown that the existence of non-trivial holomorphic families of isometries depends on the structure of the boundary of the domain.

1. The semigroup of all holomorphic isometries for $\varkappa_D$ will be denoted by Iso D. Given $g \in \mathrm{Hol}(\Delta \times D, D)$, $d_1 g(\zeta,x)$ and $d_2 g(\zeta,x)$ will indicate the partial differentials of g with respect to the first and the second variable, evaluated at the point (ζ,x).

<u>Proposition 1.1</u>. Let D be a bounded convex domain in a complex Banach space $\mathcal{E}$. Let $g : \Delta \times D \to D$ be a holomorphic map such that, for every $\zeta \in \Delta, g(\zeta,\cdot)$ is an isometry for the Kobayashi metric $\varkappa_D$. If, for every $\zeta \in \Delta$ and every $x \in D$ there is a vector $v \in \mathcal{E}\setminus\{0\}$ such that $d_2 g(\zeta,x)v$ is a compex extreme point of the ball

$$\{w \in \mathcal{E} \ : \ \varkappa_D(g(\zeta,x);w) \le 1\},$$

then g is independent of ζ.

<u>Proof</u>. First of all, $\varkappa_D(x;\cdot)$ is a norm which is equivalent to the norm $\| \ \|$ on $\mathcal{E}$. Assume that

$$\varkappa_D(g(\zeta^\circ,x^\circ);\ d_2 g(\zeta^\circ,x^\circ)v) = \varkappa_D(x^\circ;v) \tag{1.1}$$

for some $\zeta^\circ \in \Delta$, $x^\circ \in D$, $v \in \mathcal{E}\setminus\{0\}$. Setting $y^\circ = g(\zeta^\circ,x^\circ)$, the power series expansion of g at (ζ°,x°) is expressed by

$$g(\zeta,x) = y^\circ + (\zeta-\zeta^\circ)u + A(x-x^\circ) + o(|\zeta-\zeta^\circ| + \|x-x^\circ\|),$$

where $u = d_1 g(\zeta^\circ,x^\circ) \in \mathcal{E}$, $A = d_2 g(\zeta^\circ,x^\circ) \in \mathcal{L}(\mathcal{E})$ (the Banach space

---

[1] The theorem and the lemma were established in [3] for a hyperbolic domains and later on [8] the lemma was sharpened. The proofs carry over, with no substantial change, to hyperbolic manifolds.

of all bounded linear operators in $\mathcal{E}$ with the uniform norm). Let $h : \Delta\times D \to \Delta\times D$ be the holomorphic map defined by $h(\zeta,x) = (\zeta, g(\zeta,x))$. The differential $dh(\zeta^\circ,x^\circ)$ is expressed by the matrix

$$dh(\zeta^\circ,x^\circ) = \begin{pmatrix} 1 & 0 \\ u & A \end{pmatrix}.$$

Hence

$$\varkappa_{\Delta\times D}((\zeta^\circ,x^\circ);(\tau,v)) \geq \varkappa_{\Delta\times D}(h(\zeta^\circ,x^\circ);dh(\zeta^\circ,x^\circ)(\tau,v))$$
$$= \varkappa_{\Delta\times D}((\zeta^\circ,y^\circ);(\tau,\tau\cdot u+Av)),$$

i.e.,

$$\max\left\{\frac{|\tau|}{1-|\zeta_0|^2}, \varkappa_D(x^\circ;v)\right\} \geq \max\left\{\frac{|\tau|}{1-|\zeta_0|^2}, \varkappa_D(y^\circ;\tau\, u+Av)\right\} \tag{1.2}$$

for all $\tau \in \mathbb{C}$ and all $v \in \mathcal{E}$. If

$$\frac{|\tau|}{1-|\zeta_0|^2} \leq \varkappa_D(x^\circ;v),$$

the latter inequality is equivalent to

$$\varkappa_D(y^\circ\cdot\tau\, u+Av) \leq \varkappa_D(x^\circ;v),$$

which by (1.1) can be written

$$\varkappa_D(y^\circ\cdot\tau\, u+Av) \leq \varkappa_D(y^\circ;Av).$$

If Av is a complex extreme point of the ball $\{w \in \mathcal{E} : \varkappa_D(y^\circ;w) \leq 1\}$, then $u = 0$. □

If D is the open unit ball B of $\mathcal{E}$, then

$$\varkappa_B(0;v) = \|v\| \quad \text{for all} \quad v \in \mathcal{E}. \tag{1.3}$$

Hence Proposition 1.1 implies

Lemma 1.2. Let S be the set of all complex extreme points of the closure $\bar{B}$ of the open unit ball B of $\mathcal{E}$. Let $g \in \mathrm{Hol}(\Delta\times B,B)$ be such that $g(\zeta,\cdot) \in \mathrm{Iso}\, B$ for all $\zeta \in \Delta$. If for every $\zeta \in \Delta$ and every $x \in B$ there exist, $h \in \mathrm{Aut}\, B$ such that

$$h(g(\zeta,x)) = 0 \tag{1.4}$$

and

$$S \cap dh(g(\zeta,x))d_2g(\zeta,x)\mathcal{E} \neq \emptyset,$$

then g is independent of ζ.

The existence, for every $(\zeta,x) \in \Delta\times B$, of a holomorphic automorphism h satisfying (1.4) is granted provided that B is homogeneous. Furthermore, in view of (1.3), the differential at 0 of any

element of Iso B leaving 0 fixed is a linear isometry of $\mathcal{E}$. Hence Lemma 1.2 yields

<u>Theorem I</u>. Let $g \in \mathrm{Hol}(\Delta\times B, B)$ be such that $g(\zeta,\cdot) \in \mathrm{Iso}\, B$ for all $\zeta \in \Delta$. If B is homogeneous and if

$$S \cap U\mathcal{E} \neq \emptyset \tag{1.5}$$

for every linear isometry U of $\mathcal{E}$, then g is independent of ζ.

2. Condition (1.5) is fulfilled if ∂B consists entirely of complex extreme points of B. That is the case, for example, when B is the unit ball of a complex Hilbert space. Hence Theorem I yields

<u>Theorem II</u>. Let B be the open unit ball of a complex Hilbert space $\mathcal{H}$. Every holomorphic map $g : \Delta\times B \to B$ for which $g(\zeta,\cdot)$ is an isometry for the hyperbolic metric $\varkappa_B$ of B for all $\zeta \in \Delta$, is independent of ζ.

Here is a direct proof of this theorem. Let $(\,|\,)$ and $\|\ \|$ be the inner product and the associated norm in $\mathcal{H}$. There is no restriction in assuming $\zeta^\circ = 0$, $x^\circ = 0$. Choosing $v = \tau^2 u$ $(\tau\in C\setminus\{0\})$, (1.2) reads now

$$\max(1,|\tau|\|u\|) \geq \max(1,\varkappa_B(y^\circ;u+\tau\cdot Au)),$$

whence

$$|\tau|\|u\| \geq \varkappa_B(y^\circ;u+\tau\cdot Au) \tag{2.1}$$

for $|\tau| \gg 0$. Being [3, p. 154]

$$\varkappa_B(y^\circ,u+\tau\cdot Au)^2 = \frac{1}{(1-\|y^0\|^2)^2}\{|((u+\tau\cdot Au)|y^\circ)|^2+(1-\|y^\circ\|^2)\|u+\tau\cdot Au\|^2\},$$

(2.1) becomes

$$\begin{aligned}|\tau|^2(|(Au|y^\circ)|^2 + (1-\|y^\circ\|^2)(\|Au\|^2 - (1-\|y^\circ\|^2)\|u\|^2)&\\ +2\mathrm{Re}(\tau((Au|y^\circ)(y^\circ|u) + (1-\|y^\circ\|^2)(Au|u)))&\\ +|(u|y^\circ)|^2 + (1-\|y^\circ\|^2)\|u\|^2 \leq 0&\end{aligned}$$

for $|\tau| \gg 0$.

On the other hand, condition

$$\varkappa_B(0;u) = \varkappa_B(y^\circ;Au)$$

reads

$$\|u\|^2 = \frac{1}{(1-\|y^0\|^2)^2}\{|(Au|y^\circ)|^2 + (1-\|y^\circ\|^2)\|Au\|^2\}.$$

Thus (2.2) simplifies to

$$(2.3)\qquad 2\mathrm{Re}(\tau((Au|y^\circ)(y^\circ|u) + (1-\|y^\circ\|^2)(Au|u))) + |(u|y^\circ)|^2 + (1-\|y^\circ\|^2)\|u\|^2 \le 0$$

for $|\tau| \gg 0$, implying that

$$(Au|y^\circ)(y^\circ|u) + (1-\|y^\circ\|^2)(Au|u) = 0.$$

Hence (2.3) reduces to

$$|(u|y^\circ)|^2 + (1-\|y^\circ\|^2)\|u\|^2 \le 0.$$

In conclusion $u = 0$ because $\|y^\circ\| < 1$. □

The semigroup Iso B of all holomorphic isometries for the Kobayashi metric $\varkappa_B$ of the unit ball B of the complex Hilbert space $\mathcal{H}$ was fully described in [3, Theorem VI. 4.1 and Corollary VI 4.2,, 174-175]. The main tools in this description were, on one hand, the knowledge of the group Aut B of all holomorphic automorphisms of B (cf. e.g., [3, Ch. VI, 164-174]) and, on the other hand, the following statement [3, Proposition VI.1.8).

<u>Proposition 2.1</u>. If $g \in \mathrm{Hol}(B,B)$ is such that $g(0) = 0$, then g is an isometry for $\varkappa_B$ if, and only if, g is the restriction to B of a linear isometry of the Hilbert space $\mathcal{H}$.

This proposition can be established as a consequence of the following fact which may have an independent interest.

<u>Proposition 2.2</u>. Let $f \in \mathrm{Hol}(B,B)$ be such that $f(0) = 0$. If there exists an orthogonal projector P such that $df(0)\cdot P$ is a partial isometry with initial space $\mathcal{M} = P\mathcal{H}$, then the restriction of f to $\mathcal{M} \cap B$ coincides with $df(0)|_{\mathcal{M}\cap B}$.

<u>Proof</u>. Let $P_1 = df(0) \in \mathcal{L}(\mathcal{H})$ and let

$$f(x) = P_1(x) + P_2(x) + P_3(x) + \dots$$

be the power series expansion of f in B in terms of continuous polynomials P_q of degree $q = 2,3,\dots$ from $\mathcal{H}$ to $\mathcal{H}$. Being $P\circ P_1^*\circ P_1\circ P$ the orthogonal projector onto $\mathcal{M}$, then

$$P\circ P_1^*\circ f(x) = x + P\circ P_1^*\circ P_2(x) + \dots$$

for all $x \in \mathcal{M} \cap B$. For $x \in \mathcal{M} \cap B$, $x \ne 0$, let $\varphi \in \mathrm{Hol}(\Delta,B)$ be defined by

$$\varphi(\zeta) = P\circ P_1^*\, f\left(\frac{\zeta}{\|x\|}\, x\right).$$

Because $\varphi(0) = 0$ and $\varphi'(0) = \frac{1}{\|x\|}\, x$ – whence $\|\varphi'(0)\| = 1$ –

then [6, Proposition 3.2], φ is a complex geodesic for the hyperbolic metric $\varkappa_B$. Since $\mathcal{H}$ is strictly convex, then φ is linear [6, Proposition 3.4],

$$\varphi(\zeta) = \frac{\zeta}{\|x\|}\, x \quad (\zeta \in \Delta),$$

and in particular

$$P \circ P_1^* \circ f(x) = x \quad \text{for all} \quad x \in \mathcal{M} \cap B. \tag{2.4}$$

By the Schwarz lemma

$$\|f(y)\| \le \|y\| \quad \text{for all} \quad y \in B.$$

Hence (2.4) and the Cauchy inequalities yield

$$\|x\| = \|P \circ P_1^* \circ f(x)\| \le \|P\|\, \|P_1^*\|\, \|f(x)\|$$
$$= \|P_1^*\|\, \|f(x)\| \le \|f(x)\| \le \|x\|,$$

and therefore

$$\|f(x)\| = \|x\| \quad \text{for all} \quad x \in \mathcal{M} \cap B.$$

Let

$$N = \{y \in B : \|f(y)\| = \|y\|\}.$$

By Proposition VI.I.8 of [3], there exists a closed subspace $\mathcal{N} \subset \mathcal{H}$ such that $N = \mathcal{N} \cap B$, and $f|_N = df(0)|_N$. Thus $\mathcal{M}$ is a closed subspace of $\mathcal{N}$ and the conclusion follows. □

As it was shown in [3, p. 152],

$$\mathcal{N} = \{y \in \mathcal{H} : \|P_1 y\| = \|y\|\}.$$

If P_1 itself is a partial isometry, its initial space coincides with $\mathcal{N}$. Letting P be the orthogonal projector onto $\mathcal{N}$, then $P_1 = P_1 \circ P$, and Proposition 2.2 yields

<u>Corollary 2.3</u>. Let $f \in \mathrm{Hol}(B,B)$ be such that $f(0) = 0$. If $df(0)$ is a partial isometry with initial space $\mathcal{M}$, then $\mathcal{M} \cap B$ is the set of all points $x \in B$ such that $f(x) = df(0)x$.

If $df(0)$ is an isometry, Corollary 2.3 and the fact that B is homogeneous imply

<u>Corollary 2.4</u>. Let $f \in \mathrm{Hol}(B,B)$ be such that $\varkappa_B(f(x^\circ); df(x^\circ)y) = \varkappa_B(x^\circ; y)$ for some $x^\circ \in B$ and all $y \in \mathcal{H}$. Then $f \in \mathrm{Iso}\, B$.

3. Let $\mathcal{E}$ be a J^*-algebra and let B be now the open unit ball of $\mathcal{E}$. Then B is homogeneous, but the set S of all complex extreme points of the closure $\bar{B}$ of B - which coincides with the set of all real extreme points of $\bar{B}$ and is characterized by Theorem 11 of [4] -

is in general a proper subset of ∂B.

Correspondingly, examples of non-trivial holomorphic families of holomorphic isometries for $\varkappa_B$ can be constructed. A first example, exhibited by T. Franzoni in [2, p. 56] in the case of a commutative C^*-algebra, contains the extra feature that the isometries, although leaving the center 0 of B fixed are non-linear.

An example in the case of a Cartan factor of type I will now be constructed.

If F is a pure isometry in an infinite dimensional complex Hilbert space $\mathcal{H}$ with $F\mathcal{H} \neq \mathcal{H}$, setting $\mathcal{H}_0 = (F\mathcal{H})^{\perp}$, then $\mathcal{H}$ is the image of the Hilbert space direct sum $\ell_+^2(\mathcal{H}_0) = \mathcal{H}_0 \oplus \mathcal{H}_0 \oplus \ldots$ by the surjective linear isometry $T : \ell_+^2(\mathcal{H}_0) \to \mathcal{H}$ defined by $T(x_0, x_1, x_2, \ldots) = x_0 + Fx_1 + F^2x_2 + \ldots$. The linear operator T^*FT is the unilateral shift operator

$$(x_0, x_1, \ldots) \mapsto (0, x_0, x_1)$$

and T^*F^*T is the backward shift operator

$$(x_0, x_1, \ldots) \mapsto (x_1, x_2, \ldots).$$

Let $\mathcal{E}$ be the C^*-algebra $\mathcal{L}(\mathcal{H}) = (\mathcal{H}, \mathcal{H})$. For $A \in \mathcal{E}$, T^*AT has a matrix representation in $\ell_+^2(\mathcal{H}_0)$

$$T^*AT = \begin{pmatrix} A_{00} & A_{01} & \cdots \\ A_{10} & A_{11} & \cdots \\ \cdot & \cdot & \\ \cdot & \cdot & \\ \cdot & \cdot & \end{pmatrix},$$

where $A_{\alpha\beta} \in \mathcal{L}(\mathcal{H}_0, \mathcal{H}_0)$, and $\|A\| = \sup(\|A_{\alpha,\beta}\| : \alpha,\beta = 0,1,\ldots)$. Furthermore $T^*(FAF^*)T$ has the matrix representation

$$T^*(FAF^*)T = \begin{pmatrix} 0 & 0 & 0 & \cdots \\ 0 & A_{00} & A_0 & \cdots \\ 0 & A_{10} & A_{11} & \cdots \\ \cdot & \cdot & \cdot & \\ \cdot & \cdot & \cdot & \\ \cdot & \cdot & \cdot & \end{pmatrix}$$

or also the representation

$$T^*(FAF^*)T = \begin{pmatrix} 0 & 0 \\ 0 & T^*AT \end{pmatrix}$$

with respect to the decomposition $\mathcal{H}_0 \oplus \mathcal{H}_0^{\perp}$ of $\ell_+^2(\mathcal{H}_0)$.

Let P be the orthogonal projector of $\mathcal{H}$ onto $\mathcal{H}_0$. Then T^*PT

is expressed by the matrix

$$T^*PT = \begin{pmatrix} I & 0 \\ 0 & 0 \end{pmatrix}$$

(where I_0 is the identity operator on $\mathcal{H}_0$) and

$$T^*PAPT = \begin{pmatrix} A_{00} & 0 \\ 0 & 0 \end{pmatrix}.$$

Let $f \in \mathrm{Hol}(\mathbb{C}\times\mathcal{E},\mathcal{E})$ be defined by

$$f(\zeta,A) = \zeta(PAP)^2 + FAF^* \tag{3.1}$$

or, equivalently, by

$$T^*f(\zeta,A)T = \begin{pmatrix} \zeta A_{00}^2 & 0 \\ 0 & T^*AT \end{pmatrix}. \tag{3.2}$$

Note that $f(\zeta,0) = 0$, and that, being

$$\|f(\zeta,A)\| = \max(|\zeta|\,\|(PAP)^2\|,\|A\|),$$

then $f(\Delta\times B) \subset B$.

The following theorem will now be established:

<u>Theorem III</u>. There exists ℓ, with $0 < \ell < 1$, such that whenever $|\zeta| \le \ell$, $f(\zeta,\cdot)$ is an isometry for the Kobayashi metric $\varkappa_B$.

As was shown in [5], for any $A \in B$ and any $C \in \mathcal{E}$,

$$\varkappa_B(A;C) = \|(I-AA^*)^{-\frac{1}{2}}\,C(I-A^*A)^{-\frac{1}{2}}\|. \tag{3.3}$$

4. To simplify notations, $\mathcal{H}$ will henceforth be identified with $\ell_+^2(\mathcal{H}_0)$ via T, so that $f(\zeta,A)$ is expressed either by (3.1) or by the right hand side of (3.2).

For $C \in \mathcal{E}$

$$d_2f(\zeta,A)C = \frac{d}{dt}f(\zeta,A+tC)|_{t=0} = \begin{pmatrix} \zeta(A_{00}C_{00} + C_{00}A_{00} & 0 \\ 0 & C \end{pmatrix}$$

$$= \zeta(PAPCP + PCPAP) + FCF^*,$$

where $C_{00} = PCP \in \mathcal{L}(\mathcal{H}_0,\mathcal{H}_0)$.

Furthermore

$$f(\zeta,A)\,f(\zeta,A)^* = \begin{pmatrix} |\zeta|^2\,A_{00}{}^2\,A_{00}{}^{*2} & 0 \\ 0 & AA^* \end{pmatrix},$$

$$f(\zeta,A)^*f(\zeta,A) = \begin{bmatrix} |\zeta|^2 A_{00}{}^{*2} A_{00}{}^2 & 0 \\ 0 & A^*A \end{bmatrix}.$$

Then for $\|A\| < 1$,

$$(I-f(\zeta,A)f(\zeta,A)^*)^{-\frac{1}{2}} \cdot d_2 f(\zeta,A)C \cdot (I-f(\zeta,A)^*f(\zeta,A))^{-\frac{1}{2}} =$$

$$= \begin{bmatrix} \zeta(I_0-|\zeta|^2A_{00}{}^2A_{00}{}^{*2})^{-\frac{1}{2}}(A_{00}C_{00}+C_{00}A_{00})(I_0-|\zeta|^2A_{00}{}^{*2}A_{00}{}^2)^{-\frac{1}{2}} & 0 \\ 0 & (I-AA^*)^{-\frac{1}{2}}C(I-A^*A)^{-\frac{1}{2}} \end{bmatrix}$$

and therefore, by (3.3),

$$(4.1)\qquad \varkappa_B(f(\zeta,A);d_2f(\zeta,A)C) =$$

$$= \max\{|\zeta|\,\|(I_0-|\zeta|^2A_{00}{}^2A_{00}{}^{*2})^{-\frac{1}{2}}(A_{00}C_{00}+C_{00}A_{00})\cdot$$

$$(I_0 - |\zeta|^2A_{00}{}^{*2}A_{00}{}^2)^{-\frac{1}{2}}\| \;,\quad \|(I-AA^*)^{-\frac{1}{2}}C(I-A^*A)^{-\frac{1}{2}}\|\}$$

whenever $\zeta \in \Delta$, $\|A\| < 1$ and for all $C \in \mathcal{E}$.

Lemma 4.1. There exists ℓ, with $0 < \ell < 1$, such that whenever $|\zeta| \le \ell$ and $\|A\| < 1$, then

$$|\zeta|\,\|(I_0-|\zeta|^2A_{00}{}^2A_{00}{}^{*2})^{-\frac{1}{2}}(A_{00}C_{00}+C_{00}A_{00})(I_0-|\zeta|^2A_{00}{}^{*2}A_{00}{}^2)^{-\frac{1}{2}}\| \le \|C_{00}\|$$

for all $C \in \mathcal{E}$.

Proof. For any bounded linear operator X, let $\rho(X)$ be the spectral radius of X. Then

$$\|(I_0-|\zeta|^2A_{00}{}^2A_{00}{}^{*2})^{-\frac{1}{2}}(A_{00}C_{00}+C_{00}A_{00})(I_0-|\zeta|^2A_{00}{}^{*2}A_{00}{}^2)^{-\frac{1}{2}}\|^2$$

$$= \|(I_0-|\zeta|^2A_{00}{}^{*2}A_{00}{}^2)^{-\frac{1}{2}}(A_{00}C_{00}+C_{00}A_{00})^*(I_0-|\zeta|^2A_{00}{}^2A_{00}{}^{*2})^{-1}$$

$$(A_{00}C_{00}+C_{00}A_{00})(I_0-|\zeta|^2A_{00}{}^{*2}A_{00}{}^2)^{-\frac{1}{2}}\| =$$

$$= \rho((I_0-|\zeta|^2A_{00}{}^{*2}A_{00}{}^2)^{-\frac{1}{2}}(A_{00}C_{00}+C_{00}A_{00})^*(I_0-|\zeta|^2A_{00}{}^2A_{00}{}^{*2})^{-1}$$

$$(A_{00}C_{00}+C_{00}A_{00})(I_0-|\zeta|^2A_{00}{}^{*2}A_{00}{}^2)^{-\frac{1}{2}}) =$$

$$= \rho((A_{00}C_{00}+C_{00}A_{00})^*(I_0-|\zeta|^2A_{00}{}^2A_{00}{}^{*2})^{-1}(A_{00}C_{00}+C_{00}A_{00})$$

$$(I_0-|\zeta|^2A_{00}{}^{*2}A_{00}{}^2)^{-1})$$

$$\leq \|(A_{00}C_{00}+C_{00}A_{00})^*(I_0-|\zeta|^2A_{00}{}^2A_{00}{}^{*2})^{-1}$$

$$(A_{00}C_{00}+C_{00}A_{00})(I_0-|\zeta|^2A_{00}{}^{*2}A_{00}{}^2)^{-1}\| \leq$$

$$\leq \|(A_{00}C_{00}+C_{00}A_{00})^*\|\|(I_0-|\zeta|^2A_{00}{}^2A_{00}{}^{*2})^{-1}\|\|A_{00}C_{00}+C_{00}A_{00}\| \cdot$$

$$\|(I_0-|\zeta|^2A_{00}{}^{*2}A_{00}{}^2)^{-1}\|$$

$$= \|A_{00}C_{00}+C_{00}A_{00}\|^2\|(I_0-|\zeta|^2A_{00}{}^2A_{00}{}^{*2})^{-1}\| \quad \|(I_0-|\zeta|^2A_{00}{}^{*2}A_{00}{}^2)^{-1}\| \leq$$

$$\leq 4\|A_{00}\|^2\|C_{00}\|^2\|(I_0-|\zeta|^2A_{00}{}^2A_{00}{}^{*2})^{-1}\| \quad \|(I_0-|\zeta|^2A_{00}{}^{*2}A_{00}{}^2)^{-1}\|.$$

For $|\zeta| < 1$,

$$(I_0-|\zeta|^2A_{00}{}^2A_{00}{}^{*2})^{-1} = (I_0+|\zeta|^2A_{00}{}^2A_{00}{}^{*2}+|\zeta|^4(A_{00}{}^2A_{00}{}^{*2})^2+ \ldots,$$

and therefore

$$\|(I_0-|\zeta|^2A_{00}{}^2A_{00}{}^{*2})^{-1}\| < 1+|\zeta|^2\|A_{00}\|^4+|\zeta|^4\|A_{00}\|^8+ \ldots =$$

$$= \frac{1}{1-|\zeta|^2\|A_{00}\|^4} \leq \frac{1}{1-|\zeta|^2\|A_{00}\|^2} .$$

Similarly for $|\zeta| < 1$

$$\|(I_0-|\zeta|^2A_{00}{}^{*2}A_{00}{}^2)^{-1}\| \leq \frac{1}{1-|\zeta|^2\|A_{00}\|^2} .$$

Thus

$$|\zeta|\|(I_0-|\zeta|^2A_{00}{}^2A_{00}{}^{*2})^{-\frac{1}{2}}(A_{00}C_{00}+C_{00}A_{00})(I_0-|\zeta|^2A_{00}{}^{*2}A_{00}{}^2)^{-\frac{1}{2}}\| \leq$$

$$\leq \frac{2\|\zeta A_{00}\|}{1-\|\zeta A_{00}\|^2} \quad \|C_{00}\| .$$

Since the inequalities

$$0 \leq t < 1 \ , \ \frac{2t}{1-t^2} \leq 1$$

are equivalent to

$$0 \leq t \leq \sqrt{2}-1,$$

setting $\ell = \sqrt{2}-1$ The conclusion follows. □

<u>Lemma 4.2</u>. If $\|A\| < 1$, then

$$\|(I-AA^*)^{-\frac{1}{2}}C(I-A^*A)^{-\frac{1}{2}}\| \geq \|C(I-A^*A)^{-\frac{1}{2}}\|$$

for all $C \in \mathcal{E}$.

<u>Proof</u>. Being $\|AA^*\| = \|A^*\|^2 = \|A\|^2 < 1$, then

$$\|(I-AA^*)^{-\frac{1}{2}}C(I-A^*A)^{-\frac{1}{2}}x\|^2$$

$$= ((I-AA^*)^{-\frac{1}{2}}C(I-A^*A)^{-\frac{1}{2}}x|(I-AA^*)^{-\frac{1}{2}}C(I-A^*A)^{-\frac{1}{2}}x)$$

$$= ((I-AA^*)^{-1}C(I-A^*A)^{-\frac{1}{2}}x|C(I-A^*A)^{-\frac{1}{2}}x)$$

$$= \sum_{n=0}^{+\infty}((AA^*)^nC(I-A^*A)^{-\frac{1}{2}}x|C(I-A^*A)^{-\frac{1}{2}}x) =$$

$$= \|C(I-A^*A)^{-\frac{1}{2}}x\|^2 + \|A^*C(I-A^*A)^{-\frac{1}{2}}x\|^2 + \|AA^*C(I-A^*A)^{-\frac{1}{2}}x\|^2 +$$

$$+ \|A^*AA^*C(I-A^*A)^{-\frac{1}{2}}x\|^2 + \dots \geq \|C(I-A^*A)^{-\frac{1}{2}}x\|^2$$

for all $x \in \mathcal{H}$. □

<u>Lemma 4.3</u>. If $\|A\| < 1$ and $|\zeta| \leq \ell$, then

$$|\zeta|\|(I_0-|\zeta|^2A_{00}{}^2A_{00}{}^{*2})^{-\frac{1}{2}}(A_{00}C_{00}+C_{00}A_{00})(I_0-|\zeta|^2A_{00}{}^{*2}A_{00}{}^2)^{-\frac{1}{2}}\|$$

$$\leq \|(I-AA^*)^{-\frac{1}{2}}C(I-A^*A)^{-\frac{1}{2}}\|$$

for all $C \in \mathcal{E}$.

<u>Proof</u>. By Lemma 4.1 and Lemma 4.2 it suffices to show that

$$\|C_{00}\| \leq \|C(I-A^*A)^{-\frac{1}{2}}\|$$

for all $C \in \mathcal{E}$.

Let $B_{\mathcal{H}}$ be the open unit ball of the Hilbert space $\mathcal{H}$. Then

$$\|C(I-A^*A)^{-\frac{1}{2}}\| = \sup\{\|C(I-A^*A)^{-\frac{1}{2}}x\| : x \in B_{\mathcal{H}}\} =$$

$$= \sup\{\|Cx\| : (I-A^*A)^{\frac{1}{2}}x \in B_{\mathcal{H}}\}.$$

For $x \in \mathcal{E}$, $\sigma(X)$ will stand for the spectrum of X. Since

$$\|(I-A^*A)^{\frac{1}{2}}\|^2 = \|(I-A^*A)^{\frac{1}{2}}(I-A^*A)^{\frac{1}{2}}\| = \|(I-A^*A)\| = \rho(I-A^*A) =$$

$$= \sup\{t \in \mathbb{R} : t \in \sigma(I-A^*A)\} = \sup\{1-t : t \in \sigma(A^*A)\} \leq 1,$$

then

$$B_{\mathcal{H}} \subset \{x \in \mathcal{H} : \|(I-A^*A)^{-\frac{1}{2}}x\| < 1\} .$$

Let

$$C = \begin{pmatrix} C_{00} & C_{01} & \cdots \\ C_{10} & C_{11} & \cdots \\ \vdots & \vdots & \end{pmatrix}$$

be the matrix representation of C. Then for $x = (x_0,0,\dots)$

$(x_0 \in \mathcal{H}_0)$,

$$\|Cx\|^2 = \|C_{00}x_0\|^2 + C_{10}x_0\|^2 + \ldots \geq \|C_{00}x_0\|^2,$$

and therefore, by (4.2),

$$\begin{aligned} \|C_{00}\| &= \sup\{\|C_{00}x_0\| : x_0 \in \mathcal{H}_0 , \|x_0\| < 1\} \\ &\leq \sup\{\|Cx\| : x = (x_0,0,\ldots) , \|x\| < 1\} \\ &\leq \sup\{\|Cx\| : x \in B) \\ &\leq \sup\{\|Cx\| : x \in \mathcal{H} , (I-A^*A)^{-\frac{1}{2}}x \in B_{\mathcal{H}}\} = C(I-A^*A)^{-\frac{1}{2}}\| . \end{aligned}$$ □

Lemma 4.3 together with (3.3) and (4.1) yields Theorem IV. Hence for any $|\zeta| \leq \ell$, $f(\zeta,\cdot)$ defines a non-linear holomorphic isometry for $\mathcal{H}_B$ leaving 0 fixed.

Remark. Theorem III shows that H. Cartan's linearity theorem (cf. e.g. [3, Proposition III.2.2, p. 76]) does not hold for holomorphic isometries.

5. Replace now the function f expressed by (3.1) by the function $h \in \mathrm{Hol}(\mathbb{C}\times\mathcal{E},\mathcal{E})$ given by

(5.1) $$h(\zeta,A) = \zeta PAP + FAF^* .$$

Inspection of the above considerations shows that Theorem III holds also for the function h, thus providing an example of a holomorphic family of linear isometries for $\mathcal{H}_B$.

A function similar to h yields such an example also in the case in which $\mathcal{E}$ is an infinite dimensional Cartan factor of type II or of type III (cf. [4]) for the definitions.

Let $\mathcal{K}$ be another infinite dimensinal complex Hilbert space on which a pure isometry E is given. Let Q be the orthogonal projector of $\mathcal{K}$ onto $\mathcal{K}_0 = E\mathcal{K}$, let $\mathcal{E}$ be the Cartan factor of type I $\mathcal{E} = \mathcal{L}(\mathcal{H},\mathcal{K})$, and consider the function $g \in \mathrm{Hol}(\mathbb{C}\times\mathcal{E},\mathcal{E})$ defined by

$$g(\zeta,A) = \zeta QAP + EAF^* .$$

The same arguments as before show that there is some ℓ, with $0 < \ell < 1$ such that for all $|\zeta| \leq \ell$ $g(\zeta,\cdot)$ is an isometry for $\mathcal{H}_B$.

References

[1] A. Andreotti and E. Vesentini, On deformations of discontinuous groups, Acta Math., 112 (1964), 249-298.

[2] T. Franzoni, The group of holomorphic automorphisms in certain J^*-algebras, Ann. Mat. Pura Appl., (4) 127 (1981), 51-66.

[3] T. Franzoni and E. Vesentini, Holomorphic maps and invariant distances, North Holland, Amsterdam/New York/Oxford, 1980.

[4] L.A. Harris, Bounded symmetric homogeneous domains in infinite dimensional spaces, Lecture Notes in Mathematics, #364, Springer-Verlag, Berlin/Heidelberg/New York, 1973, 13-40.

[5] L.A. Harris, Analytic invariants and the Schwarz-Pick inequality, Israel J. Math. 34 (1979), 177-197.

[6] E. Vesentini, Complex geodesics and holomorphic maps, Symposia Mathematica, Vol. XXXVI, 1982, 211-230.

[7] E. Vesentini, Hyperbolic domains in Banach spaces and Banach algebras, in "Aspects of Mathematics and its Applications" J.A. Barroso Editor, Elsevier Science Publishers, New York, 1986, 859-871.

Scuola Normale Superiore
Pisa, Italy

COMPLEX MONGE-AMPÈRE EQUATION AND RELATED PROBLEMS

Pit-Mann Wong*

§1. Introduction

Geometric problems often are naturally associated to variation problems of certain functionals of geometric objects. Classically many of these variational problems are related to certain Laplace operators and can often be solved by studying heat equations of such operators. For our purpose we recall a few examples which serve as motivation for the study of the Monge-Ampère operator.

(I) Harmonic Representatives of Cohomology Classes

Given a cohomology class $\alpha \in H^p(M,\mathbb{R})$ where (M,g) is a compact Riemannian manifold, define a function $\mathcal{D}$ on the space of smooth p-forms φ representing α, i.e. $[\varphi]=\alpha$, as follows

$$(1.1) \qquad \mathcal{D}(\varphi) = \int_M |\varphi|^2 dv_g$$

where $|\ |$ is the norm induced by the metric g on p-forms and dv_g is the volume element of g.

For any small variation $\varphi + td\psi$ of φ in the same cohomology class, we have

$$\left.\frac{\partial \mathcal{D}}{\partial t}\right|_{t=0} = 2\int_M \langle d\psi, \varphi \rangle d\psi_g = 2\int_M \langle p, d^*\varphi \rangle dv_g,$$

hence the Euler-Lagrange equation is

$$d^*\varphi = 0$$

or equivalently since φ is a closed form,

$$(1.2) \qquad \Delta\varphi = dd^*\varphi + d^*d\varphi = 0.$$

*Research supported in part by an NSF grant and a Sloan Fellowship.

Namely a critical point of the functional is a harmonic form. It is a classical result of A. Milgram and P. Rosenbloom [17] that starting from any initial condition, the heat equation

$$(1.3) \qquad \begin{cases} \left(\frac{\partial}{\partial t} - \Delta\right) \varphi = 0 \\ \varphi = \varphi_o \ , \ t = 0 \end{cases}$$

is solvable for all time $t < \infty$ and the solutions φ_t converges smoothly to a harmonic form φ_∞ as $t \to \infty$.

More generally, the inhomogeneous problem

$$\Delta\varphi = \psi$$

can be analogously solved by solving the heat equation,

$$\frac{\partial \mathcal{D}}{\partial t} - \Delta\varphi + \psi = o.$$

(II) <u>Harmonic Representatives of Homotopy Classes</u>

The energy functional of mappings f between Riemannian manifolds (M,g) and (N,h) is given by,

$$(1.4) \qquad E(f) = \int_M |\nabla f|^2 dv_g$$

where

$$|\nabla f|^2(x) = g^{ij}(x)\frac{\partial f^\alpha}{\partial x^i}(x)\frac{\partial f^\beta}{\partial y^j}(x)h_{\alpha\beta}(f(x))$$

A straight forward calculation shows that the Euler-Lagrange equation is the harmonic equation,

$$(1.5) \qquad \Delta f = g^{ij}\left\{\frac{\partial^2 f^\alpha}{\partial x^i \partial x^j} - {}_M\Gamma^k_{ij} \frac{\partial f^\alpha}{\partial x^k} + {}_N\Gamma^\alpha_{\beta\gamma}\frac{\partial f^\beta}{\partial x^i}\frac{\partial f^\gamma}{\partial x^j}\right\} = 0$$

where the Γ's are the Christoffel symbols. The following results are well-known:

a) M and N are compact without boundary [11].

If the sectional curvature $K_N \leq 0$ then there is a harmonic map in every homotopy class.

b) M and N compact with boundary [13].

(i) Dirichlet Problem

Assume that $K_N \le 0$ and ∂N convex. Given a smooth map $h: \partial M \to N$ and suppose that there exists a smooth map $f_o: M \to N$ with $f_o = h$ on ∂M then there is a harmonic map in the same relative homotopy class, i.e. the Dirichlet Problem

$$(1.6)\qquad \begin{cases} \Delta f = 0 & \text{on } M \\ f = h & \text{on } \partial M \end{cases}$$

has a colution.

(ii) Neumann Problem

If $K_N \le 0$ and ∂N is convex then the ann Problem

$$(1.7)\qquad \begin{cases} \Delta f = 0 & \text{on } M \\ \nabla_\nu f = 0 & \text{on } \partial M \end{cases}$$

has a solution in every homotopy class. Here ν is the normal of ∂M.

(iii) Mixed Problem

If $K_N \le 0$ and ∂N is totally geodesic then the mixed equations,

$$(1.8)\qquad \begin{cases} \Delta f = 0 & \text{on } M \\ f(\partial M) \subset \partial N & \\ \nabla_\nu f \perp \partial N & \text{on } \partial M \end{cases}$$

has a solution in every homotopy class.

All three problems can be solved by solving the associated heat equation:

$$(1.9)\qquad \begin{cases} \frac{\partial f}{\partial t} = \Delta f & \\ f = f_o & \text{for } t = 0 \end{cases}$$

with side conditions

$$f = h \quad \text{on } \partial M \qquad \text{(Dirichlet)}$$

$$\frac{\partial f}{\partial \nu} = 0 \quad \text{on } \partial M \qquad \text{(Neumann)}$$

and

$$\begin{cases} \frac{\partial f^\alpha}{\partial \nu} = 0 & 1 \le \alpha \le n - 1 \\ f^n = 0 & \end{cases}$$

on ∂M for the mixed problem (where $y^n=0$ contains ∂Y in some appropriated embedding of Y in $\mathbb{R}^n$). Note that in the Dirichlet and Neumann Problem we may assume that ∂N is empty.

In the case of $N = \mathbb{R}^n$ we can also solve the inhomogeneous equation

$$\Delta f = g$$

with either Dirichlet or Neumann conditions, again via the heat equation:

$$\frac{\partial f}{\partial t} = \Delta f - g$$

For the Neumann problem, Green's Identity imposes the compatibility condition,

$$(1.10) \qquad \int_M g = \int_M \Delta f = \int_{\partial M} \nabla_\nu f = 0$$

In complex analysis, it is important to understand when can one embed the unit disc in a given complex manifold. For instance, given a relatively compact domain Ω in a complex manifold, with $\partial\Omega$ smooth, fix a point $p \in \Omega$. Under what conditions can we find a holomorphic map $f: \Delta \to \Omega$ such that $f(o) = p$ and $f(\partial\Delta) \subset \Omega$. Here Δ is the unit disc in $\mathbb{C}$. In some sense this a mixed problem for the $\bar{\partial}$ operator. The condition

$$f(\partial\Delta) \subset \partial\Omega$$

imposes a boundary condition which is of partially Dirichlet and partially Neumann type. If we do not fix the point p and fix the boundary value, i.e. fix a Jordan curve γ in $\partial\Omega \subset \mathbb{C}^n$, then a classical theorem of Wermer [22] says that a necessarily and sufficient condition for the Jordan curve to be the boundary of a disc is the moment condition:

$$\int_\gamma \omega = o$$

for all holomorphic 1-forms. However for applications it is more natural not to fix the boundary value but to fix an interior point. This problem perhaps should be approached from the PDE point of view as in the case of harmonic maps.

§2. The Hermitian - Einstein Equation

In recent years variational problems involving the Yang-Mills functional have been studied intensively and in some case successively solved via the heat equation approach. The Yang-Mills functional is a vector-valued version of (1.1). Let E be a complex vector bundle over a compact Riemannian manifold (M,g). Given a unitary connection A on E, denote by F_A the curvature of A, then the Yang-Mills functional is,

$$(2.1) \qquad Y(A) = \int_M |F_A|^2 dv_g$$

For any deformation $A + tB$ of A, let

$$F_{A+tB} = d(A+tB) - (A+tB)\wedge(A+tB)$$

hence

$$\frac{\partial}{\partial t} F_{A+tB} \Big|_{t=o} = dB - A\wedge B - B\wedge A = D_A B$$

where D_A is the covariant differentiation induced by A.

Thus

$$\frac{\partial V}{\partial t} \Big|_{t=o} = 2 \int_M < \frac{\partial F_A}{\partial t}, F_A > dv_g$$

$$= 2 \int_M < D_A B, F_A > dv_g$$

$$= 2 \int_M < B, D_A^* F_A > dv_g$$

and the Euler-Lagrange equation is the Yang-Mills equation,

$$(2.2) \qquad D_A^* F_A = 0$$

which because of the Bianchi identity $D_A F_A = 0$ is equivalent to

$$(2.3) \qquad \Delta_A F_A = (D_A D_A^* + D_A^* D_A) F_A = 0.$$

The natural heat (or evolution) equation for the Yang-Mills functional is then

$$\frac{\partial A}{\partial t} = - D_A^* F_A$$

so that

$$\frac{\partial Y}{\partial t} = - 2 \int_M |D_A^* F_A|^2 dv_g = - 2 \int_M |\Delta_A F_A|^2 dv_g$$

Specializing to the case of holomorphic vector bundles E over a Kähler manifold M with Kähler form ω, the covariant differentiation decomposes into

$$D_A = \partial_A + \bar\partial_A$$

according to types (1,0) and (0,1). From the Kähler identities,

$$\partial_A^* = i[\Lambda, \bar\partial_A] \ , \quad \bar\partial_A^* = - i[\Lambda, \partial_A]$$

we have (keeping in mind the Bianchi identity),

$$\frac{\partial A}{\partial t} = - 2D_A^* F_A = 2i(\bar\partial_A - \partial_A)\Lambda F_A$$

where Λ is the adjoint of the operator L_ω which takes a form φ to $\varphi\wedge\omega$. On two-forms,

$$\Lambda\varphi = -ig^{\alpha\bar\beta}\varphi_{\alpha\bar\beta}$$

where $\varphi = \varphi_{\alpha\bar\beta} dz^\alpha \wedge d\bar z^\beta$. For hermitian connections we can express the equation in terms of the hermitian metric h,

$$-2\left(\bar\partial_A - \partial_A\right)\frac{\partial h}{\partial t} h^{-1} = - 2i(\bar\partial_A - \partial_A)\Lambda F_A$$

or

$$\frac{\partial h}{\partial t} h^{-1} = - i\Lambda F_A + \eta$$

where $\eta \in \ker(\bar\partial_A - \partial_A)$. We choose for convenience $\eta = i\lambda I$, hence the equation

$$\frac{\partial h}{\partial t} h^{-1} = - i(\Lambda F - \lambda I)$$

with

(2.4) $\qquad \lambda = 2\pi \deg E/(\mathrm{vol}\ M)(\mathrm{rk}\ E)$

As usual the degree of E is defined as,

(2.5) $\qquad \deg E = \int_M c_1(E,h)\wedge\frac{\omega^{n-1}}{(n-1)!}$

which is independent of the choice of h and depends only on the class of ω. More generally the degree of a coherent sheaf is defined by taking

$$c_1(\mathcal{F}) = c_1(\det \mathcal{F})$$

where

$$\det \mathcal{F} = (\overset{r}{\Lambda}\mathcal{F})^{**}, \quad r = \mathrm{rk}\mathcal{F}$$

is a reflexive sheaf of rank one, hence a line bundle.

<u>Definition</u> (1) A holomorphic vector bundle E over a compact Kähler manifold (M,ω) is stable (resp. semi-stable) if for every proper coherent subsheaf $\mathcal{F}$,

$$\mu(\mathcal{F}) = \deg \mathcal{F}/\mathrm{rk}\mathcal{F} < \deg E/\mathrm{rk}\, E = \mu(E)$$

(resp. $\le$).

(2) A holomorphic vector bundle E over (M,ω) is Hermitian-Einstein if there exists a hermitian metric so that

$$\Lambda F - \lambda I = 0 \tag{2.6}$$

<u>Note that stability depends only on the class of ω whereas the Hermitian-Einstein condition depends on the choice of ω</u>. Note also that if ω' is cohomologous to $m\omega$ for some $m > 0$ then

$$\mu_{\omega'}(E) = m^{n-1}\mu_{\omega}(E), \quad n = \dim M$$

hence ω' - (semi)stable is equivalent to ω-(semi)stable. In terms of local coordinate the Einstein condition is,

$$g^{j\bar{k}}F^{\beta}_{\alpha j\bar{k}} = \lambda\delta^{\beta}_{\alpha}$$

where $F^{\beta}_{\alpha} = F^{\beta}_{\alpha j\bar{k}}dz^j \wedge d\bar{z}^k$ are the curvature forms.

<u>Theorem</u> (Lübke). Let (M,ω) be a compact Kähler manifold and (E,h) a holomorphic Hermitian-Einstein bundle over M then $E = \oplus E_i$ where each E_i is ω-stable with the same generalized degree $\mu_{\omega}(E_i) = \mu$.

For projective surfaces the converse is also true.

<u>Theorem</u> (Donaldson [9], [10]). Let M be a projective surface with Kähler metric ω in the hyperplane class (or more generally $m\omega$ is

in the hyperplane class for some $m > o$). Let E be a ω-stable holomorphic vector bundle then there exists hermitian metric h which is Einstein with respect to ω.

<u>Remark</u>. In the above theorem that on a stable bundle Hermitian-Einstein connection is unique, it follows that Hermitian-Einstein metric on a stable bundle is unique up to constant multiples.

Donaldson's proof goes as follows. First of all the heat equation is parabolic, hence a smooth solution exists for small time. In fact, a smooth solution exists for all finite time. This follows from the maximum principle as follows.

Suppopse H_t and K_t are two families of hermitian metrics satisfying the heat equation, define a function

$$\sigma(H,K) = \mathrm{Tr}(H^{-1}K) + \mathrm{Tr}(KH^{-1}) - 2 \text{ rank } E$$

then a direct calculation shows that

$$\frac{\partial\sigma}{\partial t} = \Delta\sigma + \text{negative term}$$

i.e. $\frac{\partial\sigma}{\partial t} \geq \Delta\sigma$. Hence maximum principle implies that $\sigma(H,K)$ is a decreasing function of time.

Since a smooth solution exists for small time t, we have for any $\varepsilon > o$, a $\delta > o$ such that

$$\sup_X \sigma(H_o, H_t) < \varepsilon$$

for all $0 \leq t \leq \delta$. Since the function $\sup_X \sigma(H_t, H_{t+\delta})$ is a decreasing function of t, that

$$\sup_X \sigma(H_{t-\delta}, H_t) \leq \sup_X \sigma(H_o, H_\delta) < \varepsilon .$$

If a solution exists for all $t < T < \infty$ then

$$\sup_X \sigma(H_t, H_{t'}) < \varepsilon$$

for all t and t' sufficiently close to T $(t,t'>T-\delta)$. This shows that H_t is a Cauchy sequence, hence converges in C^o to a continuous hermitian metric. Standard elliptic argument gives a C^∞ metric.

The argument so far is completely general, we have not used any assumption on E. The first beautiful idea of Donaldson is to use the semi-stability assumption to show that the Einstein tensor

$$\Lambda F_t - \lambda I \to 0$$

as $t \to \infty$. For this purpose it is necessary to construct another functional whose Euler-Lagrange equation is also the Einstein equation.

Consider a family of smooth hermitian metrics H_t on E and let $h_t = H_t H_o^{-1}$. The connection and curvature of H_t are related to those of H_o by,

$$\partial_t = \partial_o + \partial_o h_t \cdot h_t^{-1}$$

$$F_t = F_o + \bar{\partial}(\partial_o h_t \cdot h_t^{-1})$$

Taking trace, we have

$$\mathrm{Tr} F_t = \mathrm{Tr} F_o + \mathrm{Tr}\bar{\partial}(\partial_o h_t \cdot h_t^{-1})$$

$$\frac{\partial}{\partial t} \mathrm{Tr} F_t = \mathrm{Tr}\left[\bar{\partial}(\partial_t(\dot{h}_t \cdot h_t^{-1}))\right]$$

$$= \bar{\partial}\partial \; \mathrm{Tr}(\dot{h}_t h_t^{-1})$$

$$= \bar{\partial}\partial \frac{\partial}{\partial t} \log \det h_t$$

$$= \bar{\partial}\partial \frac{\partial}{\partial t} \log \det(H_t H_o^{-1})$$

$$= \frac{\partial}{\partial t} \bar{\partial}\partial \log \det(H_t H_o^{-1})$$

Differentiating with respect to t, we get

$$\frac{\partial F_t}{\partial t} = \dot{F}_t = \bar{\partial}(\partial_o \dot{h}_t h_t^{-1} - \partial_o h_t \cdot h_t^{-1} \dot{h}_t h_t^{-1})$$

$$= \bar{\partial}\left[\partial_o(\dot{h}_t h_t^{-1}) + \dot{h}_t h_t^{-1} \partial_o h_t h_t^{-1} - \partial_o h_t \cdot h_t^{-1} \dot{h}_t h_t^{-1}\right]$$

$$= \bar{\partial}\left[\partial_o(\dot{h}_t h_t^{-1})\right]$$

Integrating we get

$$\mathrm{Tr}F_t - \mathrm{Tr}F_o = \bar{\partial}\partial \log \det(H_t H_o^{-1}) = \bar{\partial}\partial \log \det h_t$$

the potential

$$R_1 = \log \det h_t$$

is the first Secondary Chern class, note that

$$\dot{R}_1 = \mathrm{Tr}(\dot{h}_t h_t^{-1})$$

For further secondary invariants, consider

$$F_t \wedge F_t = F_o \wedge F_o + F_o \wedge \bar{\partial}(\partial_o h_t \cdot h_t^{-1}) + \bar{\partial}(\partial_o h_t \cdot h_t^{-1}) \wedge F_o$$

$$+ \bar{\partial}(\partial_o h_t \cdot h_t^{-1}) \wedge \bar{\partial}(\partial_o h_t \cdot h_t^{-1})$$

and upon taking trace and differentiating the resulting expression, we get

$$\frac{\partial}{\partial t} \mathrm{Tr}(F_t \wedge F_t) = \mathrm{Tr}\left[F_o \wedge \bar{\partial}(\partial_t(\dot{h}_t \cdot h_t^{-1})\right] + \mathrm{Tr}\left[\bar{\partial}(\partial_t(\dot{h}_t \cdot h_t^{-1})) \wedge F_o\right]$$

$$+ 2\,\mathrm{Tr}\left[\bar{\partial}(\partial_o \dot{h}_t \cdot h_t^{-1}) \wedge \bar{\partial}\partial_t(\dot{h}_t \cdot h_t^{-1})\right]$$

$$= 2\,\mathrm{Tr}\left\{\left[F_o + \bar{\partial}(\partial_o h_t \cdot h_t^{-1})\right] \wedge \bar{\partial}\partial_t(\dot{h}_t \cdot h_t^{-1})\right\}$$

$$= 2\,\mathrm{Tr}\left\{\left[F_t \wedge \bar{\partial}\partial_t(\dot{h}_t \cdot h_t^{-1})\right\}$$

$$= 2\,\mathrm{Tr}\left[\bar{\partial}\partial_t(F_t \dot{h}_t h_t^{-1})\right]$$

$$= 2\,\bar{\partial}\partial\ \mathrm{Tr}(F_t \dot{h}_t h_t^{-1})$$

Hence we have

$$\mathrm{Tr}(F_t \wedge F_t) - \mathrm{Tr}(F_o \wedge F_o) = 2\bar{\partial}\partial \int_o^t \mathrm{Tr}(F_t \dot{h}_t h_t^{-1})dt$$

where the potential

$$R_2 = 2 \int_o^t \mathrm{Tr}(F_t \dot{h}_t h_t^{-1})dt$$

is the next Secondary Chern class.

Donaldson introduced the following functional,

$$\mathcal{M}_m = \int_M iR_2 \wedge \frac{\omega^{n-1}}{(n-1)!} - 2\lambda R_1 \frac{\omega^n}{n!}$$

and its time derivative is

$$\frac{\partial \mathcal{M}}{\partial t} = 2 \int_M \mathrm{Tr}\left[(i\Lambda F - \lambda I)\dot{h}h^{-1}\right]\frac{\omega^n}{n!}$$

which shows that the Euler-Lagrange equation is the Einstein equation.

Along the Yang-Mills flow the above expression becomes,

$$\frac{\partial \mathcal{M}}{\partial t} = -\int_M \mathrm{Tr}\left[(i\Lambda F - \lambda I)(i\Lambda F - \lambda I)\right]$$

$$= -\int_M \mathrm{Tr}\left[(i\Lambda F - \lambda I)(i\Lambda F - \lambda I)^*\right]$$

$$= -\int_M |i\Lambda F - \lambda I|^2 \frac{\omega^n}{n!}$$

By a theorem of Metha and Ramanathan:

Theorem. Let (M,ω) be a projective variety and Kähler metric ω in the hyperplane section class. Let E be an ω-(semi)stable bundle over M, then for a generic curve of sufficiently high degree, $E|_C$ is ω-(semi)stable.

Thus taking a curve as in the previous theorem, it follows that

$$\mathcal{M}_C(E|_C) = \int_C iR_2 - 2\lambda R_1 \omega$$

is bounded below, from the theory on Riemann surfaces.

However $\mathcal{M}_M(E)$ and $\mathcal{M}_C(E|_C)$ are related by the second fundamental form which is bounded,

$$\mathcal{M}_M = \frac{1}{\deg C}\mathcal{M}_C + \text{bounded term}$$

It follows that $\mathcal{M}_M$ is also bounded below if E is semi-stable, i.e.

$$\mathcal{M}_M(t) = -\int_0^t \int_M |i\Lambda F - \lambda I|^2 \, dt \geq - \text{const}$$

independent of t. This implies that

$$\int_M |i\Lambda F_t - \lambda I|^2 \to 0$$

as $t \to \infty$

By a direct computation one gets

$$\left(\frac{\partial}{\partial t} - \Delta\right) |i\Lambda F - \lambda I|^2 \le -2|i\Lambda F - \lambda I|^2$$

it follows then from standard elliptic theory that

$$\sup_M |i\Lambda F - \lambda I| \le -\| i\Lambda F - \lambda I \|_{L^2}$$

Hence the Einstein tensor

$$i\Lambda F - \lambda I \to o$$

in the C^o norm if E is semi-stable.

The above argument works in any dimension.

There is an alternative approach due to Uhlenbeck and Yau [21]. They show that the following equation,

$$i\Lambda F_h - \lambda I + s \ln h = 0$$

admits a solution for all $0 < s \le 1$, provided that E is simple, i.e. admits no holomorphic endomorphisms other than constant multiples of the identity.

Consider the reparametization,

$$t = -\ln s \ , \ 0 \le t < \infty$$

$$\varphi_t = h_{e^{-t}}$$

$$i\Lambda F_\varphi - \lambda I + e^{-t} \ln \varphi = 0$$

then

$$\frac{\partial \mathcal{M}}{\partial t} = 2 \int_M \langle \dot{\varphi}_t \varphi_t^{-1}, i\Lambda F_t - \lambda I \rangle$$

$$= -2 e^{-t} \int_M \langle \frac{\partial}{\partial t} \ln \varphi_t, \ln \varphi_t \rangle$$

$$= - 2 e^{-t} \int_M \frac{\partial}{\partial t} |\ln \varphi_t|^2$$

If we assume furthermore that E is semi-stable then

$$\mathcal{M}_t = - \int_o^t \int_M e^{-\rho} \frac{\partial}{\partial \rho} |\ln \varphi_\rho|^2 d\rho \geq - c$$

for all t.

Since

$$\int_o^t e^{-\rho} \frac{\partial}{\partial \rho} |\ln \varphi_\rho|^2 d\rho = \tfrac{1}{2} e^{-t} |\ln \varphi_t|^2 - \tfrac{1}{2} |\ln \varphi_o|^2$$

$$+ \tfrac{1}{2} \int_o^t e^{-\rho} |\ln \varphi_\rho|^2 d\rho$$

thus

$$\int_M \left(\int_o^t e^{-p} |\ln \varphi_\rho|^2 d\rho + e^{-t} |\ln \varphi_t|^2 \right) \omega^n \leq c'$$

for all t. Hence

$$\int_M |i\Lambda F - \lambda I|^2 = \int_M e^{-t} |\ln \varphi_t|^2 \rightarrow 0$$

as $t \rightarrow \infty$.

By direct computation one can show that

$$\Delta |\ln \varphi_t| \leq c |\ln \varphi_t|$$

which implies that

$$\sup_M |\ln \varphi_t| \leq \|\ln \varphi_t\|_{L^2}$$

It follows that

$$\sup_M |i\Lambda F - \lambda I|^2 = \sup_M e^{-t} |\ln \varphi_t|^2 \rightarrow 0$$

as $t \rightarrow \infty$.

From the equation we have

$$\frac{\partial}{\partial t} (i\Lambda F - \lambda I) - e^{-t} \ln \varphi_t + e^{-t} \frac{\partial}{\partial t} \ln \varphi_t = 0$$

it follows that

$$\frac{\partial}{\partial t} \int_M |i\Lambda F-\lambda I|^2 = e^{-t}\int_M |\ln \varphi_t|^2 - e^{-t}\int_M < \frac{\partial}{\partial t}\ln \varphi_t, \ln \varphi_t > \infty$$

The first term on the right tends to zero as $t \to \infty$. So does the second term = $\partial \mathcal{M}/\partial t$ because $\mathcal{M} \geq -c$. Thus for the Uhlenbeck-Yau's flow, we have

$$\begin{cases} \int_M |i\Lambda F-\lambda I|^2 \to 0 \\ \frac{\partial}{\partial t} \int_M |i\Lambda F-\lambda I|^2 \to 0 \end{cases}$$

as $t \to \infty$ provided that E is semi-stable and simple.

§3. Kähler-Einstein Metrics

Let M be a compact Kählermanifold and let E be the holomorphic tangent bundle. Fix a Kähler form ω_o, then we have the Yang-Mills functional

$$(3.1) \qquad Y = \int_M |F|^2_{\omega_o} \frac{\omega_o^n}{n!}$$

defined for say hermitian metrics along the fiber of $E=TM$. On the other hand we may restrict to Kähler metrics on $E=TM$, in which case it is more natural to take the metric on the base to be the same metric. More precisely define a functional on the space of Kähler metrics cohomologous to ω_o:

$$(3.2) \qquad K(\omega) = \int_M |F_\omega|^2_\omega \frac{\omega^n}{n!}$$

It is customary to denote the Riemannian curvature of ω by $R_{\alpha\bar\beta\mu\bar\nu}$, then

$$K(\omega) = \int_M R_{\alpha\bar\beta\mu\bar\nu} R^{\alpha\bar\beta\mu\bar\nu} \frac{\omega^n}{n!}$$

There are two other functionals closely related to $K(\omega)$:

$$L(\omega) = \int_M R_{\alpha\bar\beta} R^{\alpha\bar\beta} \frac{\omega^n}{n!} = \int_M |\mathrm{Ric}\ \omega|^2 \frac{\omega^n}{n!}$$

and

$$R(\omega) = \int_M R^2 \frac{\omega^n}{n!}$$

where $R_{\alpha\bar{\beta}} dz^\alpha \wedge d\bar{z}^\beta$ is the Ricci form and R is the scalar curvature of ω.

It is well-known (cf.[6]) that the three functionals differ from each other by a constant, hence have the same critical point. The Euler-Lagrange equation for $R(\omega)$ is

$$(2.4) \qquad \bar{\partial}(\text{grad } R) = 0, \ \text{grad } R = g^{\alpha\bar{\beta}} \frac{\partial R}{\partial \bar{z}^\beta} \frac{\partial}{\partial z^\alpha}$$

i.e. a Kähler metric is critical if the gradient of its scalar curvature function in a holomorphic vector field.

Notice that

$$\int_M R^2 \geq \frac{\left(\int_M R\right)^2}{\text{vol } M} = \sigma$$

where σ is independent of the choice the Kähler metric in the same class. The inequality above is an equality iff R = constant. Namely the lower bound σ is attained iff a Kähler metric with constant scalar curvature exists.

It is known that there is an obstruction to the existence of Kähler metric with constant scalar curvature (cf. [6], [12]). A classical theorem of Matsushima-Lichnerowiz asserts that $\text{Aut}_o M$ of such a manifold must be reductive. Calabi showed that $M = \mathbb{C}P^2$ with 1-point blown up, admits critical metrics whereas $\text{Aut}_o M$ is not reductive. Recently Calabi proved that if a critical metric g exists then $\text{Isom}_g M \cap \text{Aut}_o M$ is a maximal compact subgroup of $\text{Aut}_o M$ of positive dimension. Using this result M. Levin showed that $\mathbb{C}P^2$ with two points blown up admits no critical metric.

A Kähler metric is Kähler-Einstein if its Ricci tensor is proportional to itself, i.e.

$$R_{\alpha\bar{\beta}} = c \ g_{\alpha\bar{\beta}} \ , \ c \text{ constant.}$$

In particular $c_1(M)$ must either be zero, positive or negative according to $c = 0$, > 0 or < 0.

The converse is known as the Calabi conjecture. As is well-known the cases of $c_1(M) = 0$ and $c_1(M) < 0$ were resolved by Yau in the affirmative (and in fact the Kähler-Einstein metric is unique). The positive case $c_1(M) > 0$, as was mentioned earlier for $M = \mathbb{C}P^2$ with one or two points blown up admit no Kähler-Einstein metrics whereas $\mathbb{C}P^2$ admits many Kähler-Einstein metrics. The only Kähler surfaces with $c_1 > o$ are the del Pezzo surfaces: $\mathbb{C}P^1 \times \mathbb{C}P^1$ and $\mathbb{C}P^2$ with at most 8 points (in general position) blown up (= surfaces of degree d in $\mathbb{C}P^1$). Obviously $\mathbb{C}P^1 \times \mathbb{C}P^1$ admits a Kähler Einstein metric but uniqueness is not known. It is still open whether the remaining del Pezzo surfaces admit any Kähler-Einstein metrics.

If $c_1(M) > 0$, take any Kähler form ω with $[\omega] = c_1(M)$. Then

$$\mathrm{Ric}\omega - \omega = i\partial\bar{\partial}f$$

for some function f (determined up to an additive constant) on M.

Define a homomorphism $\mathcal{F}$ on holomorphic vector fields by,

$$\mathcal{F}(X) = \int_M (Xf)\omega^n$$

Futaki showed that $\mathcal{F}(X)$ is well-defined, i.e. independent of the choice of ω (in the anticanonical class). The Futaki obstruction is defined as,

$$\delta_M = \dim_{\mathbb{C}} \quad (\text{Holomorphic vector fields}/\ker \mathcal{F})$$

If M admits a Kähler-Einstein metric ω then $f \equiv 0$, hence $\mathcal{F} \equiv 0$ and so $\delta_M = 0$. It is also clear that $\delta_M = 0$ if M admits no holomorphic vector field. If we blow up $\mathbb{C}P^2$ at 4(or more) generic points the resulting manifold admits no holomorphic vector field (in general same is true for $\mathbb{C}P^n$ blown up at $n+2$ generic points). Thus the Futaki Obstruction for del Pezzo surfaces, $\mathbb{C}P^2$ with 4 or more points blown up vanishes. It can be shown directly that the same is true for the case with three points of $\mathbb{C}P^2$ blown up.

There is another obvious obstruction, namely since Kähler-Einstein is a priori Hermitian-Einstein, thus the tangent bundle must be a directly sum of $[c_1(M)]$-stable bundles with the same generalized degree.

<u>Question 1</u>. Can an obstruction for semi-stability be defined analogous to the Futaki obstruction? Note that stability implies in particular that there is no non-trivial holomorphic endomorphism of the tangent bundle.

<u>Question 2</u>. Are the del Pezzo surfaces semi-stable with respect to the anticanonical class?

The most natural heat equations for the Kähler-Einstein metrics are the following ones (cf. [7]):

$$(KE)_o \qquad \frac{\partial g_{\alpha\bar\beta}}{\partial t} = - R_{\alpha\bar\beta}\,, \qquad c_1(M) = 0$$

$$(KE)_- \qquad \frac{\partial g_{\alpha\bar\beta}}{\partial t} = - R_{\alpha\bar\beta} - g_{\alpha\bar\beta}\,, \quad c_1(M) < 0$$

$$(KE)_+ \qquad \frac{\partial g_{\alpha\bar\beta}}{\partial t} = - R_{\alpha\bar\beta} + g_{\alpha\bar\beta}\,, \quad c_1(M) > 0$$

Originally Calabi conjectured that given any closed (1.1) form $\frac{i}{2\pi} T_{\alpha\bar\beta} dz^{\alpha} \wedge d\bar z^{\beta} = T$ with $[T] = c_1(M)$ (with no assumption on $c_1(M)$) then there exists Kähler metric g with $\frac{i}{2\pi}$ Ric $g = T$. The corresponding equation is

$$(KE)_o' \qquad \frac{\partial g_{\alpha\bar\beta}}{\partial t} = - R_{\alpha\bar\beta} + T_{\alpha\bar\beta}$$

It turns out that this more general equation can also be solved with the $(KE)_o$ equation being a special case. These equations being parabolic admit solutions for short time.

One can also reduce the above tensor equations to scalar equations by setting

$$g_{\alpha\bar\beta} = g^o_{\alpha\bar\beta} + \partial_\alpha \partial_{\bar\beta} u$$

$$T_{\alpha\bar\beta} = R^o_{\alpha\bar\beta} + \partial_\alpha \partial_{\bar\beta} f$$

where $g^o_{\alpha\bar\beta}$ is the initial metric and $R^o_{\alpha\bar\beta}$ its Ricci curvature. The functions u and f are determined up to additive constants. For $(KE)_o'$ we have

$$\partial_\alpha \partial_{\bar\beta} \frac{\partial u}{\partial t} = - R_{\alpha\bar\beta} + R^o_{\alpha\bar\beta} + \partial_\alpha \partial_{\bar\beta} f$$

$$= \partial_\alpha \partial_{\bar\beta} \left\{ \log \det(g^o_{\alpha\bar\beta} + \partial_\alpha \partial_{\bar\beta} u) - \log \det(g^o_{\alpha\bar\beta}) \right\} + \partial_\alpha \partial_{\bar\beta} f$$

Hence,

$$= \frac{\partial u}{\partial t} \log \det(g^o_{\alpha\bar\beta} + \partial_\alpha \partial_{\bar\beta} u) - \log \det(g^o_{\alpha\bar\beta}) + f + c(t)$$

where $c(t)$ is a function of time alone.

We impose the conditions that $c(t) \equiv 0$ and the function f satisfies

$$\int_M e^{-f} \omega_o^m = \int_M \omega_o^m = \mathrm{vol}(M) \ ,$$

so we have the scalar equation

$$(KE)_o' \qquad \frac{\partial u}{\partial t} = \log \det(g^o_{\alpha\bar\beta} + \partial_\alpha \partial_{\bar\beta} u) - \log \det(g^o_{\alpha\bar\beta}) + f$$

Now for $(KE)_-$, we choose $g^o_{\alpha\bar\beta}$ the initial metric to be in the canonical class. This is possible because of Yau's solution to $(KE)_o'$. Define the functions u and f by

$$\begin{cases} g_{\alpha\bar\beta} = g^o_{\alpha\bar\beta} + \partial_\alpha \partial_{\bar\beta} u \\ -g_{\alpha\bar\beta} = R^o_{\alpha\bar\beta} + \partial_\alpha \partial_{\bar\beta} f \end{cases}$$

then we obtain

$$\partial_\alpha \partial_{\bar\beta} \frac{\partial u}{\partial t} = - R^o_{\alpha\bar\beta} + \partial_\alpha \partial_{\bar\beta} f - \partial_\alpha \partial_{\bar\beta} u$$

or

$$\frac{\partial u}{\partial t} = \log \det(g^o_{\alpha\bar\beta} + \partial_\alpha \partial_{\bar\beta} u) - \log \det(g^o_{\alpha\bar\beta}) + u + f + c(t)$$

Again we impose the conditions above so that

$$(KE)_- \qquad \frac{\partial u}{\partial t} = \log \det(g^o_{\alpha\bar\beta} + \partial_\alpha \partial_{\bar\beta} u) - \log \det(g^o_{\alpha\bar\beta}) - u + f.$$

Analogously by choosing $g^o_{\alpha\bar\beta}$ to be in the anticanonical class and defining f by

$$g^o_{\alpha\bar\beta} = R^o_{\alpha\bar\beta} + \partial_\alpha \partial_{\bar\beta} f$$

We obtain

$$(\mathrm{KE})_+ \qquad \frac{\partial u}{\partial t} = \log\det(g^o_{\alpha\bar\beta} + \partial_\alpha\partial_{\bar\beta}u) - \log\det(g^o_{\alpha\bar\beta}) - u - f.$$

§4. Monge-Ampère Equation on Domains

On a bounded (strictly pseudoconvex) domain Ω in $\mathbb{C}^n$ with C^∞ smooth boundary, consider the following functional (cf. [8], [2]) on C^2 functions on $\bar\Omega$,

$$(4.1) \qquad J(u) = \int_\Omega du\wedge d^c u\wedge (dd^c u)^{n-1}$$

where $d^c = \frac{i}{2}(\bar\partial - \partial)$. For the theory of several complex variables it is natural to consider strictly pseudoconvex domains and plurisubharmonic functions. Then the functional is non-negative and is an analogous of the energy functional in (II) of §1, however the integrand is the length of the gradient vector measured in the pseudo-metric $dd^c u \ge 0$, which unlike the harmonic case is changing as u changes. The Euler-Lagrange equation as a result is more complicated (non-linear). For any variations of the form $u+tv$ with the conditions $v = 0$ and $d^c v = 0$ on $\partial\Omega$ (or for simplicity consider only those v with compact support), we have

$$\left.\frac{\partial J}{\partial t}\right|_{t=o} = \int_\Omega dv\wedge d^c u\wedge (dd^c u)^{n-1} + du\wedge d^c v\wedge (dd^c u)^{n-1}$$
$$+ du\wedge d^c u\wedge dd^c v\wedge (dd^c u)^{n-2}$$

From type consideration and Stokes theorem we have

$$\int_\Omega du\wedge d^c v\wedge (dd^c u)^{n-1} = \int_\Omega dv\wedge d^c u\wedge (dd^c u)^{n-1}$$
$$= \int_{\partial\Omega} v d^c u\wedge (dd^c u)^{n-1} - \int_\Omega v(dd^c u)^n$$
$$= -\int_\Omega v(dd^c u)^n$$

and

$$\int_\Omega du\wedge d^c u\wedge dd^c v\wedge (dd^c u)^{n-2} - \int_\Omega du\wedge d^c v\wedge (dd^c u)^{n-1}$$

$$= \int_\Omega v(dd^c u)^{n-1}$$

Hence,

(4.2) $$\left.\frac{\partial J}{\partial t}\right|_{t=o} = - \int_\Omega v(dd^c u)^n$$

and the Euler-Lagrange equation is the homogeneous Monge-Ampère equation:

(4.3) $$(dd^c u)^n = n! \det (\partial^2 u/\partial z^\alpha \partial \bar{z}^\beta) = 0.$$

The natural heat equation associated to the Dirichlet problem is

(4.4) $$\begin{cases} \frac{\partial u}{\partial t} = \det(u_{\alpha\bar{\beta}}) \\ u|_{\partial\Omega} = \varphi \end{cases}$$

and for the inhomogeneous problem,

(4.5) $$\begin{cases} \frac{\partial u}{\partial t} = \det(u_{\alpha\bar{\beta}}) - F \ , \ F > 0 \\ u|_{\partial\Omega} = \varphi \end{cases}$$

Fix a global defining function ρ of Ω, i.e. $\Omega = \{\rho < 0\}$, $\partial\Omega = \{\rho = 0\}$ and $d\rho$ non-vanishing on $\partial\Omega$. The Dirichlet data φ can be extended smoothly to Ω so that it is strictly puri-subharmonic.

For instance extending φ smoothly but otherwise arbitrary to Ω then for some integer N, the function $\varphi + N\rho$ is strictly pluri-subharmonic. From now on such an extension will be fixed. Express $u = \varphi + v$, then

(4.6) $$\frac{\partial v}{\partial t} = \frac{\partial u}{\partial t} = \det(u_{\alpha\bar{\beta}}) = \det(\varphi_{\alpha\bar{\beta}} + v_{\alpha\bar{\beta}}) \quad \text{where } v|_{\partial\Omega} = 0,$$

For the inhomogeneous equation, we can consider instead of (4.5), the following heat equation,

(4.7) $$\frac{\partial u}{\partial t} = \log \det(u_{\alpha\bar{\beta}}) - \log F$$

which in terms of v is

(4.8) $$\frac{\partial u}{\partial t} = \log \det(\varphi_{\alpha\bar{\beta}} + v_{\alpha\bar{\beta}}) - \log F$$

$$= \log \det(\varphi_{\alpha\bar\beta}+v_{\alpha\bar\beta}) - \log \det (\varphi_{\alpha\bar\beta}) + f$$

where

$$f = \log \det(\varphi_{\alpha\bar\beta}) - \log F$$

In this form the equation is formally the same as the equation $(KE)_o'$ in the compact case.

One can also formulate the <u>Neumann problem</u>:

$$\begin{cases} \det(u_{\alpha\bar\beta}) = F & \text{on } \Omega \\ d^c u \wedge (dd^c u)^{n-1} = G dS & \text{on } \partial\Omega \end{cases}$$

where $F > 0$ for the inhomogeneous equation and $F \equiv 0$ for the homogeneous problem, dS denotes the volume element of $\partial\Omega$. We must of course impose the compatibility condition

$$\int_\Omega F dV = \int_{\partial\Omega} G dS$$

because of stokes theorem,

$$\int_\Omega (dd^c u)^n = \int_{\partial\Omega} d^c u \wedge (dd^c u)^{n-1}.$$

§5. <u>The Complex Homogeneous Monge-Ampère Equation</u>

A plurisubharmonic function u is a solution of the complex homogeneous Monge-Ampère equation if

(5.1) $\qquad (\partial\bar\partial u)^n \equiv 0.$

To avoid a trivial a situation, we impose the following rank condition.

(5.2) $\qquad \text{rank } \partial\bar\partial u = n - 1.$

where n is the complex dimension of M, a complex manifold.

With this condition, the annihilator of $\partial\bar\partial u$ then defines a complex one dimensional distribution (i.e. the zero eigen direction at each point) which is integrable, hence the manifold is foliated by Riemann surfaces. This foliation will be refered to as the Monge-

Ampère foliation associated to u.

The most interesting cases arise when the function

(5.3) $\tau = \exp u: M \to [o,c)$

is a strictly plusisubharmonic exhaustion.

In this situation, the regularity of τ at the minimum set $\tau^{-1}(o)$ becomes rather delicate. Also the function u is then singular on $\tau^{-1}(o) = u^{-1}(-\infty)$. There are a number of known results concerning manifolds with such an exhaustion, all of them exploit the situation when $\tau^{-1}(o)$ is rather simple.

In the following M will always denote a connected complex manifold of complex dimension n.

(A) <u>The unbounded case</u>

(i) <u>Stoll's Theorem [20]</u> If M admits an unbounded Monge-Ampère exhaustion $\tau: M \to [0,\infty)$ which is of class C^∞ and strictly plurisubharmonic everywhere. If $u = \log \tau$ satisfies (5.1) on $M \setminus \tau^{-1}(o)$, then $\tau^{-1}(o)$ consists of exactly one point $\{\underline{o}\}$ and the exponential map at $\underline{o}$ of the metric $\partial\bar{\partial}\tau$ is a biholomorphic map from the tangent space $T_{\underline{o}}M = \mathbb{C}^n$ onto M. Under this biholomorphism τ is identified with the standard exhaustion $\|z\|^2$ on $\mathbb{C}^n$.

If we weaken the assumption on τ slightly, one has the following theorem due to Burns:

(ii) <u>Theorem</u> [3] Same assumptions as in (i) except that τ is only assumed to be strictly plurisubharmonic on $M \setminus \tau^{-1}(o)$. Then $\tau^{-1}(o)$ can be blown down to a point, the resulting space $M^\vee$ is a normal complex space, $M^\vee$ is biholomorphic to $L^\vee$ which is obtained by blowing down the zero-section of a negative holomorphic line bundle over a $n-1$ dimensional projective manifold. Under this biholomorphism $\tau|_{M^\vee}$ is identified to $h^m|_{L^\vee}$ where m is a positive integer and h is a hermitian metric along the fibers of L with negative first chern form, $c_1(L,h) < 0$.

In the above theorem, it is the assumption that t is smooth even on the minimum set $\tau^{-1}(o)$ is crucial.

(B) <u>The bounded case</u>

First we have the characterization of balls in $\mathbb{C}^n$, due to Stoll [20].

(i) If M admits a bounded exhaustion $\tau\colon M \to [o,r^2)$ where τ is assumed to be C^∞ and strictly plurisubharmonic everywhere. If $u = \log\tau$ is plurisubharmonic and satisfies (5.1) on $M\backslash\tau^{-1}(o)$. Then $\tau^{-1}(o)$ consists of exactly one point $\underline{o}$, the exponential map of $\partial\bar\partial\tau$ at $\underline{o}$ is defined on the ball of radius r. $\mathbb{B}^n(r) \subset T_{\underline{o}}M$ and is a biholomorphic map onto M. The exhaustion τ is identified with $\|z\|^2$ on $\mathbb{B}^n(r)$.

The global regularity assumption of τ in A(i), A(ii) and B(i) forces the Monge-Ampère foliation to be holomorphic, this is one of the crucial steps in establishing these theorems. However if this assumption is weakened then in general the foliation is no longer holomorphic. If however the foliation is holomorphic, we have the following theorem (cf. Wong [24], Patrizio [19]):

(iii) <u>Theorem</u> If M admits a bounded exhaustion $\tau\colon M \to [o,c)$ which is C^o on M and C^∞ and strictly plurisubharmonic on $M\backslash\tau^{-1}(o)$. If furthermore $\tau^{-1}(o)$ consists of exactly one point $\underline{o}$ and the Monge-Ampère foliation is holomorphic. Then M is biholomorphic to a bounded circular domain in $\mathbb{C}^n$.

More specifically, under the assumptions on τ, the indicatrix of the Kobayashi metric of M at $\underline{o}$ is a bounded circular domain and the exponential map of the Kobayashi metric is well-defined and is a biholomorphic map of the indicatrix onto M. Under this biholomorphic map τ is identified with an exhaustion on $\mathbb{C}^n$ of the form $\|z\|e^{g(z)}$ where g is defined outside of the origin and is constant along each complex line through the origin. Notice that the exponential map of the Kobayashi metric (which is a Finsler metric but not a hermitian metric except for the ball $\mathbb{C}^n$) is in general not well-defined.

Examples of non-holomorphic Monge-Ampère foliations are provided by the following theorem due to Lempert [16].

(iv) <u>Theorem</u> Let $D \subset \mathbb{C}^n$ be a bounded strictly convex domain with C^∞ boundary. Given any fixed point $p \in D$, there is a unique plurisubharmonic solution u of (5.1) on $D\backslash\{p\}$ with zero boundary

values and logarithmic singularity at p, i.e.

$$u = \log |z - p|^2 + O(1)$$

on a neighborhood of p. This solution is C^∞ on $\bar{D}\setminus\{p\}$ and $\exp u$ is strictly plurisubharmonic on $\bar{D}\setminus\{p\}$.

Note that $\tau = \exp u$ in the above is of course continuous but not smooth at p, otherwise D is biholomorphic to the ball. The foliation is in general not holomorphic, unless it is biholomorphic to a circular domain.

(C) Manifolds with bounded holomorphic functions

It is quite surprising that almost nothing is known concerning intrinsic or extrinsic criterions for manifolds admitting non-trivial bounded holomorphic functions. We present here a theorem for manifolds with bounded Monge-Ampère exhaustion.

Theorem Let M be a Stein manifold and $\tau: M \to [0,1)$ is an exhaustion which of class C^0 on M and C^∞ on $M\setminus\{\tau=0\}$. We assume that $\{\tau=0\}$ consists of exactly one point $\underline{o}$, $\partial\bar{\partial}\tau > 0$ and $u = \log \tau$ satisfies the Monge-Ampère equation (5.1) on $M\setminus\{\underline{o}\}$ and has a logarithmic singularity at $\underline{o}$. Then there exists non-trivial bounded holomorphic functions on M.

Proof. We will show that under the hypothesis of the theorem, the Caratheodory metric of M can be estimated from below by the Kobayashi metric. More precisely, we will show that there exists a positive constant $\lambda > 0$ such that

$$\lambda K_M \leq C_M$$

where K_M denotes the Kobayashi metric and C_M the Canatheodory metric. It is well-known that we always have $C_M \leq K_M$.

Step I Every leaf of the Monge-Ampère foliation is a geodesic of the Kobayashi metric.

It was shown in [23] that the holomorphic sectional curvature in the direction of the complex gradient $Z = \tau^{\alpha\bar{\beta}}\tau_{\bar{\beta}}\, \partial/\partial z^\alpha$ of the Kähler metric $\partial\bar{\partial}\tau$ is zero, i.e.

$$R_{\alpha\bar{\beta}\mu\bar{\nu}}\ Z^{\alpha}\overline{Z^{\beta}}Z^{\mu}\overline{Z^{\nu}}/\|Z\|^{4} \equiv 0$$

where $R_{\alpha\bar{\beta}\mu\bar{\nu}}$ are the components of the curvature of $\partial\bar{\partial}\tau$. This implies, by a direct computation that the holomorphic sectional curvature of the metric $\bar{\partial}\partial\log(1-\tau)$, in the direction of Z is identically -1. The metric $\bar{\partial}\partial\log(1-\tau)$ is an analogue of the Bergmann metric on the unit ball. Now an argument via Schwarz lemma shows that the integral curves of Z (= leaves of the Monge-Ampère foliation) are extremal curves (or geodesic) of the Kobayashi metric.

Step II. Let L be a Kobayashi geodesic of M through the point $\underline{o}$, then $L \cap M_c$, where $M_c = \{\tau<c\}$ is a Kobayashi geodesic of M_c, where $0 < c < 1$.

This is easy to see, because the function $c^{-1}\tau$ is a Monge-Ampère exhaustion of M_c with boundary value 1. The Monge-Ampère foliation associated to $c^{-1}\tau$ on M_c is easily seen to be the restriction of the Monge-Ampère foliation on M. Step II then follows from step I.

Step III Extension of a bounded holomorphic function from a leaf of the foliation.

It is well-known that on a Stein manifold, every holomorphic function on a submanifold can be extended to a global holomorphic function; however, a bounded holomorphic function, does not in general have a bounded extension. This is only possible for instance if the submanifold is transversal to the boundary. Since we make no assumption on the ideal boundary of M, we have to work on subdomains.

From the general theory of the Monge-Ampère foliation, we know that

$$\tau^{\alpha\bar{\beta}}\tau_{\alpha}\tau_{\bar{\beta}} = \|Z\|^{2} = \tau \tag{5.4}$$

where Z is the gradient vector field, therefore $\|Z\|^2 = c$, on the level set $\tau = c$. This shows that the leaves intersects the level sets transversally. In fact (5.4) implies that the angle of intersection of a leaf with $\tau = c$, measured with respect to $\partial\bar{\partial}\tau$ is equal to c, in particular it is independent of the choice of the leaf.

Now take a leaf L and consider the curve $L \cap M_c$ in $M_c = \{\tau<c\}$, $0<c<1$. Then by carefully following the argument in

Henkin [26], one easily shows that there exists a positive constant λ_c such that for any holomorphic function f on $L \cap M_c$, there exists a holomorphic extension F on M_c satisfying

$$\sup_{M_c} |F| \le \lambda_c \sup_{L\cap M_c} |f| \tag{5.4}$$

where the constant λ_c depends on

(i) diameter of M_c

(ii) Curvature of ∂M_c

(iii) transversality of L with ∂M_c

<u>Step IV</u>

(i) diam $M_c \le 2$

(ii) Second Fundamental Form of $\partial M_c = \frac{1}{\sqrt{c}}$ Id.

(iii) $\|Z\|^2 = c$

These properties are known, cf. Wong [23]

<u>Step V</u> The estimate in Step III implies that

$$K_{M_c} \le \lambda_c C_M$$

and by Step IV, the constant λ_c remains bounded as $c \to 1$. By a limiting argument, we conclude that

$$K_M \le \lambda C_M$$

where

$$\lambda = \limsup_{c \to 1} \lambda_c < \infty$$

Since on the leaves the metric $\bar{\partial}\partial \log(1-\tau)$ is -1, the leaves are images of discs thus K_M is clearly not identically zero, the same is true for C_M. The existence of non-trivial bounded holomorphic functions follows from the definition of the Caratheodory metric.

REFERENCES

[1] Bedford, E. and B.A. Taylor, The Dirichlet problem for a complex Monge-Ampère equation, Invent. Math. 37 (1976), 1-44.

[2] Bedford, E. and B.A. Taylor, Variation properties of the complex Monge-Ampère equation, I. Dirichlet Principle, Duke Math. J. 45 (1978), 375-403 II. Intrinsic Norm. Amer. J. Math. 101 (1979), 1131-1166.

[3] Burns, D., Curvature of Monge-Ampère foliations, Ann. Math. 115 (1982), 349-373.

[4] Caffarelli, L., Nirenberg, L. and J. Spruck, The Dirichlet problem for non-linear second order equations I. Monge-Ampere equation, Comm. Pure and Appl. Math. 37 (1984), 369-402.

[5] Caffarelli, L., Kohn, J., Nirenberg, L. and J. Spruck, II. Complex Monge-Ampère equation.

[6] Calabi, E., Extremal Kähler metrics, Sem. Diff. Geom. (S.T.Yau ed.) Princeton Univ. Press (1982), 259-290; Extremal Kähler metrics II, Diff. Geom. and Complex Analysis, H.E. Rauch memorial volume (I. Chavel, H.M. Farkas ed.) Springer-Verlag (1985), 95-114.

[7] Cao, H.D., Deformation of Kähler metrics to Kähler-Einstein metrics on compact Kähler manifolds, Invent. Math., 81 (1985), 359-372.

[8] Chern, S.S., Levine, H.I. and L. Nirenberg, Intrinsic norms on a complex manifold, Global Analysis in honor of Kodaira, Princeton University Press (1969), 119-139.

[9] Donaldson, S.K., A new proof of a theorem of Narasimhan and Seshadri, J. Diff. Geom. 18 (1983), 269-278.

[10] Donaldson, S.K., Anti self-dual Yang-Mills, connections over complex algebraic surfaces and stable vector bundles, Proc. London Math. Soc. 50 (1985), 1-26.

[11] Eells, J. and J.H. Sampson, Hormonic mappings of Riemannian manifolds, Amer. J. Math. 86 (1964), 109-160.

[12] Futaki, A., An obstruction to the existence of Einstein Kähler metrics, Invent. Math. 73 (1983), 437-443.

[13] Hamilton, R., Harmonic maps of manifolds with boundary, Springer Lecture Notes, No. 471 (1975).

[14] Hildebrandt, S., Harmonic mappings of Riemannian manifolds, Springer Lecture Notes No. 1161 (1984), 1-117.

[15] Kobayashi, S., Einstein-Hermitian vector bundles and stability, Global Riem Geom (Durham Symp. 1982), Ellis Horwood Ltd. (1982), 60-64.

[16] Lempert, L., Le métrique de Kobayashi et la representation des domains sur la boule, Bull. Soc. Math. France, 109 (1981), 427-474.

[17] Milgram, A and P. Rosenbloom, Harmonic forms and heat conduction I., Proc. Nat. Acad. Sci. USA Vol 37 (1951), 180-184.

[18] Okonek, C., Schneider, M. and H. Spindler, Vector bundles on complex projective spaces, Prog. in Math. 3, Birkhäuser (1980).

[19] Patrizio, G., A characterization of complex manifolds biholomorphic to a circular domain, Math. Z., 189 (1985), 343-363.

[20] Stoll, W., The characterization of strictly parabolic manifolds, Ann. Scuola Norm. Sup. Pisa, 7 (1980), 87-154.

[21] Uhlenbeck, K. and S.T. Yau, Hermitian-Einstein metrics and stable vector bundles, preprint.

[22] Wermer, J., The hull of a curve in $\mathbb{C}^n$, Ann. Math. 68 (1958), 550-561.

[23] Wong, P.M., Geometry of the complex homogeneous Monge-Ampère equation, Invent. Math. 67 (1982), 261-274.

[24] Wong, P.M., On umbilical hypersurfaces and uniformization of circular domains, Symp. Pure Math. AMS 41 (1984), 225-252.

[25] Yau, S.T., On the Ricsi curvature of a compact Kähler manifold and the complex Monge-Ampère equation I., Comm. Pure and Appl. Math. 31 (1978), 339-411.

[26] Henkin, G.M., Continuation of bounded holomorphic functions from submanifolds in general position in a strictly pseudoconvex domain, Math. USSR Izv. 6 (1972), 536-563.

Department of Mathematics
University of Notre dame
Notre Dame, IN 46556

LIOUVILLE THEOREMS

H. Wu

One of the deeper problems in analysis on complex manifolds is the study of the relationship between the existence of complete Kähler metrics with special curvature properties on a given manifold and the "size" of the ring of bounded holomorphic functions on that manifold. On the one hand, it has been recognized for some twenty years that the existence of non-constant bounded holomorphic functions is somehow tied up with negative curvature ([KO], [W1]). On the other, it is an unfortunate fact that no precise result in this direction has ever been proved. In particular, the following simple problem, which seems to have been first proposed in a slightly different form in [W1] has been open since 1967: if M is a simply connected complete Kähler manifold with sectional curvature ≤ -1, does there exist enough bounded holomorphic functions on M to holomorphically imbed M into a high-dimensional complex euclidean space? What _is_ available are some _non-existence theorems_ of which the prototype is of course the classical theorem of Liouville. The present day geometric reformulation of Liouville's theorem is the following.

Theorem 0. If a Riemann surface admits a complete Kähler metric of non-negative Gaussian curvature, then it carries no non-constant bounded holomorphic functions.

Since on $\mathbb{C}$ it is defined the canonical complete Kähler metric of zero curvature, Theorem 0 implies that $\mathbb{C}$ can carry no non-constant bounded holomorphic function, which is precisely the original theorem of Liouville. Theorem 0 will be easily seen to be a special case of the General Schwarz Lemma of S.-T. Yau ([Y3]) to be presented below. From the viewpoint of Theorem 0, the spirit of Liuoville's theorem is that on Kähler manifolds of non-negative curvature, "bounded" holomorphic mappings must be constant. It came as a surprise to the author that no precise statement of this nature has ever been recorded in the literature. The first purpose of the present article, which is partly expository and partly original research, is to prove such a theorem. To this end, define a holomorphic mapping $f : M \longrightarrow \tilde{M}$

Work supported in part by the National Science Foundation.

between two complex manifolds to be <u>bounded</u> if $f(M)$ has compact closure in $\tilde{M}$. When $\tilde{M} = \mathbb{C}$, this recovers the usual notion of a bounded holomorphic function. Deferring the definitions of the various curvatures that enter into the ensuing discussion to § 1, we now state the theorems.

<u>Theorem 1</u>. Let M be a complete Kähler manifold of non-negative Ricci curvature. Let $F : M \longrightarrow \tilde{M}$ be any bounded holomorphic mapping into a complex manifold $\tilde{M}$ such that on $\overline{f(M)}$ is defined a strictly plurisubharmonic (abbreviation: psh) function τ. Then f is a constant.

By a result of Richberg ([RI]) there is no loss of generality in assuming that the strictly psh function τ defined on $\overline{f(M)}$ is in fact C^∞. If $\tilde{M} = \mathbb{C}$, the existence of such a τ is automatic. Hence Theorem 0 is a special case of Theorem 1. For a general $\tilde{M}$, such a τ must be postulated to assure the validity of Theorem 1. Indeed, if $\tilde{M}$ is an n-dimensional manifold obtained by blowing up a point of another complex manifold N, then $\mathbb{C}^{n-1}$ is imbedded in $\tilde{M}$. The natural inclusion $\mathbb{C}^{n-1} \subseteq \tilde{M}$ then presents a counterexample to Theorem 1 when the existence of such a τ on $\overline{f(M)}$ is not assumed. Note that there are otherwise no assumptions on $\tilde{M}$.

Theorem 1 will be seen to be an immediate consequence of the following two theorems.

<u>General Schwarz Lemma ([Y3])</u>. Let $f : M \longrightarrow \tilde{M}$ be a holomorphic mapping between a complete Kähler manifold M and an arbitrary Hermitian manifold $\tilde{M}$. Suppose the Ricci curvature of M is bounded below by $-\alpha^2$ $(\alpha \geq 0, \alpha \in \mathbb{R})$, and the bisectional curvature of $\tilde{M}$ is bounded above by $-\beta^2$ $(\beta > 0, \beta \in \mathbb{R})$. Then the metrics $d\tilde{s}^2$ and ds^2 of M and $\tilde{M}$ respectively satisfy the inequality $f^* d\tilde{s}^2 \leq (\alpha^2/\beta^2) ds^2$.

<u>Theorem 2</u>. Let M be a complex manifold and let $U \subset\subset M$ (recall: this notation means U is an open subset of M with compact closure). Suppose on $\bar{U}$ is defined a C^∞ strictly psh function. Then there exists a Hermitian metric on U whose bisectional curvature is bounded above by a negative constant.

For completeness, we will present an outline of a proof of the General Schwarz Lemma below. The proof of Theorem 2 is extremely simple and it is again a surprise that such a useful fact has never

been put in print (cf. however pp. 113-114 of [GW3] for a special case).

We next turn to a discussion of the more restricted form of the Liouville theorem, namely

(*) On a complete Kähler manifold of non-negative Ricci curvavature, there are no non-constant bounded holomorphic functions.

This is an immediate consequence of Yau's General Schwarz Lemma because a bounded holomorphic function f on M may be regarded as a holomorphic mapping $f : M \longrightarrow \mathcal{D}$ into the unit disc $\mathcal{D}$ (after normalization), and $\mathcal{D}$ has the Poincaré metric of constant curvature -1. Now we wish to point out that as a theorem about holomorphic <u>functions</u>, (*) cannot be an optimal result. Indeed, since holomorphic functions are harmonic on a Kähler manifold (cf. e.g., p. 83 of [W3]), (*) is a special case of the following theorem, also due to S.-T. Yau ([Y2]):

(**) On a complete Riemannian manifold with non-negative Ricci curvature, there are no non-constant bounded harmonic functions.

(cf. Theorem E in § 4 of [W4] for a simple proof of (**).) Thus (*) is a specialization of (**) in two respects: it assumes more (by requiring the Riemannian manifold to be Kähler) and concludes less (in being able to exclude only non-constant bounded holomorphic, rather than harmonic, functions). Suppose we keep the Kähler assumption as well as the "holomorphic" conclusion in (*), then it is altogether reasonable to conjecture that something less than non-negative Ricci curvature should suffice to guarantee the correctness of (*). In the absence of any hard evidence, we merely pose the following as a question:

If M is a complete Kähler manifold such that Ricci curvature $\geq -A/r^{2+\varepsilon}$, where r is the distance function relative to a fixed point of M and A is a positive constant, does m carry non-constant bounded holomorphic functions?

On the other hand, suppose we insist on keeping the hypothesis exactly as in (*), then we would expect to derive a strong conclusion. As a motivation of what this conclusion should be, we go back to the original Liouville theorem on $\mathbb{C}$. The latter is in fact a very special case of a general theorem:

Parabolicity of $\mathbb{C}$. On $\mathbb{C}$, there are no non-constant negative subharmonic functions.

The classical proof of this fact is analytic (cf. [AS], p. 204), and it is only quite recently that a geometric proof has become available, thanks to the work of Sibony ([SB]). Adapting K. Diederich's proof of Sibony's theorem given in [W3], p. 142, we can now give a simple and elementary geometric proof of the parabolicity of $\mathbb{C}$ using only the sub-mean-value property of subharmonic functions. This proof is presented below. However, our main interest in this theorem lies in its ability to suggest the following generalization of (*):

Conjecture. On a complete Kähler manifold with non-negative Ricci curvature, there are no non-constant negative C^∞ psh functions.

In this conjecture, the requirement that the psh functions be C^∞ is only a technical concession to the general non-smoothability of continuous psh functions on a complex manifold ([F]). Note that in general, the manifolds in this conjecture are far from being Stein, e.g., $\mathbb{C} \times P_n\mathbb{C}$ in the obvious metric is such a manifold. The latter example also points to the difficulty in the verification of the conjecture, namely, the fact that one needs to exclude negative psh functions whose Levi forms are everywhere highly degenerate. On the positive side, the following result lends considerable support to the conjecture.

Theorem 3. Let M be a complete Kähler manifold with non-negative scalar curvature, then M carries no non-constant negative C^∞ psh functions which are strictly psh at a point.

While Theorem 3 does not resolve the conjecture above it is to be remarked that Theorem 3 is an honest generalization of the theorem on the parabolicity of $\mathbb{C}$ and that Theorem 3 is not a special case of the conjecture even if the latter turns out to be true. The following consequence of Theorem 3 is of special interest because for the case $M = \mathbb{C}^n$, it is related to some earlier work of [GW1] (Corollary to Theorem 2) and [Y1] (Corollary to Theorem 5).

Corollary. Let M be any complex manifold and let τ be a C^∞ psh function defined on M. Let $c \in \mathbb{R}$ and let $M_c \equiv \{x \in M : \tau(x) < c\}$. If τ is strictly psh somewhere in M_c, then M_c does not admit any complete Kähler metric of non-negative scalar curvature.

The proof of Theorem 3 requires a mild modification in the standard L^2 technique ([H] and [AV]) in that, instead of a strictly psh function, one must now work with a psh function which is only strictly psh somewhere. A key step in the proof of Theorem 3 is the following result which is of independent interest.

Theorem 4. Let M be an n-dimensional complete Kähler manifold which admits a non-constant negative C^{∞} psh function τ which is strictly psh at one point. Then M possesses non-zero L^2 holomorphic n-forms. If τ is in addition everywhere strictly psh, then M possesses a Bergman metric.

This extends part of Theorem H in [GW3] where one can also find the relevant definitions of the various terms in Theorem 4. That the assumption of τ being strictly psh somewhere is necessary is shown by the manifold $\mathcal{D} \times \mathbb{C}$. The method of proof of Theorem 4 spawns also the following related result. It seems to be the most general known result on the existence of non-constant holomorphic functions on an abstract complex manifold.

Theorem 5. Let M be a complete Kähler manifold with non-negative Ricci curvature. If M admits a C^{∞} psh function which is strictly psh at a point $p \in M$, then the holomorphic functions on M provide local coordinates at p.

To round off the picture of this discussion, we note that the theorem on the parabolicity of $\mathbb{C}$ can also be approached from the Riemannian viewpoint. In this case the problem becomes one of deciding which Riemannian manifolds would not admit non-constant negative subharmonic functions. The theorem of Cheng-Yau ([CY]), as improved by Karp ([KA]) states that if a complete Riemannian manifold admits a negative non-constant subharmonic function, then (in self-explanatory notation) there exists a positive sequence r_i, $r_i \uparrow \infty$, such that relative to a fixed point $x_0 \in M$,

$$\frac{\text{volume}(B(x_0, r_i))}{r_1^2 \log r_i} \longrightarrow \infty$$

as $i \longrightarrow \infty$. This of course implies that $\mathbb{C}$ does not admit any non-constant negative subharmonic function since in the flat metric, $\text{volume}(B(x_0,r)) = \pi r^2$. However, it can be argued that insofar as understanding the parabolicity of $\mathbb{C}$ is concerned, the approach from the point of view of psh function on Kähler manifolds is closer to the

heart of the matter.

§ 1. In this section, we prove Theorem 2 and outline in simple terms a proof of the General Schwarz Lemma of [Y3]. Theorem 1 follows from these two results because, with notation as in Theorem 1, we may take $U \subset\subset \tilde{M}$ so that $\overline{f(M)} \subset U$ and on $\bar{U}$ is defined the given strictly psh function τ. By [RI], we may take τ to be C^{∞}. Then Theorem 2 implies U possesses a hermitian metric $d\tilde{s}^2$ with bisectional curvature $\leq -\beta^2$ $(\beta < 0)$. Consider now $f : M \longrightarrow U$. In the General Schwarz Lemma, we may let $\alpha = 0$ so that $f^* d\tilde{s}^2 = 0$ and hence f is constant.

We now outline a simple proof of the General Schwarz Lemma. The gory details of this proof are given in § 6 of [W4] so that we can afford to be as intuitive as possible. First we give the definitions of the various curvature functions which enter into the preceding theorems. For the purpose of an intuitive grasp of the arguments below, it suffices to note that from an <u>operational</u> viewpoint, the importance of these functions lies in their appearance in the various <u>Weitzenböck formulas</u> which will be pointed out in due course. More precisely, if M is a Hermitian manifold and R is the curvature tensor of the canonical Hermitian connection on M (cf. e.g., pp. 70-80 of [W3]), then one can define a function H on each ordered pair (X,Y) of the tangent vectors of type $(1,0)$ at each point $x \in M$ by:

$$H(X,Y) \equiv \langle R_{X\bar{X}} Y, \bar{Y} \rangle,$$

where $\langle , \rangle$ denotes the Hermitian inner product on the tangent space M_x. The <u>bisectional curvature</u> determined by X,Y is by definition equal to zero if X or $Y = 0$, and is $H(X,Y)/|X|^2 \cdot |Y|^2$ otherwise. Thus by definition the bisectional curvature is bounded above by $-\beta^2$ if

$$H(X,Y) \leq -\beta^2 |X|^2 \cdot |Y|^2$$

for all such X,Y. The <u>Ricci tensor</u> Ric is defined by

$$\mathrm{Ric}(X,Y) \equiv \sum_{i=1}^{n} \langle R_{X_i \bar{X}_i} X, Y \rangle$$

where $\{X_1, \ldots, X_n\}$ is an orthonormal basis of vectors of type $(1,0)$ at x and $\dim M = n$. The independence of this definition from the choice of $\{X_1, \ldots, X_n\}$ is elementary. The <u>Ricci curvature</u> determined

by a unit vector X of type $(1,0)$ is by definition the number $\mathrm{Ric}(X,X)$. In particular

$$\mathrm{Ric}(X,X) \equiv \sum_{i=1}^{n} H(X_i, X),$$

and the Ricci curvature is bounded below by $-\alpha^2$ if

$$\mathrm{Ric}\,(Y,Y) \geq -\alpha^2 |Y|^2$$

for all vectors Y of type $(1,0)$. The <u>scalar curvature</u> σ is the trace of Ric, i.e.,

$$\sigma \equiv \sum_i \mathrm{Ric}(X_i, X_i),$$

where $\{X_1,\ldots,X_n\}$ is any orthonormal basis of vectors of type $(1,0)$.

Now to the proof of the General Schwarz Lemma. Given $f : M \longrightarrow \tilde{M}$, which is holomorphic, define a function $e(f) : M \longrightarrow [0,\infty)$ as follows. If $x \in M$ and $\{X_1,\ldots,X_n\}$ is an orthonormal basis of vectors of type $(1,0)$ at x, then

$$e(f)(x) \equiv \sum_{i=1}^{n} |df(X_i)|^2.$$

It follows immediately from the definition that $f^*d\tilde{s}^2 \leq e(f)ds^2$. Therefore, it suffices to prove that $e(f) \leq \alpha^2/\beta^2$ on M. <u>Suppose</u> we have the ideal situation that $e(f)$ achieves an absolute maximum at $z \in M$. Then the second derivatives of $e(f)$ at z are all non-positive; in particular $-\Delta e(f)(z) \leq 0$, where the sign of the Laplacian Δ is (traditionally) chosen so that on $\mathbb{R}^n$, $\Delta = -\sum_i \partial^2/(\partial x^i)^2$. The Weitzenböck formula of $e(f)$, in the presence of the curvature assumptions $\overset{M}{\mathrm{Ric}} \geq -\alpha^2$ and $H^{\tilde{M}} \leq -\beta^2$, reads

$$(\S) \qquad -\Delta e(f) \geq 2|Ddf|^2 - 2\alpha^2 e(f) + 2\beta^2 e(f)^2.$$

Thus at z,

$$0 \geq -\alpha^2 e(f)(z) + \beta^2 e(f)^2(z)$$

or equivalently, $e(f)(z) \leq \alpha^2/\beta^2$. Since $e(f) \leq e(f)(z)$ we have proved $e(f) \leq \alpha^2/\beta^2$.

However, $e(f)$ does <u>not</u> in general achieve an absolute maximum on the non-compact manifold M. So, following the classical argument of Ahlfors ([A]), we "force" $e(f)$ to achieve an absolute maximum so

that the above line of reasoning remains applicable. The way to do this is as follows. Fix an arbitrary point $x_0 \in M$ and let r be the distance function on M relative to x_0. Let $B(a)$ be the ball of radius a around x_0, i.e., $B(a) \equiv \{x \in M : r(x) \leq a\}$, and define a modified version of $e(f)$, $\Phi : B(a) \longrightarrow \mathbb{R}$, by $\Phi = (a^2-r^2)^2 e(f)$. Now of course $\Phi \equiv 0$ on $\partial B(a)$ (boundary of $B(a)$) so that Φ necessarily achieves an absolute maximum at an interior point z of $B(a)$. We may clearly assume $\Phi(z) > 0$ so that $\log \Phi$ is well-defined near z. Standard tricks in a situation of this type allow us to treat r, and hence also Φ, as a C^∞ function near z. Since $\log \Phi$ also achieves an absolute maximum at z, we have $d \log \Phi(z) = 0$ and $-\Delta \log \Phi(z) \leq 0$, from which after a standard calculation using (§) we obtain:

$$e(f)(z) \leq \frac{\alpha^2}{\beta^2} + \frac{\varphi(a)}{(a^2-r^2)^2(z)} ,$$

where $\varphi(a)$ is a <u>cubic</u> polynomial in a. By the definition of Φ, this is equivalent to

$$\Phi(z) \leq \frac{\alpha^2}{\beta^2} (a^2-r^2)^2(z) + \varphi(a) \leq \frac{\alpha^2}{\beta^2} a^4 + \varphi(a).$$

Since $\Phi(x_0) \leq \Phi(z)$,

$$e(f)(x_0) = \frac{1}{a^4} \Phi(x_0) \leq \frac{\alpha^2}{\beta^2} + \frac{\varphi(a)}{a^4} .$$

Letting $a \longrightarrow \infty$, we obtain $e(f)(x_0) \leq \alpha^2/\beta^2$. Since x_0 is arbitrary, we have proved $e(f) \leq \alpha^2/\beta^2$, and the General Schwarz Lemma is proved. □

We now prove Theorem 2 by a straightforward computation. Notation as in Theorem 2, let $U \subset\subset W \subset\subset M$, such that on W we have a C^∞ strictly psh function $\varphi : W \longrightarrow \mathbb{R}$. Pick any Hermitian metric G on W and let $G_0 \equiv \varphi G$. Since we may assume $\varphi > 0$ (by replacing φ with e^φ if necessary) G_0 is also a Hermitian metric on W. Define functions $\mathcal{B}, \mathcal{B}_0, \lambda : W \longrightarrow \mathbb{R}$ by

$\mathcal{B}(x) = \max\{$bisectional curvature of G at $x\}$

$\mathcal{B}(x) = \max\{$bisectional curvature of G_0 at $x\}$

$\lambda(x) = \min\{$eigenvalues of the Levi form of $\log \varphi$ at x with respect to $G\}$.

Then using the local coordinate expression of the bisectional curvature, we shall derive the following inequality:

$$(\S\S) \qquad \mathcal{B}_0 \le \frac{1}{\varphi}\left(\mathcal{B} - \frac{1}{2}\lambda\right)$$

(compare the corresponding inequality for the holomorphic sectional curvature on p. 114 of [GW3]). Given two vectors X, Y of type $(1,0)$ at $p \in M$, we shall denote the bisectional curvature of G and G_0 determined by X,Y by $B(X,Y)$ and $B_0(X,Y)$ respectively. In the following calculations, we will assume that X and Y are linearly independent (over $\mathbb{C}$); the case that X and Y are proportional is in fact easier. let $\{z^1,\ldots,z^n\}$ be coordinate functions around p so that $z^i(p) = 0$ for $i = 1,\ldots,n$ and so that

$$X = \partial/\partial z^1(p)$$
$$Y = \partial/\partial z^2(p).$$

Write $G = \sum_{i,j} g_{ij}dz^i d\bar{z}^j$. Then $G_0 = \sum_{i,j} (\varphi g_{ij})dz^i d\bar{z}^j$. The components of the curvature tensor R of G are:

$$(1) \qquad R_{klij} = -\frac{\partial^2 g_{jl}}{\partial z^i \partial \bar{z}^j} + \sum_{m,h} g^{hm} \frac{\partial g_{kh}}{\partial z^i} \frac{\partial g_{ml}}{\partial \bar{z}^j}$$

where, as usual, $\sum_k g_{ik}g^{kj} = \delta_i^j$ (see p. 1104 of [W2]). In general, if $X = \sum_i X^i \frac{\partial}{\partial z^i}(p)$ and $Y = \sum_i Y^i \frac{\partial}{\partial z^i}(p)$, then the definition of $B(X,Y)$ gives

$$B(X,Y) = 2 \sum_{k,l,i,j} R_{klij}\, X^k \bar{X}^l Y^i \bar{Y}^j.$$

With the choice of $\{z^1,\ldots,z^n\}$ as above, we have then

$$(2) \qquad B(X,Y) = \frac{2R_{1122}}{g_{11}g_{22}}(p).$$

Similarly, with the curvature tensor of G_0 denoted by R^0, we see that

$$(3) \qquad B_0(X,Y) = \frac{2R^0_{1122}}{\varphi^2 g_{11}g_{22}}(p).$$

Now by Lemma 3 of [W2], we may assume $\{z^1,\ldots,z^n\}$ to satisfy the additional requirement that

$$\frac{\partial g_{ij}}{\partial z^2}(p) = 0$$

for all $i,j = 1,\ldots,n$. Then a direct calculation of (2) and (3) using (1) yields:

$$B(X,Y) = \frac{-2}{g_{11}g_{22}} \frac{\partial^2 g_{11}}{\partial z^2 \partial \bar{z}^2} (p),$$

$$B_0(X,Y) = \frac{1}{\varphi} B(X,Y) - \frac{2}{\varphi g_{22}} \frac{\partial^2 \log \varphi}{\partial z^2 \partial \bar{z}^2} (p).$$

Now the Levi form of $\log \varphi$ is

$$L \log \varphi \equiv 4 \sum_{ij} \frac{\partial^2 \log \varphi}{\partial z^i \partial \bar{z}^j} dz^i d\bar{z}^j.$$

Hence

$$B_0(X,Y) = \frac{1}{\varphi}\left\{B(X,Y) - \frac{1}{2} L \log \varphi \left(\frac{Y}{|Y|}, \frac{Y}{|Y|}\right)\right\},$$

from which (§§) easily follows.

Finally, let β be the maximum of $\mathcal{B}$ over $\bar{U}$. Replacing φ by $e^{k\varphi}$ for a large constant k if necessary we may assume that $\lambda|_{\bar{U}} \geq 1+2|\beta|$. Thus over $\bar{U}$, (§§) implies

$$\mathcal{B}_0 \leq -1/(\max_{\bar{U}} \varphi),$$

and this proves Theorem 2. □

§ 2. In this section, we give a simple geometric proof of the theorem on the parabolicity of $\mathbb{C}$ by specializing the argument of Sibony and Diederich to our situation ([SB] and p. 142 of [W3]). We should begin with a general remark about Sibony's theorem, which states that if τ is a negative psh function on a complex manifold M and τ is strictly psh in the neighborhood of a point $p \in M$, then for some positive constant A, the inequality $F(p,X) \geq A|X|$ is valid for all vectors X in a neighborhood of p, where F is the Kobayashi-Royden differential metric of M ([RO]). This theorem immediately implies the parabolicity of $\mathbb{C}$, in the following way. In complex dimension 1, psh functions are precisely the subharmonic functions and strict plurisubharmonicity is the same as strict subharmonicity. Suppose on $\mathbb{C}$ is defined a negative non-constant subharmonic function τ. By convolution with a fixed C_0^∞ kernel, τ may be assumed to be C^∞. Since $d\tau(p) \neq 0$ for some $p \in \mathbb{C}$, the function $\varphi \equiv e^\tau - 1$ is C^∞, negative and $-\Delta\varphi(p) > 0$ (for uniformity with the rest of this paper, we are forced to adopt the monstrous sign convention that $\Delta = -\left(\frac{\partial^2}{\partial x^2} + \frac{\partial^2}{\partial y^2}\right)$ in $\mathbb{C}$). By Sibony's theorem, $F(p,X) > 0$ for all non- zero vectors near p. But it is trivial (and well known) that $F \equiv 0$ on $\mathbb{C}$. This contradiction completes the proof. What we

do now is to adapt K. Diederich's simple proof of the preceding theorem to directly prove the parabolicity of $\mathbb{C}$ without ever mentioning the Kobayashi-Royden metric. This argument is hopefully more digestible to the geo- meters than the classical argument.

So suppose τ is a negative non-constant subharmonic function on $\mathbb{C}$. We have already observed that it is permissible to assume τ is C^∞. Moreover, the reasoning in the preceding paragraph shows that we may further assume that $-\Delta\tau(0) > 0$ (use a translation if necessary). Thus τ is strictly subharmonic in a neighborhood of the origin 0, and after a dilation of $\mathbb{C}$, we may even assume that $-\Delta\tau > 0$ on the disc $\mathcal{D}_2$ of radius 2. We now show that there exists a positive constant Γ,Γ depending only on τ, such that if $f : \bar{\mathcal{D}} \longrightarrow \mathbb{C}$ ($\bar{\mathcal{D}}$ = the closed unit disc) is any holomorphic mapping satisfying $f(0) = 0$, then $|f'(0)| < \Gamma$. This is of course absurd and the parabolicity of $\mathbb{C}$ would follow. The constant Γ is defined as follows. Let h be a fixed function on $\mathbb{C}$ such that $h \le 0$ and is C^∞ in $\mathbb{C} - \{0\}$ such that

$$h = \begin{cases} \log |z|^2 & \text{on the disc of radius } 1/2, \\ 0 & \text{on } \mathbb{C} - \mathcal{D}. \end{cases}$$

Let c be a positive constant, so small that $\tau_0 \equiv \tau + ch$ is subharmonic on $\mathbb{C}$ (this makes use of the fact that $-\Delta\tau > 0$ on $\mathcal{D}_2$). Note that $\tau_0 < 0$. Now the definition of Γ is: $\Gamma \equiv \exp \frac{|\tau(0)|}{2c}$. To show that Γ has the property stated above, let $f : \bar{\mathcal{D}} \longrightarrow \mathbb{C}$ be holomorphic and satisfy $f(0) = 0$, $f'(0) \neq 0$. The composite function $\tau_0(f)$ is of course subharmonic on $\bar{\mathcal{D}}$. Hence the function $g \equiv \tau_0(f) - c \log |z|^2$ is also subharmonic on $\bar{\mathcal{D}}$. Near 0,

$$g(z) = \tau(f(z)) + c \log \left|\frac{f(z)}{z}\right|^2,$$

so that $g(0) = \tau(0) + 2c \log |f'(0)|$. Since g is subharmonic the sub-mean-value theorem gives:

$$g(0) \le \frac{1}{2\pi} \int_0^{2\pi} g(e^{i\theta})d\theta = \frac{1}{2\pi} \int_0^{2\pi} \tau_0(f(e^{i\theta}))d\theta < 0.$$

This implies $\tau(0) + 2c \log |f'(0)| < 0$, whence $|f'(0)| < \Gamma$, as claimed. □

§ 3. This section proves Theorems 3 and 4. The proof of Theorem 5 is a straightforward modification of the proof of Theorem 4 and will therefore be omitted.

We first show how to deduce Theorem 3 from Theorem 4. Assumption as in Theorem 3, suppose M carries a negative non-constant C^∞ psh function τ which is strictly psh at one point. By Theorem 4, M then carries a non-zero L^2 holomorphic n-form ψ ($n = \dim M$). It has been observed in [GW4], Theorem 2, that under the assumption of non-negative scalar curvature (so that the sum of all the eigenvalues of the Ricci tensor at each point is non-negative), $|\psi|$ is constant. The simple reasoning behind this assertion is as follows. Since ψ is an L^2 n-form, the function $|\psi|$ is in L^2 with respect to the canonical Borel measure associated with the given Kähler metric on M. By a standard Weitzenböck formula (cf. pp. 72-72 of [GW4] or Theorem 3 in § 4 of [W4]), M having non-negative scalar curvature implies $|\psi|$ is subharmonic. Thus $|\psi|$ is a non-negative subharmonic function in L^2. By a theorem of Yau ([Y1], see also Theorem B in § 4 of [W4] for a simplified exposition), on a complete Riemannian manifold, a non-constant non-negative subharmonic function is never in L^q for $1 < q < \infty$. Thus $|\psi|$ is constant, as claimed. Let $|\psi| \equiv A$, $A \in \mathbb{R}$. Since ψ is non-zero, $A > 0$. Hence

$$\text{volume}(M) = \frac{1}{A^2}\int_M A^2 = \frac{1}{A^2}\int_M |\psi|^2 < \infty.$$

Now recall that we also have a negative non-constant psh function τ on M. Then e^τ is a non-negative psh (and hence subharmonic) function on M. Since $e^{2\tau} < 1$, $\int_M e^{2\tau} < \int_M 1 = \text{volume}(M) < \infty$. By Yau's theorem again, e^τ must be constant, which contradicts the non-constancy of τ. □

It remains to prove Theorem 4. This is nothing more than a collage of ideas of [GW3] (on pp. 149-150), of Siu and Yau (p. 508 of [SI]), and above all, of the standard L^2 technique as presented in Chapter IV of [H]. Let τ be the given C^∞ psh function on M which is negative and is strictly psh at a point. Then there is a C^∞ non-negative function $\alpha : M \longrightarrow [0,\infty)$ such that $\partial\bar\partial\tau \geq \alpha$ (in the sense that for all vector fields X of type $(1,0)$, $\partial\bar\partial\tau(X,\bar X) \geq \alpha|X|^2$), and such that $\alpha(p) > 0$ for some $p \in M$. Let $\{z^1,\ldots,z^n\}$ be complex coordinate functions around p so that $z^i(p) = 0$ for

$i = 1,\ldots,n$. Let f be a C^∞ form of type $(n,0)$ such that $f = dz^1 \wedge \ldots \wedge dz^n$ in a neighborhood of p and such that

$$\overline{\text{supp } f} \subset (\text{supp } \alpha)^0,$$

where supp stands for support, and the right hand side means the interior of supp α. Define a C^∞ psh function $\psi : M \longrightarrow \mathbb{R}$ with the property that $\psi \le k$ for a positive constant k, and such that

$$\psi = (2n+1) \log \left(\sum_{i=1}^{n} |z^i|^2 \right) + (C^\infty \text{ function})$$

in a neighborhood of p. Such a ψ can be constructed as follows. Let h be a C^∞ function, $h \equiv 1$ in $\left\{ \sum_{i=1}^{n} |z^i|^2 < \varepsilon \right\}$ and $h \equiv 0$ in $\left\{ \sum_{i=1}^{n} |z^i|^2 \ge 2\varepsilon \right\}$ where ε is chosen to be so small that $\alpha > 0$ on the set $\left\{ \sum_{i=1}^{n} |z^i|^2 \le 3\varepsilon \right\}$. Now by definition

$$\psi = c_0 \tau + (2n+1) h \log \sum_{i} |z^i|^2,$$

where c_0 is a positive constant chosen to be so large that $\partial\bar\partial\psi > 0$ in $\left\{ \sum_{i=1}^{n} |z^i|^2 \le 2\varepsilon \right\}$. With this choice of c_0 fixed, ψ clearly has the requisite properties. By the usual reasoning in standard L^2 theory, we try to find an $(n,0)$-form u such that $\bar\partial u = \bar\partial f$ and such that $u(p) \ne f(p)$. This will guarantee that $u - f$ is a non-zero $(n,0)$-form and that $\bar\partial(u-f) = 0$ so that $u - f$ is holomorphic. Finally, by making good use of the special properties of τ and ψ, we will also show that $u - f$ is also in L^2.

To find such a u, let $\lambda = \tau + \psi$ and for any form φ, define

$$\|\varphi\|_\lambda^2 \equiv \int_M |\varphi|^2 e^{-\lambda}$$

and let $L^2_{(n,q)}(M,\lambda)$ be the Hilbert space which is the completion of C_0^∞ (n,q)-forms ($C_0^\infty \equiv C^\infty$ with compact support) with respect to this norm $\|\cdot\|_\lambda^2$. As in [H], Chapter IV, we denote by T,S the closures of the densely defined operator $\bar\partial$ on $L^2_{(n,0)}(M,\lambda)$ and $L^2_{(n,1)}(M,\lambda)$ respectively:

$$T : L^2_{(n,0)}(M,\lambda) \longrightarrow L^2_{(n,1)}(M,\lambda)$$

$$S : L^2_{(n,1)}(M,\lambda) \longrightarrow L^2_{(n,2)}(M,\lambda).$$

Also, denote the formal adjoints of T and S by T^*, S^*, and their domains of definition by $\mathcal{D}_T, \mathcal{D}_{T^*}$, etc. For now, we will ignore the fact that λ has a singularity at p and will treat it as though it were a C^∞ psh function on M. Then $\partial\bar{\partial}\lambda \geq \alpha$, and the usual Weitzenböck formula (e.g., WF VI in § 2 of [W4]) on $(n,1)$-forms implies that for every C_0^∞ $(n,1)$-form ξ,

$$(\#) \qquad \|S\xi\|_\lambda^2 + \|T^*\xi\|_\lambda^2 \geq \|D\xi\|_\lambda^2 + \int_M \alpha|\xi|^2 e^{-\lambda},$$

where $D\xi$ denote the covariant differential of ξ. By the density of C_0^∞ forms in $\mathcal{D}_{T^*} \cap \mathcal{D}_S$ with respect to the graph norm $\|\xi\|_\lambda^2 + \|S\xi\|_\lambda^2 + \|T^*\xi\|_\lambda^2$, (cf. Lemma 5.2.1. of [H], or [V]; cf. also the discussion on p. 89 of [GW2]), (#) is actually valid for every $\xi \in \mathcal{D}_{T^*} \cap \mathcal{D}_S$. In particular

$$(\#\#) \qquad \|T^*\xi\|_\lambda^2 \geq \int_M \alpha|\xi|^2 e^{-\lambda}$$

for all $(n,1)$-forms ξ such that $\xi \in D_{T^*}$ and $S\xi = 0$.

Now let $\zeta \equiv \bar{\partial}f$, f being the $(n,0)$-form defined above. Then

$$\overline{\operatorname{supp}\zeta} \subset (\operatorname{supp}\alpha)^0 - \left\{\sum_i |z^i|^2 < \varepsilon\right\},$$

where we have also harmlessly assumed that in fact $f = dz^1 \wedge \ldots \wedge dz^n$ in $\left\{\sum_i |z^i|^2 < \varepsilon\right\}$. this implies that if we define

$$A^2 \equiv \int_{\operatorname{supp}\alpha} \frac{1}{\alpha}|\zeta|^2 e^{-\lambda},$$

then $A^2 < \infty$. We claim:

$$(\dagger) \qquad |\langle\langle\zeta,\xi\rangle\rangle_\lambda| \leq A\|T^*\xi\|_\lambda^2$$

for all $\xi \in \mathcal{D}_{T^*}$, where $\langle\langle\xi,\xi\rangle\rangle_\lambda \equiv \|\xi\|_\lambda^2$. To prove $(\dagger)$, note that if $\xi \in \ker S$ (the kernel of S), then by the Schwarz inequality,

$$\begin{aligned}
|\langle\langle\zeta,\xi\rangle\rangle_\lambda|^2 &= \left|\int_{\operatorname{supp}\alpha} \langle\zeta,\xi\rangle e^{-\lambda}\right|^2 \\
&\leq \int_{\operatorname{supp}\alpha} \frac{1}{\alpha}|\zeta|^2 e^{-\lambda} \cdot \int_{\operatorname{supp}\alpha} \alpha\,|\xi|^2 e^{-\lambda} \\
&= A^2 \int_M \alpha\,|\xi|^2 e^{-\lambda} \\
&\leq A^2\|T^*\xi\|_\lambda^2,
\end{aligned}$$

by (#). Thus (†) holds for $\mathcal{D}_{T^*} \cap \ker S$. The fact that (†) also holds for al $\xi \in \mathcal{D}_{T^*}$ such that $\xi \perp \ker S$ is trivial (because $\zeta \in \ker S$). Now, let ξ be an arbitrary element of $\mathcal{D}_{T^*}$. Write $\xi = \xi' + \xi^{\perp}$, where $\xi' \in \ker S$ and $\xi^{\perp} \perp \ker S$ (this is possible because $\ker S$ is always closed). By general operator theory, the orthogonal complement $(\ker S)^{\perp}$ of $\ker S$ satisfies $(\ker S)^{\perp} \subset \ker T^*$, so that $T^*\xi^{\perp} = 0$ and in particular $\xi^{\perp} \in \mathcal{D}_{T^*}$. Thus also $\xi' \in \mathcal{D}_{T^*}$. Hence

$$\begin{aligned} |\langle\langle\zeta,\xi\rangle\rangle_\lambda|^2 &= |\langle\langle\zeta,\xi'\rangle\rangle_\lambda| \\ &\le A\|T^*\xi'\|_\lambda^2 \\ &= A\|T^*(\xi'+\xi^{\perp})\|_\lambda^2 \\ &= A\|T^*\xi\|_\lambda^2, \end{aligned}$$

and (†) is completely proved.

Let R_{T^*} be the image of T^* in $L^2_{(n,0)}(M,\lambda)$. Define a linear functional $F : \bar{R}_{T^*} \longrightarrow \mathbb{C}$ by $F(T^*\xi) \equiv \langle\langle\zeta,\xi\rangle\rangle_\lambda$ for every $\xi \in \mathcal{D}_{T^*}$. We must first show F is well-defined, i.e., if $T^*\xi = 0$ for some $\xi \in \mathcal{D}_{T^*}$, then necessarily $\langle\langle\zeta,\xi\rangle\rangle_\lambda = 0$. Indeed, if $\xi \in \mathcal{D}_{T^*} \cap \ker S$, then $T^*\xi = 0 \Rightarrow \int_M \alpha|\xi|^2 e^{-\lambda} = 0, \Rightarrow \operatorname{supp}\xi \subset M - \operatorname{supp}\alpha \subset M - \operatorname{supp}\zeta$, $\Rightarrow \langle\langle\zeta,\xi\rangle\rangle_\lambda = \int_M \langle\zeta,\xi\rangle e^{-\lambda} = 0$, as desired. The case of a general $\xi \in \mathcal{D}_{T^*}$ can be handled in exactly the same way as in the proof of (†). So F is well defined. Now by (†), we have

(††) $$|F(T^*\xi)| \le A\|T^*\xi\|_\lambda$$

for every $\xi \in \mathcal{D}_{T^*}$. Thus F is a bounded linear functional and the Riesz representation theorem implies that there exists a $u \in \bar{R}_{T^*}$ such that $F(T^*\xi) = \langle\langle u,T^*\xi\rangle\rangle_\lambda$ for all $\xi \in \mathcal{D}_{T^*}$. By the definition of F, $\langle u,T^*\xi\rangle\rangle = \langle\langle\zeta,\xi\rangle\rangle_\lambda$ for every $\xi \in \mathcal{D}_{T^*}$, so that $\langle\langle Tu,\xi\rangle\rangle_\lambda = \langle\langle\zeta,\xi\rangle\rangle_\lambda$ for all $\xi \in \mathcal{D}_{T^*}$ and hence $Tu = \zeta$. Changing notation, we get $\bar{\partial}u = \bar{\partial}f$. Note that since $\|u\|_\lambda = \|F\|$, (††) implies that

(‡) $$\int_M |u|^2 e^{-\lambda} \le \int_{\operatorname{supp}\alpha} \frac{1}{\alpha}|\zeta|^2 e^{-\lambda} < \infty.$$

Now we pause to take care of the singularity of λ at the point

p. The argument is based on the proof of Theorem 4.4.2 in [H]. We define the regularized version of ψ by letting

$$\psi_\varepsilon \equiv c_0\tau + (2n+1)h \log (\varepsilon + \sum_i |z^i|^2)$$

for each $\varepsilon > 0$. Note that each ψ_ε is C^∞, psh and most importantly, $\psi_\varepsilon \downarrow \psi$ as $\varepsilon \downarrow 0$. The preceding argument with ψ replaced by ψ_ε now becomes completely rigorous. In other words, with $\lambda_\varepsilon \equiv \tau + \psi_\varepsilon$ for each ε, there is a $u_\varepsilon \in L^2_{(n,0)}(M, \lambda_\varepsilon)$ such that $\bar\partial u_\varepsilon = \bar\partial f$ and

$$(\ddagger\ddagger) \qquad \begin{aligned} \int_M |u_\varepsilon|^2 e^{-\lambda_\varepsilon} &\le \int_{\text{supp }\alpha} \frac{1}{\alpha} |\zeta|^2 e^{-\lambda_\varepsilon} \\ &\le \int_{\text{supp }\alpha} \frac{1}{\alpha} |\zeta|^2 e^{-\lambda}, \end{aligned}$$

where the last inequality makes use of the fact that $\lambda \le \lambda_\varepsilon$. By the definition of λ and λ_ε, there is a $k_1 > 0$ so that for all $\varepsilon < 1$, $\lambda_\varepsilon < k_1$ on M. Hence on any compact set K of M,

$$\begin{aligned} \int_K |u_\varepsilon|^2 = e^{k_1} \int_K |u_\varepsilon|^2 e^{-k_1} &\le e^{k_1} \int_M |u_\varepsilon|^2 e^{-\lambda_\varepsilon} \\ &\le e^{k_1} \int_{\text{supp}} \frac{1}{\alpha} |\zeta|^2 e^{-\lambda}, \end{aligned}$$

where we have made use of $(\ddagger\ddagger)$. Thus on each compact set $K \subset M$, $\{u_\varepsilon\}$ is a bounded set in $L^2(K)$ for all $\varepsilon < 1$ so that on each K, there is a subsequence which also can be denoted by u_ε so that $u_\varepsilon \longrightarrow u$ <u>weakly</u> to a $u \in L^2(K)$. By letting K exhaust M and by the diagonal process, we obtain a function u on M which is locally in L^2. Since $\bar\partial u_\varepsilon \longrightarrow \bar\partial u$, we see that $\bar\partial u = \bar\partial f$. Moreover, on each compact set K, $u_\varepsilon e^{-\lambda_\varepsilon/2} \longrightarrow u e^{-\lambda/2}$ weakly, so that $\limsup_{\varepsilon \to 0} \int_K |u_\varepsilon|^2 e^{-\lambda_\varepsilon} \ge \int_K |u|^2 e^{-\lambda}$. By $(\ddagger\ddagger)$,

$$\int_K |u|^2 e^{-\lambda} \le \int_{\text{supp }\alpha} \frac{1}{\alpha} |\zeta|^2 e^{-\lambda}.$$

Letting K exhaust M we see that $(\ddagger)$ is now valid for λ as defined above, i.e., $\lambda = \tau + \psi$ and $\psi = c_0\tau + (2n+1)h \log \sum_i |z^i|^2$.

Thus we have $\bar\partial(u-f) = 0$ and $u - f$ must be a holomorphic n-form. It is not identically zero because $f(p) = (dz^1 \wedge \ldots \wedge dz^n)(p) \neq$

0 while $u(p) = 0$, and the latter is seen as follows (standard argument): by the definition of ψ, near the point p,

$$e^{-\lambda} = \frac{C^{\infty} \text{ function}}{\left(\sum_i |z^i|^2\right)^{2n+1}}$$

which blows up at p to the order r^{4n+2} (where $r^2 \equiv \sum_i |z^i|^2$), so that (‡) implies that u must vanish at p at least to order $(2n+3)$. Finally, to see that $(u-f)$ is in L^2, note that f has compact support so that it suffices to prove that u is in L^2. Recall that $\psi < k$ on M so that $\lambda = \psi + \tau < k$, which implies

$$\int_M |u|^2 < e^k \int_M |u|^2 e^{-\lambda} < \infty,$$

again by (‡). This then completes the proof of the first part of Theorem 4. The proof of the second assertion follows from the argument above together with the argument in the proofs of Propositions 8.9 and 8.14 in [GW3]. Briefly, replacing the coefficient $(2n+1)$ in the definition of ψ by m, we can insure that u must vanish at p at elast in order $2m - (2n-1)$. Replacing f by a form which near p equals $(z^{m_1} z^{m_2} \ldots z^{m_n}) dz^1 \wedge \ldots \wedge dz^n$ (m_i being non-negative integers), we can then insure that the L^2 holomorphic n-form $u - f$ vanishes at p to any prescribed order. In particular, if K denotes the Bergman kernel restricted to the diagonal of $M \times M$, this guarantees that $\partial\bar\partial \log K(p) \neq 0$. If τ is everywhere strictly psh, $\partial\bar\partial \log K$ is nowhere zero, which is equivalent to saying that the Bergman metric exists.

References

[A] L.V. Ahlfors, An extension of Schwarz's lemma, Trans. Amer. Math. Soc. 43 (1938), 359-364.

[AS] L.V. Ahlfors and L. Sario, Riemann Surfaces, Princeton University Press, 1960.

[AV] A. Andreotti and E. Vesentini, Carleman estimates for the Laplace-Beltrami equation on complex manifolds, Inst. Hautes Etudes Sci. Publ. Math. 25 (1965), 81-138.

[CY] S.Y. Cheng and S.T. Yau, Differential equations on Riemannian manifolds and their geometric applications, Comm. Pure Appl. Math. 28 (1975), 333-354.

[F] J.-E. Fornaess, Regularization of plurisubharmonic functions, Math. Ann. 259 (1982), 119-123.

[GW1] R.E. Greene and H. Wu, Curvature and complex analysis II, Bull. Amer. Math. Soc. 78 (1972), 866-870.

[GW2] ————————, Analysis on non-compact Kähler manifolds, Proc. Symp. Pure Math. Vol. 30, Part II (Several Complex Variables), Amer. Math. Soc., 1977., 69-100.

[GW3] ————————, Function Theory on Manifolds Which Possess a Pole, Springer-Verlag Lecture Notes in Mathematics 699, 1979.

[GW4] ————————, Harmonic forms on non-compact Riemannian and Kähler manifolds, Mich. Math. J. 28 (1981), 63-81.

[H] L. Hörmander, An Introduction to Complex Analysis in Several Variables, Van Nostrand, 1966.

[KA] L. Karp, Subharmonic functions on real and complex manifolds, Math. Z. 179 (1982), 535-554.

[KO] S. Kobayashi, Hyperbolic Manifolds and Holomorphic Mappings, Marcel Dekken, New York, 1970.

[RI] R. Richberg, Stetige streng pseudoconvexe Funktionen, Math. Ann. 175 (1968), 257-286.

[RO] H.L. Royden, Remarks on the Kobayashi metric, Several Complex Variables II, Springer-Verlag Lecture Notes in Mathematics 185, 1971, 125-137.

[SB] N. Sibony, A class of hyperbolic manifolds, Recent Developments in Several Complex Variables, Annals of Mathematics Studies 100, Princeton University Press, 1981, 357-372.

[SI] Y.-T. Siu, Pseudoconvexity and the problem of Levi, Bull. Amer. Math. Soc. 84 (1978), 481-512.

[V] E. Vesenti, Lectures on Levi Convexity of Complex Manifolds and Cohomology Vanishing Theorems, Tata Inst. Fund. Res. Lectures on Math. 39, Bombay, India, 1967.

[W1] H. Wu, Normal families of holomorphic mappings, Acta. Math. 119 (1967), 193-233.

[W2] ———————, A remark on holomorphic sectional curvature, Indiana Univ. Math. J. 22 (1973), 1103-1108.

[W3] ———————, Function theory on non-compact Kähler manifolds, Complex Differential Geometry, DMV Seminar Band 3, Birkhäuser, 1983, 67-158.

[W4] ———————, The Bochner Technique in Differential Geometry, Mathematical Reports, Hardwood Academic Publishers, London, to appear.

[Y1] S.-T. Yau, Some function-theoretic properties of complete Riemannian manifolds and their applications to geometry, Indiana Univ. Math. J. 25 (1976), 659-670.

[Y2] ———————, Harmonic function on complete Riemannian manifolds, Comm. Pure Appl. Math. 28 (1975), 201-228.

[Y3] ———————, A general Schwarz lemma for Kähler manifolds, Amer. J. Math. 100 (1978), 197-203.

University of California
Berkeley, California 94720
U.S.A.

A FINAL WORD

Now that this third volume of the Proceedings of the Special Year in Complex Analysis is really complete I would like once more to thank the technical typists of our department, Stephanie Smith, Virginia Sauber and Jaya Nagendra for the excellent job they have done. In so doing I reflect also the praise they received in many of the letters from the contributors to these Proceedings.

I would also like to add that due to the generous grant from the Argonne Universities' Association Trust Fund and the support of our Department, the first semester of the academic year 1986-87 saw some more activities of the Special Year. Two were specially noteworthy.

On November 13 & 14, a beautiful mini-conference on the role of Complex Analysis in Geometry and Analysis, was held to mark the retirement of Professor Maurice Heins. The speakers were A. Baernstein II, F. W. Gehring, B. Rodin and A. Weitsman.

As a fitting conclusion to a special year in Complex Analysis, the University of Maryland, to commemorate the fiftieth anniversary of the Fields Medal award, decided to confer a Doctorate Honoris Causa on December 23, 1986 to a giant in the subject, Lars V. Ahlfors, whose work has deeply influenced all the areas covered in these Proceedings.

Carlos Berenstein

LECTURE NOTES IN MATHEMATICS
Edited by A. Dold and B. Eckmann

Some general remarks on the publication of proceedings of congresses and symposia

Lecture Notes aim to report new developments - quickly, informally and at a high level. The following describes criteria and procedures which apply to proceedings volumes.

1. One (or more) expert participant(s) of the meeting should act as the responsible editor(s) of the proceedings. They select the papers which are suitable (cf. points 2, 3) for inclusion in the proceedings, and have them individually refereed (as for a journal). It should not be assumed that the published proceedings must reflect conference events faithfully and in their entirety. Contributions to the meeting which are not included in the proceedings can be listed by title. The series editors will normally not interfere with the editing of a particular proceedings volume - except in fairly obvious cases, or on technical matters, such as described in points 2, 3. The names of the responsible editors appear on the title page of the volume.

2. The proceedings should be reasonably homogeneous (concerned with a limited area). For instance, the proceedings of a congress on "Analysis" or "Mathematics in Wonderland" would normally not be sufficiently homogeneous.

   One or two longer survey articles on recent developments in the field are often very useful additions to such proceedings - even if they do not correspond to actual lectures at the congress. An extensive introduction on the subject of the congress would be desirable.

3. The contributions should be of a high mathematical standard and of current interest. Research articles should present new material and not duplicate other papers already published or due to be published. They should contain sufficient information and motivation and they should present proofs, or at least outlines of such, in sufficient detail to enable an expert to complete them. Thus resumes and mere announcements of papers appearing elsewhere cannot be included, although more detailed versions of a contribution may well be published in other places later.

   Surveys, if included, should cover a sufficiently broad topic, and should in general not simply review the author's own recent research. In the case of surveys, exceptionally, proofs of results may not be necessary.

   The editors of a volume are strongly advised to inform contributors about these points at an early stage.

.../...

4. Proceedings should appear soon after the meeeting. The publisher should, therefore, receive the complete manuscript within nine months of the date of the meeting at the latest.

5. Plans or proposals for proceedings volumes should be sent to one of the editors of the series or to Springer-Verlag Heidelberg. They should give sufficient information on the conference or symposium, and on the proposed proceedings. In particular, they should contain a list of the expected contributions with their prospective length. Abstracts or early versions (drafts) of some of the contributions are very helpful.

6. Lecture Notes are printed by photo-offset from camera-ready typed copy provided by the editors. For this purpose Springer-Verlag provides editors with technical instructions for the preparation of manuscripts and these should be distributed to all contributing authors. Springer-Verlag can also, on request, supply stationery on which the prescribed typing area is outlined. Some homogeneity in the presentation of the contributions is desirable.

   Careful preparation of manuscripts will help keep production time short and ensure a satisfactory appearance of the finished book. The actual production of a Lecture Notes volume normally takes 6 -8 weeks.

   Manuscripts should be at least 100 pages long. The final version should include a table of contents.

7. Editors receive a total of 50 free copies of their volume for distribution to the contributing authors, but no royalties. (Unfortunately, no reprints of individual contributions can be supplied.) They are entitled to purchase further copies of their book for their personal use at a discount of 33 1/3%, other Springer mathematics books at a discount of 20% directly from Springer-Verlag.

   Commitment to publish is made by letter of intent rather than by signing a formal contract. Springer-Verlag secures the copyright for each volume.

LECTURE NOTES

ESSENTIALS FOR THE PREPARATION OF CAMERA-READY MANUSCRIPTS

Springer
Springer-Verlag
Berlin Heidelberg New York
London Paris Tokyo

The preparation of manuscripts which are to be reproduced by photo-offset requires special care. Manuscripts which are submitted in technically unsuitable form will be returned to the author for retyping. There is normally no possibility of carrying out further corrections after a manuscript is given to production. Hence it is crucial that the following instructions be adhered to closely. If in doubt, please send us 1 - 2 sample pages for examination.

Typing area. On request, Springer-Verlag will supply special paper with the typing area outlined.

The CORRECT TYPING AREA is 18 x 26 1/2 cm (7,5 x 11 inches).

Make sure the TYPING AREA IS COMPLETELY FILLED. Set the margins so that they precisely match the outline and type right from the top to the bottom line. (Note that the page-number will lie outside this area). Lines of text should not end more than three spaces inside or outside the right margin (see example on page 4).

Type on one side of the paper only.

Type. Use an electric typewriter if at all possible. CLEAN THE TYPE before use and always use a BLACK ribbon (a carbon ribbon is best).

Choose a type size large enough to stand reduction to 75%.

Word Processors. Authors using word-processing or computer-typesetting facilities should follow these instructions with obvious modifications. Please note with respect to your printout that
i) the characters should be sharp and sufficiently black;
ii) if the size of your characters is significantly larger or smaller than normal typescript characters, you should adapt the length and breadth of the text area proportionally keeping the proportions 1:0.68.
iii) it is not necessary to use Springer's special typing paper. Any white paper of reasonable quality is acceptable.
IF IN DOUBT, PLEASE SEND US 1-2 SAMPLE PAGES FOR EXAMINATION. We will be glad to give advice.

Spacing and Headings (Monographs). Use ONE-AND-A-HALF line spacing in the text. Please leave sufficient space for the title to stand out clearly and do NOT use a new page for the beginning of subdivisions of chapters. Leave THREE LINES blank above and TWO below headings of such subdivisions.

Spacing and Headings (Proceedings). Use ONE-AND-A-HALF line spacing in the text. Start each paper on a NEW PAGE and leave sufficient space for the title to stand out clearly. However, do NOT use a new page for the beginning of subdivisions of a paper. Leave THREE LINES blank above and TWO below headings of such subdivisions. Make sure headings of equal importance are in the same form.

The first page of each contribution should be prepared in the same way. Therefore, we recommend that the editor prepares a sample page and passes it on to the authors together with these ESSENTIALS. Please take

.../...

the following as an example.

MATHEMATICAL STRUCTURE IN QUANTUM FIELD THEORY

John E. Robert
Fachbereich Physik, Universität Osnabrück
Postfach 44 69, D-4500 Osnabrück

Please leave THREE LINES blank below heading and address of the author. THEN START THE ACTUAL TEXT OF YOUR CONTRIBUTION.

Footnotes. These should be avoided. If they cannot be avoided, place them at the foot of the page, separated from the text by a line 4 cm long, and type them in SINGLE LINE SPACING to finish exactly on the outline.

Symbols. Anything which cannot be typed may be entered by hand in BLACK AND ONLY BLACK ink. (A fine-tipped rapidograph is suitable for this purpose; a good black ball-point will do, but a pencil will not). Do not draw straight lines by hand without a ruler (not even in fractions).

Equations and Computer Programs. Equations and computer programs should begin four spaces inside the left margin. Should the equations be numbered, then each number should be in brackets at the right-hand edge of the typing area.

Pagination. Number pages in the upper right-hand corner in LIGHT BLUE OR GREEN PENCIL ONLY. The final page numbers will be inserted by the printer.

There should normally be NO BLANK PAGES in the manuscript (between chapters or between contributions) unless the book is divided into Part A, Part B for example, which should then begin on a right-hand page.

It is much safer to number pages AFTER the text has been typed and corrected. Page 1 (Arabic) should be THE FIRST PAGE OF THE ACTUAL TEXT. The Roman pagination (table of contents, preface, abstract, acknowledgements, brief introductions, etc.) will be done by Springer-Verlag.

Corrections. When corrections have to be made, cut the new text to fit and PASTE it over the old. White correction fluid may also be used.

Never make corrections or insertions in the text by hand.

If the typescript has to be marked for any reason, e.g. for TEMPORARY page numbers or to mark corrections for the typist, this can be done VERY FAINTLY with BLUE or GREEN PENCIL but NO OTHER COLOR: these colors do not appear after reproduction.

Table of Contents. It is advisable to type the table of contents later, copying the titles from the text and inserting page numbers.

Literature References. These should be placed at the end of each paper or chapter, or at the end of the work, as desired. Type them with single line spacing and start each reference on a new line.
Please ensure that all references are COMPLETE and PRECISE.

Vol. 1117: D.J. Aldous, J.A. Ibragimov, J. Jacod, Ecole d'Été de Probabilités de Saint-Flour XIII – 1983. Édité par P.L. Hennequin. IX, 409 pages. 1985.

Vol. 1118: Grossissements de filtrations: exemples et applications. Seminaire, 1982/83. Edité par Th. Jeulin et M. Yor. V, 315 pages. 1985.

Vol. 1119: Recent Mathematical Methods in Dynamic Programming. Proceedings, 1984. Edited by I. Capuzzo Dolcetta, W.H. Fleming and T. Zolezzi. VI, 202 pages. 1985.

Vol. 1120: K. Jarosz, Perturbations of Banach Algebras. V, 118 pages. 1985.

Vol. 1121: Singularities and Constructive Methods for Their Treatment. Proceedings, 1983. Edited by P. Grisvard, W. Wendland and J.R. Whiteman. IX, 346 pages. 1985.

Vol. 1122: Number Theory. Proceedings, 1984. Edited by K. Alladi. VII, 217 pages. 1985.

Vol. 1123: Séminaire de Probabilités XIX 1983/84. Proceedings. Edité par J. Azéma et M. Yor. IV, 504 pages. 1985.

Vol. 1124: Algebraic Geometry, Sitges (Barcelona) 1983. Proceedings. Edited by E. Casas-Alvero, G.E. Welters and S. Xambó-Descamps. XI, 416 pages. 1985.

Vol. 1125: Dynamical Systems and Bifurcations. Proceedings, 1984. Edited by B.L.J. Braaksma, H.W. Broer and F. Takens. V, 129 pages. 1985.

Vol. 1126: Algebraic and Geometric Topology. Proceedings, 1983. Edited by A. Ranicki, N. Levitt and F. Quinn. V, 423 pages. 1985.

Vol. 1127: Numerical Methods in Fluid Dynamics. Seminar. Edited by F. Brezzi, VII, 333 pages. 1985.

Vol. 1128: J. Elschner, Singular Ordinary Differential Operators and Pseudodifferential Equations. 200 pages. 1985.

Vol. 1129: Numerical Analysis, Lancaster 1984. Proceedings. Edited by P.R. Turner. XIV, 179 pages. 1985.

Vol. 1130: Methods in Mathematical Logic. Proceedings, 1983. Edited by C.A. Di Prisco. VII, 407 pages. 1985.

Vol. 1131: K. Sundaresan, S. Swaminathan, Geometry and Nonlinear Analysis in Banach Spaces. III, 116 pages. 1985.

Vol. 1132: Operator Algebras and their Connections with Topology and Ergodic Theory. Proceedings, 1983. Edited by H. Araki, C.C. Moore, Ş. Strătilă and C. Voiculescu. VI, 594 pages. 1985.

Vol. 1133: K.C. Kiwiel, Methods of Descent for Nondifferentiable Optimization. VI, 362 pages. 1985.

Vol. 1134: G.P. Galdi, S. Rionero, Weighted Energy Methods in Fluid Dynamics and Elasticity. VII, 126 pages. 1985.

Vol. 1135: Number Theory, New York 1983–84. Seminar. Edited by D.V. Chudnovsky, G.V. Chudnovsky, H. Cohn and M.B. Nathanson. V, 283 pages. 1985.

Vol. 1136: Quantum Probability and Applications II. Proceedings, 1984. Edited by L. Accardi and W. von Waldenfels. VI, 534 pages. 1985.

Vol. 1137: Xiao G., Surfaces fibrées en courbes de genre deux. IX, 103 pages. 1985.

Vol. 1138: A. Ocneanu, Actions of Discrete Amenable Groups on von Neumann Algebras. V, 115 pages. 1985.

Vol. 1139: Differential Geometric Methods in Mathematical Physics. Proceedings, 1983. Edited by H. D. Doebner and J. D. Hennig. VI, 337 pages. 1985.

Vol. 1140: S. Donkin, Rational Representations of Algebraic Groups. VII, 254 pages. 1985.

Vol. 1141: Recursion Theory Week. Proceedings, 1984. Edited by H.-D. Ebbinghaus, G.H. Müller and G.E. Sacks. IX, 418 pages. 1985.

Vol. 1142: Orders and their Applications. Proceedings, 1984. Edited by I. Reiner and K. W. Roggenkamp. X, 306 pages. 1985.

Vol. 1143: A. Krieg, Modular Forms on Half-Spaces of Quaternions. XIII, 203 pages. 1985.

Vol. 1144: Knot Theory and Manifolds. Proceedings, 1983. Edited by D. Rolfsen. V, 163 pages. 1985.

Vol. 1145: G. Winkler, Choquet Order and Simplices. VI, 143 pages. 1985.

Vol. 1146: Séminaire d'Algèbre Paul Dubreil et Marie-Paule Malliavin. Proceedings, 1983–1984. Edité par M.-P. Malliavin. IV, 420 pages. 1985.

Vol. 1147: M. Wschebor, Surfaces Aléatoires. VII, 111 pages. 1985.

Vol. 1148: Mark A. Kon, Probability Distributions in Quantum Statistical Mechanics. V, 121 pages. 1985.

Vol. 1149: Universal Algebra and Lattice Theory. Proceedings, 1984. Edited by S. D. Comer. VI, 282 pages. 1985.

Vol. 1150: B. Kawohl, Rearrangements and Convexity of Level Sets in PDE. V, 136 pages. 1985.

Vol 1151: Ordinary and Partial Differential Equations. Proceedings, 1984. Edited by B.D. Sleeman and R.J. Jarvis. XIV, 357 pages. 1985.

Vol. 1152: H. Widom, Asymptotic Expansions for Pseudodifferential Operators on Bounded Domains. V, 150 pages. 1985.

Vol. 1153: Probability in Banach Spaces V. Proceedings, 1984. Edited by A. Beck, R. Dudley, M. Hahn, J. Kuelbs and M. Marcus. VI, 457 pages. 1985.

Vol. 1154: D.S. Naidu, A.K. Rao, Singular Pertubation Analysis of Discrete Control Systems. IX, 195 pages. 1985.

Vol. 1155: Stability Problems for Stochastic Models. Proceedings, 1984. Edited by V.V. Kalashnikov and V.M. Zolotarev. VI, 447 pages. 1985.

Vol. 1156: Global Differential Geometry and Global Analysis 1984. Proceedings, 1984. Edited by D. Ferus, R.B. Gardner, S. Helgason and U. Simon. V, 339 pages. 1985.

Vol. 1157: H. Levine, Classifying Immersions into $\mathbb{R}^4$ over Stable Maps of 3-Manifolds into $\mathbb{R}^2$. V, 163 pages. 1985.

Vol. 1158: Stochastic Processes – Mathematics and Physics. Proceedings, 1984. Edited by S. Albeverio, Ph. Blanchard and L. Streit. VI, 230 pages. 1986.

Vol. 1159: Schrödinger Operators, Como 1984. Seminar. Edited by S. Graffi. VIII, 272 pages. 1986.

Vol. 1160: J.-C. van der Meer, The Hamiltonian Hopf Bifurcation. VI, 115 pages. 1985.

Vol. 1161: Harmonic Mappings and Minimal Immersions, Montecatini 1984. Seminar. Edited by E. Giusti. VII, 285 pages. 1985.

Vol. 1162: S.J.L. van Eijndhoven, J. de Graaf, Trajectory Spaces, Generalized Functions and Unbounded Operators. IV, 272 pages. 1985.

Vol. 1163: Iteration Theory and its Functional Equations. Proceedings, 1984. Edited by R. Liedl, L. Reich and Gy. Targonski. VIII, 231 pages. 1985.

Vol. 1164: M. Meschiari, J.H. Rawnsley, S. Salamon, Geometry Seminar "Luigi Bianchi" II – 1984. Edited by E. Vesentini. VI, 224 pages. 1985.

Vol. 1165: Seminar on Deformations. Proceedings, 1982/84. Edited by J. Ławrynowicz. IX, 331 pages. 1985.

Vol. 1166: Banach Spaces. Proceedings, 1984. Edited by N. Kalton and E. Saab. VI, 199 pages. 1985.

Vol. 1167: Geometry and Topology. Proceedings, 1983–84. Edited by J. Alexander and J. Harer. VI, 292 pages. 1985.

Vol. 1168: S.S. Agaian, Hadamard Matrices and their Applications. III, 227 pages. 1985.

Vol. 1169: W.A. Light, E.W. Cheney, Approximation Theory in Tensor Product Spaces. VII, 157 pages. 1985.

Vol. 1170: B.S. Thomson, Real Functions. VII, 229 pages. 1985.

Vol. 1171: Polynômes Orthogonaux et Applications. Proceedings, 1984. Edité par C. Brezinski, A. Draux, A.P. Magnus, P. Maroni et A. Ronveaux. XXXVII, 584 pages. 1985.

Vol. 1172: Algebraic Topology, Göttingen 1984. Proceedings. Edited by L. Smith. VI, 209 pages. 1985.

Vol. 1173: H. Delfs, M. Knebusch, Locally Semialgebraic Spaces. XVI, 329 pages. 1985.

Vol. 1174: Categories in Continuum Physics, Buffalo 1982. Seminar. Edited by F.W. Lawvere and S.H. Schanuel. V, 126 pages. 1986.

Vol. 1175: K. Mathiak, Valuations of Skew Fields and Projective Hjelmslev Spaces. VII, 116 pages. 1986.

Vol. 1176: R.R. Bruner, J.P. May, J.E. McClure, M. Steinberger, H_∞ Ring Spectra and their Applications. VII, 388 pages. 1986.

Vol. 1177: Representation Theory I. Finite Dimensional Algebras. Proceedings, 1984. Edited by V. Dlab, P. Gabriel and G. Michler. XV, 340 pages. 1986.

Vol. 1178: Representation Theory II. Groups and Orders. Proceedings, 1984. Edited by V. Dlab, P. Gabriel and G. Michler. XV, 370 pages. 1986.

Vol. 1179: Shi J.-Y. The Kazhdan-Lusztig Cells in Certain Affine Weyl Groups. X, 307 pages. 1986.

Vol. 1180: R. Carmona, H. Kesten, J.B. Walsh, École d'Été de Probabilités de Saint-Flour XIV – 1984. Édité par P.L. Hennequin. X, 438 pages. 1986.

Vol. 1181: Buildings and the Geometry of Diagrams, Como 1984. Seminar. Edited by L. Rosati. VII, 277 pages. 1986.

Vol. 1182: S. Shelah, Around Classification Theory of Models. VII, 279 pages. 1986.

Vol. 1183: Algebra, Algebraic Topology and their Interactions. Proceedings, 1983. Edited by J.-E. Roos. XI, 396 pages. 1986.

Vol. 1184: W. Arendt, A. Grabosch, G. Greiner, U. Groh, H.P. Lotz, U. Moustakas, R. Nagel, F. Neubrander, U. Schlotterbeck, One-parameter Semigroups of Positive Operators. Edited by R. Nagel. X, 460 pages. 1986.

Vol. 1185: Group Theory, Beijing 1984. Proceedings. Edited by Tuan H.F. V, 403 pages. 1986.

Vol. 1186: Lyapunov Exponents. Proceedings, 1984. Edited by L. Arnold and V. Wihstutz. VI, 374 pages. 1986.

Vol. 1187: Y. Diers, Categories of Boolean Sheaves of Simple Algebras. VI, 168 pages. 1986.

Vol. 1188: Fonctions de Plusieurs Variables Complexes V. Séminaire, 1979–85. Edité par François Norguet. VI, 306 pages. 1986.

Vol. 1189: J. Lukeš, J. Malý, L. Zajíček, Fine Topology Methods in Real Analysis and Potential Theory. X, 472 pages. 1986.

Vol. 1190: Optimization and Related Fields. Proceedings, 1984. Edited by R. Conti, E. De Giorgi and F. Giannessi. VIII, 419 pages. 1986.

Vol. 1191: A.R. Its, V.Yu. Novokshenov, The Isomonodromic Deformation Method in the Theory of Painlevé Equations. IV, 313 pages. 1986.

Vol. 1192: Equadiff 6. Proceedings, 1985. Edited by J. Vosmansky and M. Zlámal. XXIII, 404 pages. 1986.

Vol. 1193: Geometrical and Statistical Aspects of Probability in Banach Spaces. Proceedings, 1985. Edited by X. Fernique, B. Heinkel, M.B. Marcus and P.A. Meyer. IV, 128 pages. 1986.

Vol. 1194: Complex Analysis and Algebraic Geometry. Proceedings, 1985. Edited by H. Grauert. VI, 235 pages. 1986.

Vol.1195: J.M. Barbosa, A.G. Colares, Minimal Surfaces in $\mathbb{R}^3$. X, 124 pages. 1986.

Vol. 1196: E. Casas-Alvero, S. Xambó-Descamps, The Enumerative Theory of Conics after Halphen. IX, 130 pages. 1986.

Vol. 1197: Ring Theory. Proceedings, 1985. Edited by F.M.J. van Oystaeyen. V, 231 pages. 1986.

Vol. 1198: Séminaire d'Analyse, P. Lelong – P. Dolbeault – H. Skoda. Seminar 1983/84. X, 260 pages. 1986.

Vol. 1199: Analytic Theory of Continued Fractions II. Proceedings, 1985. Edited by W.J. Thron. VI, 299 pages. 1986.

Vol. 1200: V.D. Milman, G. Schechtman, Asymptotic Theory of Finite Dimensional Normed Spaces. With an Appendix by M. Gromov. VIII, 156 pages. 1986.

Vol. 1201: Curvature and Topology of Riemannian Manifolds. Proceedings, 1985. Edited by K. Shiohama, T. Sakai and T. Sunada. VII, 336 pages. 1986.

Vol. 1202: A. Dür, Möbius Functions, Incidence Algebras and Power Series Representations. XI, 134 pages. 1986.

Vol. 1203: Stochastic Processes and Their Applications. Proceedings, 1985. Edited by K. Itô and T. Hida. VI, 222 pages. 1986.

Vol. 1204: Séminaire de Probabilités XX, 1984/85. Proceedings. Edité par J. Azéma et M. Yor. V, 639 pages. 1986.

Vol. 1205: B.Z. Moroz, Analytic Arithmetic in Algebraic Number Fields. VII, 177 pages. 1986.

Vol. 1206: Probability and Analysis, Varenna (Como) 1985. Seminar. Edited by G. Letta and M. Pratelli. VIII, 280 pages. 1986.

Vol. 1207: P.H. Bérard, Spectral Geometry: Direct and Inverse Problems. With an Appendix by G. Besson. XIII, 272 pages. 1986.

Vol. 1208: S. Kaijser, J.W. Pelletier, Interpolation Functors and Duality. IV, 167 pages. 1986.

Vol. 1209: Differential Geometry, Peñíscola 1985. Proceedings. Edited by A.M. Naveira, A. Ferrández and F. Mascaró. VIII, 306 pages. 1986.

Vol. 1210: Probability Measures on Groups VIII. Proceedings, 1985. Edited by H. Heyer. X, 386 pages. 1986.

Vol. 1211: M.B. Sevryuk, Reversible Systems. V, 319 pages. 1986.

Vol. 1212: Stochastic Spatial Processes. Proceedings, 1984. Edited by P. Tautu. VIII, 311 pages. 1986.

Vol. 1213: L.G. Lewis, Jr., J.P. May, M. Steinberger, Equivariant Stable Homotopy Theory. IX, 538 pages. 1986.

Vol. 1214: Global Analysis – Studies and Applications II. Edited by Yu.G. Borisovich and Yu.E. Gliklikh. V, 275 pages. 1986.

Vol. 1215: Lectures in Probability and Statistics. Edited by G. del Pino and R. Rebolledo. V, 491 pages. 1986.

Vol. 1216: J. Kogan, Bifurcation of Extremals in Optimal Control. VIII, 106 pages. 1986.

Vol. 1217: Transformation Groups. Proceedings, 1985. Edited by S. Jackowski and K. Pawalowski. X, 396 pages. 1986.

Vol. 1218: Schrödinger Operators, Aarhus 1985. Seminar. Edited by E. Balslev. V, 222 pages. 1986.

Vol. 1219: R. Weissauer, Stabile Modulformen und Eisensteinreihen. III, 147 Seiten. 1986.

Vol. 1220: Séminaire d'Algèbre Paul Dubreil et Marie-Paule Malliavin. Proceedings, 1985. Edité par M.-P. Malliavin. IV, 200 pages. 1986.

Vol. 1221: Probability and Banach Spaces. Proceedings, 1985. Edited by J. Bastero and M. San Miguel. XI, 222 pages. 1986.

Vol. 1222: A. Katok, J.-M. Strelcyn, with the collaboration of F. Ledrappier and F. Przytycki, Invariant Manifolds, Entropy and Billiards; Smooth Maps with Singularities. VIII, 283 pages. 1986.

Vol. 1223: Differential Equations in Banach Spaces. Proceedings, 1985. Edited by A. Favini and E. Obrecht. VIII, 299 pages. 1986.

Vol. 1224: Nonlinear Diffusion Problems, Montecatini Terme 1985. Seminar. Edited by A. Fasano and M. Primicerio. VIII, 188 pages. 1986.

Vol. 1225: Inverse Problems, Montecatini Terme 1986. Seminar. Edited by G. Talenti. VIII, 204 pages. 1986.

Vol. 1226: A. Buium, Differential Function Fields and Moduli of Algebraic Varieties. IX, 146 pages. 1986.

Vol. 1227: H. Helson, The Spectral Theorem. VI, 104 pages. 1986.

Vol. 1228: Multigrid Methods II. Proceedings, 1985. Edited by W. Hackbusch and U. Trottenberg. VI, 336 pages. 1986.

Vol. 1229: O. Bratteli, Derivations, Dissipations and Group Actions on C*-algebras. IV, 277 pages. 1986.

Vol. 1230: Numerical Analysis. Proceedings, 1984. Edited by J.-P. Hennart. X, 234 pages. 1986.

Vol. 1231: E.-U. Gekeler, Drinfeld Modular Curves. XIV, 107 pages. 1986.